Studies in Weak Arithmetics

CSLI *Lecture Notes Number 217*

Studies in Weak Arithmetics

Volume 3

Edited by
Patrick CÉGIELSKI
Ali ENAYAT
Roman KOSSAK

CSLI
PUBLICATIONS
Center for the Study of
Language and Information
STANFORD, CALIFORNIA

Copyright © 2016
CSLI Publications
Center for the Study of Language and Information
Leland Stanford Junior University
Printed in the United States
20 19 18 17 16   1 2 3 4 5

Library of Congress Cataloging-in-Publication Data

Studies in weak arithmetics / edited by Patrick Cégielski.

p. cm. – (CSLI lecture notes ; no. 196)

ISBN-13 978-1-57586-953-7 (alk. paper)
ISBN-10 1-57586-953-5   (alk. paper)

1. Number theory. 2. Mathematics–Philosophy. 3. Logic, Symbolic
and mathematical. I. Cegielski, Patrick, 1954-

QA241.S824 2010
512.7–dc22   2009048021
CIP

∞ The acid-free paper used in this book meets the minimum requirements
of the American National Standard for Information Sciences—Permanence
of Paper for Printed Library Materials, ANSI Z39.48-1984.

CSLI was founded in 1983 by researchers from Stanford University, SRI
International, and Xerox PARC to further the research and development of
integrated theories of language, information, and computation. CSLI headquarters
and CSLI Publications are located on the campus of Stanford University.

CSLI Publications reports new developments in the study of language,
information, and computation. Please visit our web site at
http://cslipublications.stanford.edu/
for comments on this and other titles, as well as for changes
and corrections by the author and publisher.

Contents

Contributors

RASMUS BLANCK : Department of Philosophy, Linguistics and Theory of Science, University of Gothenburg, Olof Wijksgatan 6, P.O. Box 200, SE-405 30 Göteborg, Sweden.
rasmus.blanck@gu.se

PATRICK CÉGIELSKI : Université Paris – Est Créteil, IUT de Sénart-Fontainebleau, Département Informatique, route forestière Hurtault, F-77300 Fontainebleau, France.
cegielski@u-pec.fr

PIETRO CORVAJA : University of Udine, DIMI, Via delle Scienze 206, I-33100 Udine, Italy.
pietro.corvaja@uniud.it

ALI ENAYAT : Department of Philosophy, Linguistics and Theory of Science, University of Gothenburg, Olof Wijksgatan 6, P.O. Box 200, SE-405 30 Göteborg, Sweden.
ali.enayat@gu.se

HENRI-ALEX ESBELIN : Clermont Université, CNRS UMR 6158, LIMOS, Campus Universitaire des Czeaux 1 rue de la Chebarde, F-63178 Aubière cedex, France.
alex.esbelin@univ-bpclermont.fr

JOOST J. JOOSTEN : Department of Philosophy, Carrer Montalegre 6, 08001 Barcelona, Spain.
jjoosten@ub.edu

ROMAN KOSSAK : City University of New York, The Graduate Center, CUNY 365 Fifth Avenue, New York, NY 10016, USA.
rkossak@gc.cuny.edu

YOANN MARQUER : Université Paris 7 – Denis Diderot, 6, avenue de la Brèche, F-94000 Créteil, France.
yoann.apeiron.marquer@gmail.com

MOJTABA MONIRI : Department of Mathematics, Western Illinois University, 1 University Circle, Macomb, IL 61455, U.S.A.
m-moniri@wiu.edu

EUGENIO G. OMODEO : DMG, via Alfonso Valerio, 12/1, I-34127 Trieste, Italy.
eomodeo@units.it

JAMES H. SCHMERL : Department of Mathematics, University of Connecticut, Storrs, CT 06269, U.S.A.
james.schmerl@uconn.edu

HENRY TOWSNER : Department of Mathematics, University of Philadelphia, 209 South 33rd. Street, Philadelphia, PA 19104-6395, U.S.A.
htowsner@math.upenn.edu

PIERRE VALARCHER : Université Paris – Est Crteil, IUT de Sénart-Fontainebleau, Département Informatique, route forestière Hurtault, F-77300 Fontainebleau, France.
pierre.valarcher@u-pec.fr

LUCA VALLATA : DMG, via Alfonso Valerio, 12/1, I-34127 Trieste, Italy.
luca.vallata@gmail.com

KEITA YOKOYAMA : Japan Advanced Institute of Science and Technology, 1-1 Asahidai, Nomi, Ishikawa, 923-1292 Japan.
y-keita@jaist.ac.jp

Introduction

PATRICK CÉGIELSKI, ALI ENAYAT, ROMAN KOSSAK

When the first author of this preface founded the conference series, *Journées sur les Arithmétiques Faibles* (*Weak Arithmetics Days*, JAF), launched in June 1990 with a first meeting held at École Normale Supérieure in Lyon, France, he didn't imagine that he would be able to program thirty-seven meetings. During first biannual and then annual meetings, researchers from all over the world presented work and discussed research ideas concerning *Weak Arithmetics*, that is, the research area concerning, roughly speaking, the application of logical methods to Number Theory and related topics.

The latest meetings of the series, JAF33 & JAF 34, respectively took place in June 16-18, 2014, at the University of Gothenburg, Sweden; and in July, 7-9, 2015 at the CUNY Graduate Center, New York, USA. The Steering Committee of the JAF series decided that a volume be published, containing substantive papers based on lectures delivered during these meetings.

All papers published in the present volume were refereed by (at least) two referees each. We are grateful to the following colleagues for their invaluable assistance in the refereeing process: Jeremy Avigad, Lev Beklemishev, Alexis Bès, Henri-Alex Esbelin, Isaac Goldbring, Yuri Gurevich, Emil Jeřábek, Dave Marker, Yuri Matiyasevich, Volodya Shavrukov, Albert Visser, and Tin Lok Wong.

We would like to express our deep gratitude to the project TAR-MAC (*Théorie des algorithmes : machines, complétude, axiomatisation et contraintes physiques*), agreement ANR-12-BS02-0007 from French

Studies in Weak Arithmetics, Volume 3.
Patrick Cégielski, Ali Enayat,
Roman Kossak.
Copyright © 2016, CSLI Publications.

ANR (*Agence Nationale de la Recherche*), for having generously covered the cost of production of the present volume, and to Marie-Annick Le Traon (Université Paris Est Créteil–IUT de Sénart-Fontainebleau) for the design of the cover.

The JAF website:

http://lacl.fr/jaf/

contains further information on the conference series "Journées sur les Arithmétiques Faibles".

1

Flexibility in Fragments of Peano Arithmetic

RASMUS BLANCK

Abstract: This paper concerns flexible formulae of arithmetic: formulae whose "extensions as sets are left undetermined by the formal
system". Formally, this means that a formula $\gamma(x)$ is flexible for a class
of formulae X if, for each $\xi(x) \in X$, the theory $T + \forall x(\gamma(x) \leftrightarrow \xi(x))$ is
consistent. We compare different kinds of flexibility results, and gauge
the amount of induction needed for their proofs. By formalising these
arguments, we are also able to derive their model-theoretic counterparts, assuring the existence of certain kinds of end-extensions of models of fragments of Peano arithmetic.

1 Introduction

This paper is centered around the notion of flexibility, giving a survey
of a number of known flexibility results, as well as gauging the amount
of induction needed for different flavours of flexibility. Flexible (or independent, free etc.) formulae were introduced almost simultaneously by
Kripke [13] and Mostowski [17].[1] The existence of independent formulae
is often presented as a generalisation of the Gödel-Rosser incompleteness theorem: For every reasonable, consistent theory of arithmetic T,

---

[1]Kripke's paper was initially submitted on March 11, 1960 (with subsequent
revisions submitted on November 30, 1960 and April 21, 1961) while Mostowski's
paper was submitted on April 21, 1960. However, Kripke's paper appeared in print
about a year after Mostowski's, and it includes a postscript dated June 4, 1962
concerning the overlap between his results and Mostowski's.

Studies in Weak Arithmetics, Volume 3.
Patrick Cégielski, Ali Enayat,
Roman Kossak.

there is a sentence γ such that

$$T \nvdash \gamma, \neg\gamma.^2$$

The connection is straightforwardly seen by noting that if $\gamma(x)$ is an independent formula, then the only propositional combinations of sentences of the form $\gamma(n)$ provable in T are the tautologies, so in particular, for each $n \in \omega$,

$$T \nvdash \gamma(n), \neg\gamma(n).$$

The notion of independent formulae is in turn generalised by the notion of flexible formulae, and Kripke describes these latter ones by saying that "their extensions as sets are left undetermined by the formal system". This description can be made precise by defining $\gamma(x)$ to be flexible (for a prescribed class of formulae X) whenever the theory

$$T + \forall x(\gamma(x) \leftrightarrow \xi(x))$$

is consistent for every formula $\xi(x)$ in X. Taking a model-theoretic perspective, this definition states that there is a model of T in which the *extensions* of $\gamma(x)$ and $\xi(x)$ coincide, not only on the standard numbers, but also on the non-standard ones.

Flexible formulae of the kind defined above also appear in a number of later texts, for example Lindström [14, 15], Montagna [16], Sommaruga-Rosolemos [23], and Visser [24]. In all of these treatments, the relevant existence theorems are proved by making indispensable use of partial truth definitions (or partial satisfaction predicates), which suggests that these theorems do not apply to theories as weak as Robinson's Q. The theories under consideration in Kripke's paper, however, are defined by a number of very modest conditions (satisfied by all extensions of Q), even if Kripke's example of such a theory is essentially equivalent to PA. These observations, taken together with the fact that there is a slight discrepancy between Kripke's notion of flexibility and the (implicit) one used in later treatments, give rise to some confusion about what is actually proved in Kripke's paper.

Hence, in Section 3, we give an introduction to the kind of flexibility available to theories extending Q, by examining Kripke's theorem with a focus on methods and ideas. We also introduce the notion of independent formulae, compare them with Kripke's flexible formulae, and rederive Mostowski's existence result using Kripke's method. Section 4 contains a proof of Kripke's theorem that lends itself to many generalisations and extensions. In Section 5 we take a model-theoretic approach, using formalisations of the results in the earlier sections to construct end-extensions of models of fragments of arithmetic in which

---

[2]Gödel [5], Rosser [19].

flexible formulae obtain a prescribed extension. In the final section, we discuss some recent developments and also (re-)state an open problem.

2 Preliminaries

Let $\mathscr{L}_A$ be the language of arithmetic, and fix some reasonable Gödel numbering of syntactical objects. We freely confuse natural numbers with formal numerals, but write $\ulcorner \phi \urcorner$ for the Gödel number of the formula ϕ, and $\ulcorner \phi(\dot{x}) \urcorner$ for the Gödel number of the sentence obtained by replacing the free variable in ϕ by the value of x. For each sentence ϕ, let $\phi^0 := \phi$ and $\phi^1 := \neg\phi$.

Let $\Delta_0 = \Sigma_0 = \Pi_0$ be the set of arithmetical formulae, all of whose quantifiers are bounded by terms in $\mathscr{L}_A$, and let Σ_n and Π_n be defined from Δ_0 in some standard way. We demand that these classes satisfy $\phi \in \Sigma_n$ iff $\neg\phi \in \Pi_n$, together with the closure of Σ_n under existential quantification, and the closure of Π_n under universal quantification. B_n is the set of Boolean combinations of Σ_n formulae, and a formula $\phi(x)$ is Δ_n^T if $\phi(x)$ is Σ_n and provably equivalent in T to a Π_n-formula. When writing Σ_n, Π_n, Δ_n, or B_n, we always assume that $n > 0$.

Q is the finitely axiomatisable theory known as Robinson's arithmetic, and PA (Peano arithmetic) is the strengthening of Q obtained by adding the scheme of induction for all formulae in the language of arithmetic. We always assume that T is some consistent, r.e. extension of Q. $I\Delta_0 + \exp$ is the fragment of PA obtained by restricting the induction axiom to formulae in Δ_0, plus adding an axiom asserting the totality of the exponential function. $I\Sigma_1$ is similarly obtained from PA by restricting the induction scheme to formulae in Σ_1.

A formula $\xi(x)$ *numerates* a set $X \subseteq \omega$ in T, if for all $k \in \omega$,

$$k \in X \text{ iff } T \vdash \xi(k).$$

If we also have

$$k \notin X \text{ iff } T \vdash \neg\xi(k),$$

then $\xi(x)$ *binumerates* X in T.

Let $\sigma(z)$ be any formula. We let $\mathrm{Prf}_\sigma(x, y)$ be a binumeration of the relation "y is a proof of x from the sentences satisfying $\sigma(z)$". Let $\mathrm{Pr}_\sigma(x) := \exists y \mathrm{Prf}_\sigma(x, y)$, and let $\mathrm{Con}_\sigma := \neg\mathrm{Pr}_\sigma(0 = 1)$. Whenever $\sigma(z)$ is Σ_n, then $\mathrm{Pr}_\sigma(x)$ is Σ_n, and Con_σ is Π_n. We further demand that

$$T \vdash \forall x(\sigma(x) \to \sigma'(x)) \to \forall y(\mathrm{Pr}_\sigma(y) \to \mathrm{Pr}_{\sigma'}(y)),$$

and

$$T \vdash \forall x(\sigma(x) \to \sigma'(x)) \to (\mathrm{Con}_{\sigma'} \to \mathrm{Con}_\sigma).$$

We ambiguously let $\text{Pr}_T(x)$ and Con_T denote $\text{Pr}_\sigma(x)$ and Con_σ, respectively, where $\sigma(z)$ is any Σ_1 numeration of T such that $\text{Pr}_\sigma(x)$ satisfies Löb's derivability conditions: For all formulae ϕ, ψ, possibly with free variables,

L1. If $T \vdash \phi$, then $I\Delta_0 + \exp \vdash \text{Pr}_\sigma(\ulcorner\phi\urcorner)$,
L2. $I\Delta_0 + \exp \vdash \text{Pr}_\sigma(\ulcorner\phi\urcorner) \wedge \text{Pr}_\sigma(\ulcorner\phi \rightarrow \psi\urcorner) \rightarrow \text{Pr}_\sigma(\ulcorner\psi\urcorner)$,
L3. $I\Delta_0 + \exp \vdash \text{Pr}_\sigma(\ulcorner\phi\urcorner) \rightarrow \text{Pr}_\sigma(\ulcorner\text{Pr}_\sigma(\phi)\urcorner)$.

Proofs of the following well-known facts can be found in e.g. [8], [11], or [22].

Fact 1 (Formalised Σ_1-completeness). For each $\phi(x_1, \ldots, x_k) \in \Sigma_1$,

$$I\Delta_0 + \exp \vdash \forall x_1, \ldots x_k(\phi(x_1, \ldots, x_k) \rightarrow \text{Pr}_T(\ulcorner\phi(\dot{x}_1, \ldots, \dot{x}_k)\urcorner)).$$

Fact 2 (Existence of partial satisfaction predicates). Given $k > 0$, there is a $k + 1$-ary Σ_n-formula $\text{Sat}^k_{\Sigma_n}(x, x_1, \ldots, x_k)$ such that for all Σ_n-formulae $\phi(x_1, \ldots, x_k)$ with exactly the variables $x_1, \ldots, x_k$ free,

$$I\Delta_0 + \exp \vdash \forall x_1, \ldots, x_k(\phi(x_1, \ldots, x_k) \leftrightarrow \text{Sat}^k_{\Sigma_n}(\ulcorner\phi\urcorner, x_1, \ldots, x_k)).$$

Similar formulae can be found for Π_n, and there are also Δ_{n+1} formulae acting as partial satisfaction predicates for formulae in the class B_n of Boolean combinations of Σ_n formulae.

Fact 3 (The selection theorem). For each Σ_1-formula $\phi(x_1, \ldots, x_k)$, there is a Σ_1-formula $\text{Sel}\{\phi\}$ with exactly the same free variables, such that

1. $I\Delta_0 \vdash \forall x_1, \ldots, x_k(\text{Sel}\{\phi\}(x_1, \ldots, x_k) \rightarrow \phi(x_1, \ldots, x_k))$,
2. $I\Delta_0 \vdash \forall x_1, \ldots, x_k, y(\text{Sel}\{\phi\}(x_1, \ldots, x_k) \wedge$
$$\text{Sel}\{\phi\}(x_1, \ldots, x_{k-1}, y) \rightarrow x_k = y),$$
3. $I\Delta_0 \vdash \forall x_1, \ldots, x_{k-1}(\exists x_k \phi(x_1, \ldots, x_k) \rightarrow \exists x_k \text{Sel}\{\phi\}(x_1, \ldots, x_k))$.

We use these formulae as a device to talk about partial recursive functions within T. Let the k-ary partial recursive function φ_i with index i be the function whose graph is defined by

$$\text{Sel}\{\text{Sat}^{k+1}_{\Sigma_1}\}(i, x_1, \ldots, x_{k+1}).$$

For each $k > 0$, let $R^k(x, y_1, \ldots, y_k, z)$ be the Σ_1-formula

$$\text{Sel}\{\text{Sat}^{k+1}_{\Sigma_1}\}(x, y_1, \ldots, y_k, z).$$

This formula acts as a *formalisation* of a function that is universal for k-ary partial recursive functions, in the sense that $\varphi(n_1, \ldots, n_k) = m$ iff $T \vdash R^k(\ulcorner\varphi\urcorner, n_1, \ldots, n_k, m)$. Note also that by clause 2 of Fact 3, we have $T \vdash R^k(\ulcorner\varphi\urcorner, n_1, \ldots, n_k, m) \rightarrow \exists! z R^k(\ulcorner\varphi\urcorner, n_1, \ldots, n_k, z)$. Henceforth, we omit the superscript k, as the arity is always clear from the context.

Remark 2.1. When T is an extension of Q, we still have access to formulae $R(i, n_1, \ldots, n_k, m)$ such that $\varphi_i(n_1, \ldots, n_k) = m$ iff Q $\vdash$ $R(i, n_1, \ldots, n_k, m) \wedge \exists! z R(i, n_1, \ldots, n_k, z)$, although the indexing of the partial recursive functions might differ from the one described above. The fact that R can be defined in terms of Sat_{Σ_1} whenever T extends $\mathrm{I}\Delta_0 + \exp$ will be used in Section 5.

Occasionally we need to distinguish between formulae in the object language and the corresponding relations in the metalanguage. Hence, we write $\mathbf{\Sigma}_n$ and $\mathbf{\Pi}_n$ for the collection of relations on ω that are definable by Σ_n and Π_n formulae, respectively.

Whenever $\phi(x_1, \ldots, x_k, y_1, \ldots, y_n)$ is a formula numerating a relation $P(x_1, \ldots, x_k, y_1, \ldots, y_n)$ in T, we say that

$$Qy_1, \ldots Qy_n \phi(x_1, \ldots, x_k, y_0, \ldots, y_n)$$

is a formalisation of $Qy_1, \ldots Qy_n P(x_1, \ldots, x_k, y_1, \ldots, y_n)$ in T, where each Q is either $\exists$ or $\forall$.

Fact 4 (Kleene's enumeration theorem, [12] Theorem IV). Let $n > 0$. Every k-ary $\mathbf{\Sigma}_n$ relation can be defined by a formula of the form

$$\exists y_1 \forall y_2 \ldots Qy_n T(m, x_1, \ldots, x_k, y_1, \ldots, y_n)$$

for a suitable choice of m. In the above formula, T is Kleene's primitive recursive T-predicate, Q is $\exists$ or $\forall$ depending on whether n is odd or even, and in this latter case T is prefixed with a negation.

Similarily, every k-ary $\mathbf{\Pi}_n$ relation can be defined by a formula of the form

$$\forall y_1 \exists y_2 \ldots Qy_n \neg T(m, x_1, \ldots, x_k, y_1, \ldots, y_n)$$

for a suitable choice of m. Whenever n is even, Q is $\exists$ and in this case the negation is dropped.

Since, for each m, and for each $n > 0$, there is a binumeration of the relation $T(m, x_1, \ldots, x_k, y_1, \ldots, y_n)$ in Q, it follows that every arithmetical relation has a formalisation in one of the above forms in Q. In particular, every unary relation in $\mathbf{\Sigma}_1$ is numerated in Q by a formula of the form $\exists y T(m, x, y)$.

For any formula $\sigma(z)$, let $(\sigma + y)(z) := \sigma(z) \vee z = y$. The following fact then follows from the definition of Con_σ, together with the formalised deduction theorem.

Fact 5. For each formula $\sigma(z)$,

$$\mathrm{I}\Delta_0 + \exp \vdash \forall x (\mathrm{Pr}_\sigma(x) \leftrightarrow \neg \mathrm{Con}_{\sigma + \neg x}).$$

3   Flexibility and Independence in Q

This section is devoted to an overview of the results of [13], and their respective relation to the main result of [17] and the central lemma of [20]. We also provide a rederivation of Mostowski's theorem by Kripke's method as suggested in the final remark of Kripke's paper.

The terminology used by different authors to discuss the concepts and results of this section is remarkably inconsistent. In [17], Mostowski defines a formula $\phi(x)$ to be *free* for T if, for each n, the formulae $\phi(0), \ldots, \phi(n)$ are *completely independent*, which is taken to mean that every conjunction formed of some of these formulae and the negation of the others is consistent with T. When Feferman, Kreisel & Orey [4] refers to Mostowski's (then unpublished, but privately communicated) result, the *formula* $\phi(x)$ is called *completely independent*. Scott [20] calls such a formula *independent*, a terminology which reappears in Lindström [14, 15]. Hájek & Pudlák [8] instead call these formulae *flexible*. To complicate things further, Kripke has a notion of an *independent set* which is closely related to the complete independence of Mostowski. Myhill [18] calls such a set *absolutely independent*, and attributes the definition to Harary [7], although their definitions differ.

In an attempt to unify the terminology (at least for the rest of the present paper) we define the following three central concepts. The first is essentially equivalent to Mostowski's original definition of a free formula.

Definition 1. A formula $\phi(x)$ is independent over T, if for each function f from ω to $\{0, 1\}$, the theory $T + \{\phi(n)^{f(n)} : n \in \omega\}$ is consistent.

The concept of a flexible formula first appears in Kripke's [13] with the following definition.

Definition 2. A formula $\gamma(x)$ is flexible for Σ_n (or, equivalently, Σ_n-flexible) over T if every relation in Σ_n has a formalisation $\sigma(x)$ in T such that $T + \forall x(\gamma(x) \leftrightarrow \sigma(x))$ is consistent.

This definition is to be contrasted with the more modern concept of flexibility used in later treatments such as Lindström [14, 15], Montagna [16], Sommaruga-Rosolemos [23], and Visser [24]. Although none of these authors explicitly define the notion, it is possible to extract the following salient definition from their writings.[3]

Definition 3. A formula $\gamma(x)$ is flexible for Σ_n (or, equivalently, Σ_n-flexible) over T if, for every formula $\sigma(x) \in \Sigma_n$, the theory $T + \forall x(\gamma(x) \leftrightarrow \sigma(x))$ is consistent.

---

[3]Yet another notion of Σ_n-flexibility appears in Hájek [9], but this concept can be readily formulated in terms of independent formulae instead.

As we will see in the sequel, every Σ_n-flexible formula is also independent. In general we can construct Σ_n-flexible, but not Σ_n-flexible, formulae in extensions of Q, while in extensions of $I\Delta_0 + \exp$ the two notions coincide.

3.1 Kripke's Theorem

The following theorem is the main result of Kripke's paper, and is also the theorem that this paper is centered around.

Theorem 3.1 (Kripke [13], Theorem 1). *Suppose that* T *is a consistent, r.e. extension of* Q. *Then there is a* Σ_n *formula* $\gamma(x)$ *which is flexible for* Σ_n *over* T.

Kripke derives his theorem from the existence of a flexible function: a partial recursive function f from ω to ω such that if T is consistent, then there is an e such that for all k, the theory T $+$ "$f(e) = k$" is also consistent. The ingenious idea is to take advantage of Kleene's enumeration theorem: every relation in (say) Σ_1 can be formalised, using Kleene's T-predicate, as $\exists y T(k, x, y)$ for a suitable choice of k. Then

$$\text{T} + \text{``}f(e) = k\text{''} \vdash \text{``}\exists y T(f(e), x, y)\text{''} \leftrightarrow \exists y T(k, x, y),$$

which means that "$\exists y T(f(e), x, y)$" is Σ_1-flexible in Kripke's sense. What remains is to prove the existence of a flexible function, and to make the quoted formulas formal. Let $\Phi(x, y)$ be a recursive function which is universal for (unary) partial recursive functions, and let $R(x, y, z)$ be a Σ_1 numeration of $\Phi(x, y) = z$.

Lemma 3.2 (Kripke [13], Lemma 1). *Suppose that* T *is a consistent, r.e. extension of* Q. *Then there is a partial recursive function with index* e *(that depends on* T*) such that for each* k, *the theory* T $+ R(e, e, k) + \exists! z R(e, e, z)$ *is consistent.*

Proof. Let T be a consistent, r.e. extension of Q. Let a partial recursive function f be defined by $f(m) = k$ iff

$$\text{T} \vdash \neg(R(m, m, k) \land \exists! z R(m, m, z)).$$

Should more than one sentence of this form be provable, choose the one whose proof has the least Gödel number. Let e be an index of f. We claim that for each k, T $+ R(e, e, k) + \exists! z R(e, e, z)$ is consistent.

Suppose, for a contradiction, that T $\vdash \neg(R(e, e, k) \land \exists! z R(e, e, z))$ for some k. Then $f(e) = k$ is defined, and by Remark 2.1 we have T $\vdash R(e, e, k) \land \exists! z R(e, e, z)$. This contradicts the consistency of T, so it follows that T $\vdash R(e, e, k) + \exists! z R(e, e, z)$ is consistent. $\dashv$

Proof of Theorem 3.1. Let T be a consistent, r.e. extension of Q, and let e be as in the proof of Lemma 3.2. Recall that e depends on the actual choice of the theory T. For brevity, we only state the proof for the case $n = 1$. Let $\gamma(x)$ be the formula $\exists z(R(e, e, z) \wedge \exists y T(z, x, y))$, where T is Kleene's T-predicate. The proof can then easily be modified to arbitrary n by using the corresponding normal forms.

If $P(x)$ is any relation in Σ_1, then by Fact 4 there is a k such that $\exists y T(k, x, y)$ is a formalisation of $P(x)$ in Q. By the Lemma, the theory $T + R(e, e, k) + \exists! z R(e, e, z)$ is consistent. We have

$$T + R(e, e, k) + \exists! z R(e, e, z) \vdash \exists z(R(e, e, z) \wedge \exists y T(z, x, y)) \leftrightarrow \exists y T(k, x, y)$$

and the left-hand side of the equivalence is definitionally equivalent to $\gamma(x)$, so it follows that

$$T + \forall x(\gamma(x) \leftrightarrow \exists y T(k, x, y))$$

is consistent. Hence $\gamma(x)$ is flexible for Σ_n. ⊣

Remark 3.3. A slight modification of the proof yields a Π_n formula which is flexible for $\mathbf{\Pi}_n$ in Kripke's sense. For Π_1, this is accomplished by letting $\gamma'(x) := \forall z(R(e, e, z) \rightarrow \forall y \neg T(z, x, y))$, and for higher complexities by using the corresponding normal forms.

Remark 3.4. Note that if T is strong enough to formalise the proof of Kleene's enumeration theorem, then the formula $\exists y T(x, y, z)$ acts as a Σ_1-satisfaction predicate in T. Hence, if T is an extension of Q, but not as strong as, say, $I\Delta_0 + \exp$, it is not likely that we can assure that $T \vdash \forall x(\sigma(x) \leftrightarrow \exists y T(k, x, y))$, even though this equivalence holds in the standard model.

3.2 Mostowski's Theorem

We now turn to the independent formulae of Mostowski. Recall that $\phi(x)$ is independent over T if the theory

$$T + \{\phi(n)^{f(n)} : n \in \omega\}$$

is consistent for each function f from ω to $\{0, 1\}$. The main existence result is the following theorem:

Theorem 3.5 (Mostowski [17], Theorem 4). *If $\{T_i : i \in \omega\}$ is an r.e. sequence of consistent, r.e. extensions of Q, then there is a formula which is simultaneously independent over each T_i.*

Kripke defines a set A to be independent over T if, for each subset B of A, the theory obtained from T by adding the elements of B plus the negations of elements of $A \setminus B$ is consistent. It is easy to see that $\phi(x)$ is independent over T in Mostowski's sense iff $\{\phi(n) : n \in \omega\}$ is an independent set in Kripke's sense. As is noted in the concluding remark of Kripke's paper, Mostowski's "generalisation [...] can be proved by the present methods also", and we give this Kripke-style proof of Mostowski's theorem by way of two lemmas.

Lemma 3.6 (Suggested in [13], Remark 3). *Suppose that $\{T_i : i \in \omega\}$ is an r.e. sequence of consistent, r.e. theories extending* Q. *Then there is a partial recursive function f with index e, such that for each $i, k \in \omega$, the theory $T_i + R(e, e, k) + \exists!zR(e, e, z)$ is consistent.*

Proof. Let f be a partial recursive function defined by the stipulation that $f(m) = k$ iff there is an $i \in \omega$ such that T_i proves

$$\neg(R(m, m, k) \wedge \exists!zR(m, m, z)). \tag{$*$}$$

If more than one T_i proves sentences of this form, consider only the theory with the least index. Should this theory prove more than one sentence of this form, choose the one whose proof has the least Gödel number.

Let e be an index of f. Then e is as desired as long as no T_i proves a sentence of the form $(*)$ with $m = e$. Suppose, for a contradiction, that for some $i, k \in \omega$, we have

$$T_i \vdash \neg(R(e, e, k) \wedge \exists!zR(e, e, z)).$$

We may assume that i, and the (Gödel number of the) proof of this sentence are the minimal such numbers. Then $f(e) = k$, so

$$Q \vdash R(e, e, k) \wedge \exists!zR(e, e, z).$$

Since $T_i \vdash Q$, it follows that T_i is inconsistent. $\dashv$

Lemma 3.7 (Kripke [13], Corollary 1.1). *Suppose that* T *is a consistent extension of* Q. *If $\gamma(x)$ is Σ_1-flexible over* T, *then $\{\gamma(n) : n \in \omega\}$ is an independent set over* T.

Proof. Suppose that $\gamma(x)$ is Σ_1-flexible over T, and let $A = \{\gamma(n) : n \in \omega\}$. We want to show that for each $B \subseteq A$, the theory

$$T + B + \{\neg\phi : \phi \in A \setminus B\}$$

is consistent. By compactness, it suffices to show this for each finite subset of A, so let B be any finite subset of A, and let $X = \{n : \gamma(n) \in B\}$. Since X is finite, it is recursive, so by Fact 4 there is a k such that $\exists yT(k, x, y)$ binumerates X in Q. Since $\gamma(x)$ is Σ_1-flexible over T, the theory

$$T + \forall x(\gamma(x) \leftrightarrow \exists yT(k, x, y))$$

is consistent.

For each $n \in \omega$ we have

1. if $\gamma(n) \in B$, then $n \in X$, so $Q \vdash \exists y T(k, n, y)$. But then $T + \forall x(\gamma(x) \leftrightarrow \exists y T(k, x, y)) \vdash \gamma(n)$, so $T + \gamma(n)$ is consistent.
2. if $\gamma(n) \in A \setminus B$, then $n \notin X$, so $Q \vdash \neg \exists y T(k, n, y)$, which in a similar manner implies that $T + \neg \gamma(n)$ is consistent.

Hence we have shown that for each finite $B \subseteq A$, the theory $T + B + \{\neg \phi : \phi \in A \setminus B\}$ is consistent, so $\{\gamma(n) : n \in \omega\}$ is an independent set. $\dashv$

Proof of Theorem 3.5. Let $\{T_i : i \in \omega\}$ be an r.e. sequence of consistent, r.e. theories extending Q, and let e be as in the proof of Lemma 3.6. It follows that the theory $T_i + R(e, e, k) + \exists! z R(e, e, z)$ is consistent for each $i, k \in \omega$. Let $\gamma(x) := \exists z(R(e, e, z) \wedge \exists y T(z, x, y))$; then $\gamma(x)$ is Σ_1-flexible over T_i. By Lemma 3.7, $\{\gamma(n) : n \in \omega\}$ is an independent set over T_i, so $\gamma(x)$ is independent over T_i. $\dashv$

3.3 Scott's Lemma

No survey of flexibility and independence is complete without mentioning the following result, which was used by Scott to prove that every countable Scott set can be realised as the standard system of some prime model of PA (i.e., a model $\mathcal{M}$ of PA which is elementarily embeddable into any model that is elementary equivalent to $\mathcal{M}$).

Theorem 3.8 (Scott's Lemma [20]). *Suppose that* T *is a consistent, r.e. extension of* Q, *and that* $\sigma(x)$ *is any* Σ_n *formula. Then there is a* Σ_{n+1} *formula* $\gamma(x)$ *which is simultaneously independent over each consistent* $T_f^\sigma = T + \{\sigma(n)^{f(n)} : n \in \omega\}$, *where* f *is any function from* ω *to* $\{0, 1\}$.

Scott's Lemma and its proof has a slightly different flavour than the others discussed in this paper. Partly, this is due to the fact that we are not constructing a formula from a set of consistent theories, as in Mostowski's theorem, but rather *first* constructing a formula and *then* proving that if T_f^σ is consistent, then the augmented theory is also consistent. Because of this, the diagonal trap used in the previous proofs does not work, since the supposedly flexible function gets defined in the possibly inconsistent theories T_f^σ. Neither can we pick out the consistent T_f^σ's in advance (in an attempt to use Mostowski's theorem as it stands) since the collection of such theories is not r.e. On the other hand, given an *independent* formula $\sigma(x)$ of any complexity (i.e. such that every T_f^σ is assured to be consistent), it is straightforward to adjust Kripke's method to construct a Σ_1 formula $\gamma(x)$ which is simultaneously independent over each T_f^σ.

The reader interested in the proof of Scott's lemma is directed to Scott's original paper [20], or Lindström's [15], Theorem 2.10. Another proof, which makes indispensable use of partial satisfaction predicates is found as Proposition 2 in Lindström's [14]. Hájek [9] also attributes a similar result, with a different proof, to Jensen & Ehrenfeucht [10].

4   The Streamlined Case

In this section, we prove a more general version of Kripke's theorem for extensions of $I\Delta_0 + \exp$. The presence of a slight amount of induction allows for the use of partial truth definitions, which makes the proofs somewhat more transparent, as well as allowing further generalisations. The flexible formulae are defined using a version of Kripke's trick: by using a flexible function f and a suitable partial truth definition, we define our $\gamma(x)$ as a formula expressing "x satisfies the formula whose Gödel number is output by $f(e)$".

Theorem 4.1 (Essentially Kripke). *If* T *is a consistent, r.e. extension of* $I\Delta_0 + \exp$, *then there is a* Σ_n *formula that is flexible for* Σ_n *over* T.[4]

Proof. Let e be as defined in Lemma 3.2, and let $\gamma(x)$ be the Σ_n formula

$$\exists z(R(e, e, z) \wedge \mathrm{Sat}_{\Sigma_n}(z, x)).$$

Pick any Σ_n-formula $\sigma(x)$. By Lemma 3.2, the theory

$$T + R(e, e, \ulcorner \sigma \urcorner) + \exists ! z R(e, e, z)$$

is consistent. Hence we have that $T + R(e, e, \ulcorner \sigma \urcorner) + \exists ! z R(e, e, z)$ proves the sentence

$$\forall x \big(\exists z (R(e, e, z) \wedge \mathrm{Sat}_{\Sigma_n}(z, x)) \leftrightarrow \mathrm{Sat}_{\Sigma_n}(\ulcorner \sigma \urcorner, x) \big).$$

Then the left-hand side of the equivalence is definitionally identical to $\gamma(x)$, and by Fact 2, the right-hand side is equivalent to $\sigma(x)$, which allows us to conclude that

$$T + \forall x(\gamma(x) \leftrightarrow \sigma(x))$$

is consistent.   ⊣

Remark 4.2. The existence of Π_n-flexible Π_n formulae follows immediately from the theorem above. Let $\gamma(x)$ be Σ_n-flexible over T, and let $\pi(x)$ be any Π_n formula. Then $\neg\pi(x) \in \Sigma_n$, so $T + \forall x(\gamma(x) \leftrightarrow \neg\pi(x))$

---

[4]The theorem appears as Corollary 6.2.7 of [23], where it is stated for extensions of a peculiar version of primitive recursive arithmetic (essentially equivalent to $I\Sigma_1$), with a different proof. It can also be obtained as a special case of Theorem 3.6 of [24], Proposition 1 of [14], and Theorem 2.11 of [15]. All three results are stated for extensions of PA, and the proofs are different from ours.

is consistent. Since $T \vdash \forall x(\gamma(x) \leftrightarrow \neg\pi(x)) \leftrightarrow \forall x(\neg\gamma(x) \leftrightarrow \pi(x))$, it follows that $\neg\gamma(x)$ is (a Π_n formula that is) Π_n-flexible over T.

Remark 4.3. Montagna [16] constructs a Δ_{n+1} formula which is flexible for B_n formulae over extensions of PA. It is possible to prove this result for extensions of $I\Delta_0 + \exp$ by using the present method and the corresponding satisfaction predicates.

A strong version of flexibility is provided by the Gödel-Rosser-Mostowski-Myhill-Kripke-Visser theorem, encompassing many of the possible generalisations. The name stems from the fact that Mostowski, Myhill, and Kripke independently proved generalisations of the Gödel-Rosser incompleteness theorem. To the best of our knowledge, Visser was the first to put all these generalisations together, stated as Theorem 3.6 of [24] for extensions of PA.[5]

Theorem 4.4 (The Gödel-Rosser-Mostowski-Myhill-Kripke-Visser theorem). *Suppose that $\{T_i : i \in \omega\}$ is an r.e. sequence of consistent, r.e. theories extending $I\Delta_0 + \exp$, and that $\{U_i : i \in \omega\}$ is an r.e. sequence of sets of sentences such that $T_i \nvdash B$ for all $B \in U_i$. Then there is a Σ_n-formula $\gamma(x)$ such that for all $\sigma(x) \in \Sigma_n$, for all $i \in \omega$, and for all $B \in U_i$, $T_i + \forall x(\gamma(x) \leftrightarrow \sigma(x)) \nvdash B$.*

Proof. Let the T_i's and U_i's be as in the statement of the theorem. Define a partial recursive function f by $f(m) = k$ iff $i \in \omega$ and (the Gödel number of) $B \in U_i$ are the least numbers such that

$$T_i + R(m, m, k) + \exists! z R(m, m, z) \vdash B.$$

Let e be an index of f, and suppose that for some $i, k \in \omega$, and $B \in U_i$, we have

$$T_i + R(e, e, k) + \exists! z R(e, e, z) \vdash B.$$

Then $f(e) = k$, so $I\Delta_0 + \exp \vdash R(e, e, k) \wedge \exists! z R(e, e, z)$. But then $T_i \vdash B$, contradicting our assumptions. Hence for each $i, k \in \omega$, and $B \in U_i$, $T_i + R(e, e, k) \wedge \exists! z R(e, e, z) \nvdash B$.

Let $\sigma(x)$ be any Σ_n formula, and let $\gamma(x)$ be the formula

$$\exists z (R(e, e, z) \wedge \mathrm{Sat}_{\Sigma_n}(z, x)).$$

Suppose, for a contradiction, that for some $i \in \omega$ and $B \in U_i$,

$$T_i + \forall x(\gamma(x) \leftrightarrow \sigma(x)) \vdash B.$$

---

[5]A weaker result appears in Smoryński [21] as the Gödel-Rosser-Mostowski-Myhill-Kripke theorem. Visser [24] uses the same name for his Theorem 3.6. Sommaruga-Rosolemos [23] states, without a proof, a version of Visser's Theorem 3.6, essentially for extensions of $I\Sigma_1$.

By the same argument as in the proof of Theorem 3.1, we have

$$T_i + R(e, e, \ulcorner \sigma \urcorner) + \exists! z R(e, e, z) \vdash \forall x (\gamma(x) \leftrightarrow \sigma(x))$$

so

$$T_i + R(e, e, \ulcorner \sigma \urcorner) \wedge \exists! z R(e, e, z) \vdash B,$$

which is a contradiction. Hence, for each $i \in \omega$, and for each $B \in U_i$,

$$T_i + \forall x (\gamma(x) \leftrightarrow \sigma(x)) \nvdash B. \qquad \dashv$$

Remark 4.5. By using the methods of Section 3, it is possible to prove a version of the result above with $I\Delta_0 + \exp$ replaced by Q and $\sigma(x)$ restricted to formulae on Kleene normal form. It is also possible to obtain a Δ_{n+1} formula which is flexible in the above sense for B_n, by using the corresponding satisfaction predicates.

For the final variation of this section, recall that a set X is mono-consistent with T if $T + \phi$ is consistent for each $\phi \in X$. A number of generalisations of the kind below can be found in [15], with the original results following from the special case $X = \text{Th}(T)$. This theorem is curious in that it does not seem possible to prove using the methods we have used so far.

Theorem 4.6 (Lindström [14], Proposition 1). *Suppose that* T *is a consistent, r.e. extension of* $I\Delta_0 + \exp$, *and that* X *is an r.e. set that is monoconsistent with* T. *Then there is a formula* $\gamma(x) \in \Sigma_n$ *such that for each* $\sigma(x) \in \Sigma_n$, $\neg \forall x (\gamma(x) \leftrightarrow \sigma(x)) \notin X$.

If we attempt to prove Theorem 4.6 using the strategy that we have employed earlier, we can straightforwardly construct a function f such that "$f(e) \neq k$" $\notin X$ for each k. For this to imply $\neg \forall x (\gamma(x) \leftrightarrow \sigma(x)) \notin X$, however, we have to assume that X is a deductively closed set containing $I\Delta_0 + \exp$, in which case the conclusion is no stronger than that of Theorem 4.1. Hence, we give Lindström's original proof, which instead uses the diagonalisation lemma to construct the formula $\gamma(x)$.

Proof of Theorem 4.6. Let $P(k, m, n)$ be a primitive recursive relation such that

$$\sigma(x) \in \Sigma_n \ \& \ \neg \forall x (\eta(x) \leftrightarrow \sigma(x)) \in X \text{ iff } \exists n P(\eta, \sigma, n).$$

Let $\psi(x, y, z)$ be a Δ_1^T binumeration of $P(k, m, n)$, and let

$$\psi^*(x, y, z) = \psi(x, y, z) \wedge \forall y' z' (\langle y', z' \rangle < \langle y, z \rangle \rightarrow \neg \psi(x, y', z')).$$

Then $\psi^*(x, y, z)$ is also Δ_1^T. By the diagonalisation lemma, let $\gamma(x)$ be a Σ_n formula such that

$$T \vdash \forall x (\gamma(x) \leftrightarrow \exists y \exists z (\psi^*(\ulcorner \gamma \urcorner, y, z) \wedge \text{Sat}_{\Sigma_n}(x, y))).$$

Suppose, for a contradiction, that for some $\sigma(x) \in \Sigma_n$, $\neg\forall x(\gamma(x) \leftrightarrow \sigma(x)) \in X$. For each such formula there is a n such that $P(\gamma, \sigma, n)$. Pick $\sigma(x)$ and n such that $\langle \ulcorner \sigma \urcorner, n \rangle$ is minimal. Then

$$\text{T} \vdash \forall y \forall z (\psi^*(\ulcorner \gamma \urcorner, y, z) \leftrightarrow y = \ulcorner \sigma \urcorner \wedge z = n),$$

so $\text{T} \vdash \forall x(\gamma(x) \leftrightarrow \sigma(x))$, which contradicts the fact that X is mono-consistent with T. ⊣

5 End-Extensions

Recall that Theorem 3.1 gives rise to consistent theories of the form $\text{T} + \forall x(\gamma(x) \leftrightarrow \sigma(x))$. By the completeness theorem, this means that there is some model $\mathcal{M} \models \text{T} + \forall x(\gamma(x) \leftrightarrow \sigma(x))$, and trivially, each such model is an end-extension of the standard model of arithmetic $\mathbb{N}$. The question now arises: For which models $\mathcal{M} \models \text{T}$ and formulae $\sigma(x)$ can we find an end-extension $\mathcal{K}$ of $\mathcal{M}$ such that $\mathcal{K} \models \text{T} + \forall x(\gamma(x) \leftrightarrow \sigma(x))$?

By formalising Lemma 3.2, we are able to prove that for each Σ_n, there is a $\gamma(x) \in \Sigma_n$ such that for each $\sigma(x) \in \Sigma_n$, every model of $\text{I}\Sigma_1 + \text{Con}_\text{T}$ can be end-extended to a model of $\text{T} + \forall x(\gamma(x) \leftrightarrow \sigma(x))$. From this point on, we assume that the formula $R(x, y, z)$ is defined in terms of partial satisfaction predicates and the selection theorem. We summarise the formalisation procedure in the following lemma:

Lemma 5.1 (Formalisation of Kripke's lemma). *If* T *is a consistent, r.e. extension of* Q, *then there is an index e such that*

1. $\text{I}\Delta_0 + \exp \vdash \forall z(\text{Con}_\text{T} \rightarrow \neg R(e, e, z))$, *and*
2. $\text{I}\Delta_0 + \exp \vdash \forall z(\text{Con}_\text{T} \rightarrow \text{Con}_{\text{T}+R(e,e,z)})$.[6]

Proof. Let $\phi(x, z)$ be the formula $\text{Pr}_\text{T}(\ulcorner \neg R(\dot{x}, \dot{x}, \dot{z}) \urcorner)$, and let $e = \ulcorner \phi \urcorner$. We show that e has the desired properties. By definition of R, we have

$$\text{I}\Delta_0 + \exp \vdash \forall z(R(e, e, z) \leftrightarrow \text{Sel}\{\text{Sat}_{\Sigma_1}\}(e, e, z)), \tag{1}$$

which by Fact 3 gives

$$\text{I}\Delta_0 + \exp \vdash \forall z(R(e, e, z) \rightarrow \text{Sat}_{\Sigma_1}(e, e, z)). \tag{2}$$

Fact 2 gives

$$\text{I}\Delta_0 + \exp \vdash \forall z(R(e, e, z) \rightarrow \phi(e, z)), \tag{3}$$

so by construction of ϕ,

$$\text{I}\Delta_0 + \exp \vdash \forall z(R(e, e, z) \rightarrow \text{Pr}_\text{T}(\ulcorner \neg R(e, e, \dot{z}) \urcorner)). \tag{4}$$

---

[6]This Lemma and the Theorem below were announced in [2], and although the conclusions are correct, the proof of the Lemma contained a small mistake. The full proof is presented here, both as an example of methodology, and as a correction of that erroneous proof.

Fact 1 gives

$$\text{I}\Delta_0 + \exp \vdash \forall z(R(e,e,z) \to \Pr_T(\ulcorner R(e,e,\dot{z})\urcorner)), \qquad (5)$$

so (4) and (5) together with the derivability conditions give

$$\text{I}\Delta_0 + \exp \vdash \forall z(\text{Con}_T \to \neg R(e,e,z)). \qquad (6)$$

This concludes the proof of the first claim. For the second part, note that by Fact 5,

$$\text{I}\Delta_0 + \exp \vdash \exists z \neg \text{Con}_{T+R(e,e,z)} \leftrightarrow \exists z \Pr_T(\ulcorner \neg R(e,e,\dot{z})\urcorner). \qquad (7)$$

By construction of ϕ, this implies

$$\text{I}\Delta_0 + \exp \vdash \exists z \neg \text{Con}_{T+R(e,e,z)} \leftrightarrow \exists z \phi(e,z), \qquad (8)$$

so Fact 2 gives

$$\text{I}\Delta_0 + \exp \vdash \exists z \neg \text{Con}_{T+R(e,e,z)} \leftrightarrow \exists z \text{Sat}_{\Sigma_1}(e,e,z). \qquad (9)$$

Then by Fact 3,

$$\text{I}\Delta_0 + \exp \vdash \exists z \neg \text{Con}_{T+R(e,e,z)} \to \exists z \text{Sel}\{\text{Sat}_{\Sigma_1}\}(e,e,z), \qquad (10)$$

so (10) together with the definition of R give

$$\text{I}\Delta_0 + \exp \vdash \exists z \neg \text{Con}_{T+R(e,e,z)} \to \exists z R(e,e,z). \qquad (11)$$

By (6) and (11), we can conclude

$$\text{I}\Delta_0 + \exp \vdash \exists z \neg \text{Con}_{T+R(e,e,z)} \to \neg \text{Con}_T. \qquad (12)$$

By contrapositioning, and moving the universal quantifier to the left, we obtain

$$\text{I}\Delta_0 + \exp \vdash \forall z(\text{Con}_T \to \text{Con}_{T+R(e,e,z)}). \qquad \dashv$$

Theorem 5.2 (Formalisation of Kripke's theorem). *If* T *is a consistent, r.e. extension of* Q, *then there is a* Σ_n-*formula* $\gamma(x)$ *such that*

1. *if* $\mathcal{M} \models \text{I}\Sigma_1 + \text{Con}_T$, *then* $\mathcal{M} \models \forall x \neg \gamma(x)$, *and*
2. *if* $\sigma(x) \in \Sigma_n$, *then every model of* $\text{I}\Sigma_1 + \text{Con}_T$ *can be end-extended to a model of* $T + \forall x(\gamma(x) \leftrightarrow \sigma(x))$.

Proof. Let T be a consistent, r.e. extension of Q, and let $\mathcal{M} \models \text{I}\Sigma_1 + \text{Con}_T$. Let e be as in Lemma 5.1, let $\gamma(x) := \exists z(R(e,e,z) \wedge \text{Sat}_{\Sigma_n}(z,x))$, and let $\sigma(x)$ be any Σ_n-formula.

By part 1 of Lemma 5.1, $\mathcal{M} \models \forall z \neg R(e,e,z)$, so it follows that $\mathcal{M} \models \forall x \neg \gamma(x)$). By part 2 we have that, for each k, $\mathcal{M} \models \text{Con}_{T+R(e,e,k)}$, and since $\mathcal{M} \models \text{I}\Sigma_1$ it is possible to construct an end-extension $\mathcal{K}$ of $\mathcal{M}$ with $\mathcal{K} \models T + R(e,e,\ulcorner\sigma\urcorner)$.[7] By the same reasoning as in the proof of Theorem 4.1, it follows that $\mathcal{K} \models T + \forall x(\gamma(x) \leftrightarrow \sigma(x))$. $\qquad \dashv$

---

[7]The details of how this end-extension is constructed can be found in Theorem 2.6 of [1].

Remark 5.3. Note that the end-extension obtained above necessarily is such that $\mathcal{K} \models \neg\mathrm{Con}_T$. Since in $\mathcal{K}$, the function $f(e)$ is defined, we have added a non-standard object that codes a proof of $\neg R(e, e, k)$ in $\mathcal{K}$. By the Lemma, we have $\mathcal{K} \models R(e, e, k) \wedge \mathrm{Pr}_T(\ulcorner \neg R(e, e, k) \urcorner)$, and since $R(e, e, k)$ is a Σ_1-formula, we obtain

$$\mathcal{K} \models \mathrm{Pr}_T(\ulcorner R(e, e, k) \urcorner) \wedge \mathrm{Pr}_T(\ulcorner \neg R(e, e, k) \urcorner).$$

Remark 5.4. Recall the model-theoretic characterisation of Π_1-conservativity: If $T \vdash PA$, then $T + \phi$ is Π_1-conservative over T iff every model of T can be end-extended to a model of $T + \phi$ (Guaspari [6], Theorem 6.5(i)). It is easy to extend the notion of Π_1-conservativity to: $T + \theta$ is Π_1-conservative over $T + \phi$ if, for each $\pi \in \Pi_1$, $T + \phi \vdash \pi$ implies $T + \theta \vdash \pi$. Guaspari's characterisation result can be similarily expanded, see [1] for details, so for theories extending PA, we can rephrase Theorem 5.2 as follows:

Theorem 5.5. *Suppose that* T *and* S *are consistent, r.e. extensions of* PA *and* Q*, respectively. Then there is a* Σ_n *formula* $\gamma(x)$ *such that for each* $\sigma(x) \in \Sigma_n$*, the theory* $T + \forall x(\gamma(x) \leftrightarrow \sigma(x))$ *is* Π_1*-conservative over* $T + \mathrm{Con}_S$.

The final result of this section is a version of the Gödel-Rosser-Mostowski-Myhill-Kripke-Visser theorem, formalised in the style above.

Theorem 5.6 (Formalisation of the GRMMKV theorem). *Suppose that* $\{T_i : i \in \omega\}$ *is an r.e. sequence of consistent, r.e. theories extending* Q*, and that* $\{U_i : i \in \omega\}$ *is an r.e. sequence of sets of sentences such that* $T_i \nvdash B$ *for all* $B \in U_i$*. Then there is a* Σ_n*-formula* $\gamma(x, u, y)$ *such that for all* $\sigma(x) \in \Sigma_n$*, for all* $i \in \omega$*, and for all* $B \in U_i$,

1. *If* $\mathcal{M} \models I\Sigma_1 + \mathrm{Con}_{T_i + \neg B}$*, then* $\mathcal{M} \models \forall x \neg\gamma(x, i, B))$*, and*
2. *every model of* $I\Sigma_1 + \mathrm{Con}_{T_i + \neg B}$ *can be end-extended to a model of* $T_i + \forall x(\gamma(x, i, B) \leftrightarrow \sigma(x)) + \neg B$.

The method is similar to the one used in proving Lemma 5.1.

Lemma 5.7. *If* $\{T_i : i \in \omega\}$ *is an r.e. sequence of consistent, r.e. theories extending* Q*, and* $\{U_i : i \in \omega\}$ *is an r.e. sequence of sets of sentences such that* $T_i \nvdash B$ *for each* $B \in U_i$*, then there is an index* e *such that for all* $i \in \omega$ *and* $B \in U_i$,

1. $I\Delta_0 + \exp \vdash \forall z(\mathrm{Con}_{T_i + \neg B} \to \neg R(e, e, i, B, z))$*, and*
2. $I\Delta_0 + \exp \vdash \forall z(\mathrm{Con}_{T_i + \neg B} \to \mathrm{Con}_{T_i + R(e,e,i,B,z) + \neg B})$.

Proof. Let the T_i's and U_i's be as in the statement of the lemma. Let $\phi(x, u, y, z)$ be the formula $\mathrm{Pr}_{T_u}(\ulcorner R(\dot{x}, \dot{x}, \dot{u}, \dot{y}, \dot{z}) \to \dot{y} \urcorner)$, and let

$e = \ulcorner \phi \urcorner$. Pick an arbitrary $i \in \omega$, and an arbitrary $B \in U_i$. By definition of R, we have

$$\mathrm{I}\Delta_0 + \exp \vdash \forall z \big(R(e, e, i, B, z) \to \phi(e, i, B, z) \big) \qquad (13)$$

so by construction of ϕ,

$$\mathrm{I}\Delta_0 + \exp \vdash \forall z \big(R(e, e, i, B, z) \to \mathrm{Pr}_{\mathrm{T}_i}(\ulcorner R(e, e, i, B, \dot{z}) \to B \urcorner) \big). \qquad (14)$$

By Fact 1,

$$\mathrm{I}\Delta_0 + \exp \vdash \forall z \big(R(e, e, i, B, z) \to \mathrm{Pr}_{\mathrm{T}_i}(\ulcorner R(e, e, i, B, \dot{z}) \urcorner) \big), \qquad (15)$$

which together with (14) and the derivability conditions give

$$\mathrm{I}\Delta_0 + \exp \vdash \forall z \big(R(e, e, i, B, z) \to \mathrm{Pr}_{\mathrm{T}_i}(\ulcorner B \urcorner) \big), \qquad (16)$$

so by Fact 5,

$$\mathrm{I}\Delta_0 + \exp \vdash \forall z \big(\mathrm{Con}_{\mathrm{T}_i + \neg B} \to \neg R(e, e, i, B, z) \big). \qquad (17)$$

For the latter part, recall that by definition of ϕ,

$$\mathrm{I}\Delta_0 + \exp \vdash \exists z \mathrm{Pr}_{\mathrm{T}_i}(\ulcorner R(e, e, i, B, \dot{z}) \to B \urcorner) \to \exists z \phi(e, \overset{i}{i}, B, z), \qquad (18)$$

so by Fact 2

$$\mathrm{I}\Delta_0 + \exp \vdash \exists z \mathrm{Pr}_{\mathrm{T}_i}(\ulcorner R(e, e, i, B, \dot{z}) \to B \urcorner) \to \exists z \mathrm{Sat}_{\Sigma_1}(e, e, i, B, z), \qquad (19)$$

and by Fact 3.3

$$\mathrm{I}\Delta_0 + \exp \vdash \exists z \mathrm{Pr}_{\mathrm{T}_i}(\ulcorner R(e, e, i, B, \dot{z}) \to B \urcorner) \to \exists z R(e, e, i, B, z). \qquad (20)$$

But then by (16) and (20),

$$\mathrm{I}\Delta_0 + \exp \vdash \exists z \mathrm{Pr}_{\mathrm{T}_i}(\ulcorner R(e, e, i, B, \dot{z}) \to B \urcorner) \to \mathrm{Pr}_{\mathrm{T}_i}(\ulcorner B \urcorner), \qquad (21)$$

so by Fact 5,

$$\mathrm{I}\Delta_0 + \exp \vdash \forall z \big(\mathrm{Con}_{\mathrm{T}_i + \neg B} \to \mathrm{Con}_{\mathrm{T}_i + R(e, e, i, B, z) + \neg B} \big). \qquad \dashv$$

Proof of Theorem 5.6. Let the T_i's and U_i's be as in the statement of the theorem. Pick any $i \in \omega$ and $B \in U_i$, and let $\mathcal{M} \models \mathrm{I}\Sigma_1 + \mathrm{Con}_{\mathrm{T}_i + \neg B}$. Let e be as in Lemma 5.7, let $\sigma(x)$ be any Σ_n formula, and let $\gamma(x, u, y)$ be $\exists z \big(R(e, e, u, y, z) \wedge \mathrm{Sat}_{\Sigma_n}(z, x) \big)$.

By part 1 of Lemma 5.7, we have that $\mathcal{M} \models \forall x \neg \gamma(x, i, B)$. By part 2, it follows that $\mathcal{M} \models \mathrm{Con}_{\mathrm{T}_i + R(e, e, i, B, \ulcorner \sigma \urcorner) + \neg B}$, and since $\mathcal{M} \models \mathrm{I}\Sigma_1$, there is an end-extension $\mathcal{K}$ of $\mathcal{M}$ such that $\mathcal{K} \models \mathrm{T}_i + R(e, e, i, B, \ulcorner \sigma \urcorner) + \neg B$. By a reasoning similar to the the one in the proof of Theorem 4.1, we conclude that the model is as desired. $\dashv$

Remark 5.8. Using a version of the arithmetised completeness theorem due to Enayat and Wong [3], the same proof works for countable models of $\mathrm{I}\Delta_0 + \exp + \mathrm{B}\Sigma_1$.

6 Recent Developments and Open Problems

The history of the following theorem starts with [25], in which Woodin proves the result for completions T of PA, and for countable models $\mathcal{M}$ of T. Later, in unpublished work, Enayat and Shavrukov together showed that the countability assumption of $\mathcal{M}$ can be removed. The following refinement of Woodin's result is established in [1].

Theorem 6.1. *Suppose that* T *is a consistent, r.e. extension of* $I\Sigma_1$, *and that* $\mathcal{M}$ *is a countable model of* T. *Then there is a formula* $\gamma(x)$ *such that for each* $\sigma(x) \in \Sigma_n$, *if the* $\mathcal{M}$-*extension of* $\sigma(x)$ *is* $\mathcal{M}$-*finite, and* $\mathcal{M} \models \forall x(\gamma(x) \rightarrow \sigma(x))$, *then there is an end-extension* $\mathcal{K}$ *of* $\mathcal{M}$ *such that* $\mathcal{K} \models T + \forall x(\gamma(x) \leftrightarrow \sigma(x))$.

The proof is more involved than the ones seen earlier in the paper, and is similar to Japaridze's proof of the "hard" direction of the Hájek-Montagna completeness theorem for the modal logic of Π_1-conservativity.[8] This, in combination with the strong form of flexibility provided by the formalisation of the GRMMKV theorem, suggests that the real problem lies in handling models of $T + \neg\mathrm{Con}_T$.

By Remark 5.3, we see that in the transition from the original model $\mathcal{M}$ to the end-extension $\mathcal{K}$ in Theorems 5.2 and 5.6, one consistency statement is lost in the process, since $\mathcal{M} \models \mathrm{Con}_T$ and $\mathcal{K} \models \neg\mathrm{Con}_T$. Since the proofs are obtained by using formalisations of partial recursive functions, which are by nature (equivalent to) Σ_1 formulae, Σ_1-persistence implies that we can not change the value of our flexible function once it has been defined. This is due to the fact that some minimalisation ("pick the least proof") is required to make the definition of the flexible function admissible. Combining the above theorem with the formalisation of the GRMMKV theorem, the following question arises:

Is there a formula $\gamma(x) \in \Sigma_n$ such that for each $\sigma(x) \in \Sigma_n$, every countable model of $T + \forall x(\gamma(x) \rightarrow \sigma(x))$ can be end-extended to a model of $T + \forall x(\gamma(x) \leftrightarrow \sigma(x))$?

Acknowledgements

I am indebted to Ali Enayat and Taishi Kurahashi for carefully reading an earlier version of this paper and suggesting numerous improvements, and also to the anonymous referee who has further helped me improve the paper. During the preparation of this paper, I was partially supported by grants from *Kungliga och Hvitfeldtska stiftelsen* and *Stiftelsen Paul och Marie Berghaus donationsfond*.

---

[8]This observation is due to Volodya Shavrukov.

References

[1] Rasmus Blanck and Ali Enayat, Marginalia on a theorem of Woodin, to appear in *The Journal of Symbolic Logic*.

[2] Rasmus Blanck, Two consequences of Kripke's lemma, *Idées fixes*, Martin Kaså (ed.), Philosophical Communications, Web Series, No. 61, University of Gothenburg, 2014, pp. 45–53.

[3] Ali Enayat and Tin Lok Wong, Model theory of WKL_0^*, to appear in *The Annals of Pure and Applied Logic*.

[4] S. Feferman, G. Kreisel, and S. Orey, 1-consistency and faithful interpretations, *Archive for Mathematical Logic*, vol. 6, 1960, pp. 52–63.

[5] Kurt Gödel, Über formal unentscheidbare sätze der Principia Mathematica und verwandter systeme, I, *Monatshefte für Mathematik*, vol. 38, 1931, pp. 173–198.

[6] David Guaspari, Partially conservative extensions of arithmetic, *Transactions of the American Mathematical Society*, vol. 254, 1979, pp. 47–68.

[7] Frank Harary, A very independent axiom system, *The American Mathematical Monthly*, vol. 68 (2), 1961, pp. 159–162.

[8] Petr Hájek and Pavel Pudlák, *Metamathematics of First Order Arithmetic*, Perspectives in Mathematical Logic, Springer-Verlag, 1993.

[9] Petr Hájek, Completion closed algebras and models of Peano arithmetic, *Commentationes Mathematicae Universitatis Carolinae*, vol. 22 (3), 1981, pp. 585–594.

[10] D. Jensen and A. Ehrenfeucht, Some problem in elementary arithmetics, *Fundamenta Mathematicae*, vol. 92 (3), 1976, pp. 223–245.

[11] Richard Kaye, *Models of Peano Arithmetic*, Oxford University Press, 1991.

[12] Stephen Cole Kleene, *Introduction to Metamathematics*, North-Holland, 1952.

[13] Saul A. Kripke, "Flexible" predicates of formal number theory, *Proceedings of the American Mathematical Society*, vol. 13 (4), 1962, pp. 647–650.

[14] Per Lindström, A note on independent formulas, *Notes on formulas with prescribed properties in arithmetical theories*, Philosophical Communications, Red Series, No. 5, Göteborgs Universitet, 1984, pp. 1–5.

[15] Per Lindström, *Aspects of Incompleteness*, Lecture Notes in Logic, No. 10, 2nd edition, A. K. Peters, 2003.

[16] Franco Montagna, Relatively precomplete numerations and arithmetic, *Journal of Philosophical Logic*, vol. 11 (4), 1982, pp. 419–430.

[17] A. Mostowski, A generalization of the incompleteness theorem, *Fundamenta Mathematicae*, vol. 49 (2), 1961, pp. 205–232.

[18] John Myhill, An absolutely independent set of Σ_1^0-sentences, *Zeitschrift für mathematische Logik und Grundlagen der Mathematik*, vol. 18 (3), 1972, pp. 107–109.

[19] J. Barkley Rosser, Extensions of some theorems of Gödel and Church, *The Journal of Symbolic Logic*, vol. 1 (3), 1936, pp. 87–91.

[20] Dana Scott, Algebras of sets binumerable in complete extensions of arithmetic, *Recursive Function Theory*, J. C. E. Dekker (ed.), Proceedings of Symposia in Pure Mathematics, vol. 5, American Mathematical Society, 1962, pp. 171–121.

[21] C. A. Smoryński, Applications of Kripke models, *Metamathematical Investigation of Intuitionistic Arithmetic and Analysis*, A. S. Troelstra (ed.), Lecture Notes in Mathematics, No. 344, Springer-Verlag, 1973, pp. 324–391.

[22] C. Smoryński, *Self-Reference and Modal Logic*, Springer-Verlag, 1985.

[23] Giovanni Sommaruga-Rosolemos, *Fixed Point Constructions in Various Theories of Mathematical Logic*, Bibliopolis, 1991.

[24] Albert Visser, Numerations, λ-calculus & arithmetic, *To H. B. Curry: Essays on Combinatory Logic, Lambda Calculus and Formalism*, J. P. Seldin and J. R. Hindly (eds.), Academic Press, 1980, pp. 259–284.

[25] W. Hugh Woodin, A potential subtlety concerning the distinction between determininsm and nondeterminism, *Infinity, New Research Frontiers*, Heller and Woodin (eds.), Cambridge University Press, 2011, pp. 119–129.

2

On the Diophantine Complexity
of the Set of Prime Numbers

Pietro Corvaja, Eugenio Omodeo,
Luca Vallata

Abstract: The '*positive aspects of a negative solution*' (the recursive unsolvability of Hilbert's 10th problem) include the discovery of Diophantine representations of the set $\mathcal{P}$ of primes. What is the *rank* of $\mathcal{P}$, namely the smallest possible number, r, of unknowns in a polynomial representing $\mathcal{P}$?

Siegel's theorem on integral points on curves (1929) hands us a revealing characterisation of the Diophantine subsets of $\mathbf{Z}$ which can be represented in terms of a single unknown; thereby, since 19th century results about the density of $\mathcal{P}$ entail that $\mathcal{P}$ does not meet that characterisation, we get the lower rank bound $r \geqslant 2$.

We also show that the Diophantine set consisting of those integers $\kappa > 3$ which satisfy the congruence $\binom{2\,\kappa}{\kappa} \equiv 2 \bmod \kappa^3$ has rank not exceeding 7. As a consequence, the least known upper rank bound for $\mathcal{P}$, namely $r \leqslant 9$ as found by Yu. V. Matiyasevich in 1977, can be lowered to $r \leqslant 7$ if the converse of Wolstenholme's theorem (1862) holds, as conjectured by J. P. Jones.

Key words. Diophantine representations, DPRM theorem, Hilbert's 10th problem, Siegel's theorem on integral points, Wolstenholme's theorem.

1 Introduction

The combination of the DPR-theorem of 1961, [4], with Matiyasevich's decisive result of 1970, [13], yields that every recursively enumerable

Studies in Weak Arithmetics, Volume 3.
Patrick Cégielski, Ali Enayat,
Roman Kossak.

set is Diophantine. As a direct consequence, we have that the set $\mathcal{P}$ of all prime numbers is Diophantine; and, in fact, the history of explicit Diophantine polynomial representations of the primes begins in 1971.

In consideration of the central role primes play in arithmetic, it is natural to wonder what is the *rank* of $\mathcal{P}$—i.e., the least number of unknowns that must appear in a polynomial defining $\mathcal{P}$. Matiyasevich's important result about a universal Diophantine equation, announced in 1975 and published in 1982 by Jones [9], tells us that for every Diophantine set $\mathcal{S}$ the upper rank bound rank $\mathcal{S} \leq 9$ holds; hence, in particular, that bound applies to the set $\mathcal{P}$.

As observed in [7], p. 449, at least two variables must occur in a prime-representing polynomial: otherwise stated, rank $\mathcal{P} \geq 1$. Any improvement, no matter how slight, of this lower estimate of rank $\mathcal{P}$ must rely on comparatively advanced results in number theory: we will resort to the celebrated Siegel's Theorem on integral points on curves [25] to increase this estimate by 1.

A primary aim of this paper is, in fact, that of offering a useful characterisation of the Diophantine sets of rank 1: this will have an independent mathematical significance, and will be obtained via Siegel's theorem just mentioned. After classifying certain sequences of integers as being of exponential type and certain others as being of polynomial type, we will prove that the rank-1 Diophantine sets of integers are, precisely, the finite unions of sequences of exponential or polynomial type. The set of primes cannot result from such a union, else—as we will explain in due course—it would turn out to have either a logarithmic or a polynomial density, which would contradict long-established facts about the density of primes. To avoid density considerations and reach the same conclusion, one can argue that any non-constant sequence of polynomial or exponential type takes infinitely many non-prime values.

Another aim of this paper is to show that the least known upper rank bound for $\mathcal{P}$, which is 9 as recalled above, can be lowered to rank $\mathcal{P} \leq 7$ if the converse of Wolstenholme's theorem [30] holds, as conjectured by J. P. Jones.

In Sec. 2 we outline some prerequisite notions: we recall first the definitions of *Diophantine set* and of *rank* of a Diophantine set, and then some major propositions about Pell's and Pell-type equations; we also state, without proof, Siegel's Theorem on integral points on curves, which subsequently will play an essential role in the proof of a key theorem. Our key theorem, stating that the Diophantine sets of integers of rank 1 are the finite unions of sequences of exponential or polynomial type, is stated in Sec. 3 and proved in Sec. 5; as a corollary of that characterisation, we draw important information (stated in Sec. 3 and proved in Sec. 6) about the density of rank-one Diophantine sets, whence we get, in Sec. 4:

Theorem. rank $\mathcal{P} \geq 2$.

Sec. 7 provides a quick historical survey on prime-representing polynomials. Then, in Sec. 8, we prove:

Theorem. rank $\mathcal{P} \leq 7$ holds if, as conjectured by J. P. Jones, the primes are precisely the numbers $2, 3$, and those integers $\kappa > 3$ which satisfy the congruence

$$\binom{2\,\kappa}{\kappa} \equiv 2 \mod \kappa^3 .$$

2 Preliminaries

2.1 Diophantine Sets and Their Ranks

Before entering into a deeper technical discussion, we must define the notion of *rank* of a Diophantine set. Recall first:

Definition 1. A relation $\mathcal{D}$ among n positive integers is said to be DIOPHANTINE if a parametric Diophantine equation

$$D(\overbrace{\underbrace{a_1, \ldots, a_n}_{\text{parameters}}, \underbrace{x_1, \ldots, x_m}_{\text{unknowns}}}^{\text{variables}}) = 0 \tag{1}$$

exists such that the n-tuples $(a_1, \ldots, a_n)$ of positive integers belonging to $\mathcal{D}$ are precisely the ones to which there correspond m-tuples $(x_1, \ldots, x_m)$ over $\mathbf{N} \setminus \{0\}$ satisfying (1). One says that D is a defining polynomial for $\mathcal{D}$. (Note: Unlike values for the variables, the coefficients of D may be arbitrary elements of $\mathbf{Z}$.)

Such a $\mathcal{D}$, as seen from a geometrical viewpoint, is the projection on the first n coordinates of the set of integral points of a variety in affine space of dimension $n + m$.

Occasionally we must refer the notion of Diophantine relation to subsets of $\mathbf{Z}^n$, leaving the analogue of the definition just seen as understood; Def. 2, to be given next, should also be altered accordingly when we focus on signed integers.

After Matiyasevich ([17], Chap. 8), for $n = 1$, we adopt the following terminology:

Definition 2. The RANK of a Diophantine subset $\mathcal{D}$ of $\mathbf{N} \setminus \{0\}$ is the number

$$\min \left\{ m \mid D(a, x_1, \ldots, x_m) \text{ is a polynomial defining } \mathcal{D} \right\},$$

henceforth denoted as rank $\mathcal{D}$.

Notice that the rank-0 Diophantine sets are simply the finite sets; therefore, the rank of the set $\mathcal{P}$ of primes must exceed 0.

2.2 Pell's Equation and Conics

Let $\mathbb{K}$ be a real quadratic field. The multiplicative group of units of norm 1 is generated by the fundamental unit ϵ along with -1. Letting α_1, α_2 be a $\mathbf{Z}$-module basis of the ring of integers of $\mathbb{K}$, consider the quadratic form

$$q(x, y) := N_{\mathbb{K}/\mathbf{Q}}(\alpha_1 x + \alpha_2 y) = (\alpha_1 x + \alpha_2 y) \cdot (\alpha_1' x + \alpha_2' y),$$

where the symbol $'$ denotes Galois conjugation. Note that if $\mathbb{K} = \mathbf{Q}(\sqrt{\delta})$ and $(\alpha_1, \alpha_2) = (1, \sqrt{\delta})$, then $q(x, y) = x^2 - \delta y^2$ and the corresponding equation $q(x, y) = 1$, i.e.

$$x^2 - \delta y^2 = 1, \tag{2}$$

is the classical Pell equation.

Let $G = \mathrm{SO}(q)$ be the orthogonal group of the quadratic form q: it can be viewed as the group of 2×2 matrices T of determinant 1 such that the multiplication by T preserves the quadratic form q. Its sub-group $G(\mathbf{Z})$ of the integral-valued matrices of G is generated by $\pm I$ and a single matrix of infinite order with eigenvalues ϵ, ϵ^{-1}. In the case of the quadratic form $x^2 - \delta y^2$ (arising in the Pell equation), the group G consists of matrices of the form

$$\begin{pmatrix} a & \delta b \\ b & a \end{pmatrix},$$

where $(a, b) \in \mathbf{Z}^2$ are the solutions to Pell's equation $q(a, b) = 1$.

Letting (a_0, b_0) be the minimal solution to this Pell equation, it is known (see, e.g., [12], pp. 216–217) that

$$G(\mathbf{Z}) = \left\{ \pm \begin{pmatrix} a_0 & \delta b_0 \\ b_0 & a_0 \end{pmatrix}^n : n \in \mathbf{Z} \right\},$$

so every solution (x, y) in positive integers of the Pell equation (2) is of the form

$$\begin{pmatrix} x \\ y \end{pmatrix} = \begin{pmatrix} a_0 & \delta b_0 \\ b_0 & a_0 \end{pmatrix}^n \begin{pmatrix} 1 \\ 0 \end{pmatrix}.$$

In particular, Pell equations of the special form

$$x^2 - (a^2 - 1) y^2 = 1, \tag{3}$$

which occur extensively below, are solved over $\mathbf{N}$ by the pairs $(\chi_a(n), \psi_a(n))$ such that

$$\begin{pmatrix} \chi_a(n) \\ \psi_a(n) \end{pmatrix} = \begin{pmatrix} a & a^2-1 \\ 1 & a \end{pmatrix}^n \begin{pmatrix} 1 \\ 0 \end{pmatrix},$$

with $n = 0, 1, 2, \ldots$

This description of the sequence of positive solutions to the Pell's equation of special type (3) immediately leads us to the following congruence properties satisfied by such sequences:

Lemma 1. *If $a \equiv b \bmod m$ then, for every n,*

$$\psi_a(n) \equiv \psi_b(n) \pmod m \qquad and \qquad \chi_a(n) \equiv \chi_b(n) \pmod m.$$

In particular, since $\psi_1(n) = n$, if $m \mid (a-1)$ then $\psi_a(n) \equiv n \bmod m$; thus:

$$\psi_a(n) \equiv n \pmod{(a-1)}.$$

Proof. It suffices to notice that if $a \equiv b \bmod m$ then the matrix $\begin{pmatrix} a & a^2-1 \\ 1 & a \end{pmatrix}$ is congruent to the matrix $\begin{pmatrix} b & b^2-1 \\ 1 & b \end{pmatrix}$ modulo m. ⊣

The following two propositions recall useful (elementary) inequalities of the sequences $\chi_a(n)$ and $\psi_a(n)$; the proofs of these and related popular results can be found, for instance: in [5], in [15] Part I, Sec. 2, and in [29].

Lemma 2. *The inequalities $n \leq \psi_a(n) < \psi_a(n+1)$ hold for each n.*

Lemma 3. *If $a \geq 2$, then $(2a-1)^n \leq \psi_a(n+1) \leq (2a)^n$.*

For the general equation $q(x, y) = c$, we will use the following proposition, proved e.g. in [1] Chap. 9, Lemma 6.1:

Lemma 4. *Given a non-degenerate quadratic form $q(x, y) \in \mathbf{Q}[x, y]$ and a non zero integer c, the equation $q(x, y) = c$ can have infinitely many solutions in integers x, y only if the splitting field of $q(x, y)$ is a real quadratic field. In that case, letting $G = \mathrm{SO}(q)$ be the corresponding orthogonal group, the set of integral solutions to the equation $q(x, y) = c$ is formed by finitely many orbits for the group $G(\mathbf{Z})$.*

2.3 Siegel's Theorem on Integral Points on Curves

Siegel's Theorem on integral points on curves, which we now state in terms convenient for our purpose, is a principal tool that we use in the following.[1]

Theorem 1. *Let $C \subset \mathbb{A}^2$ be a plane irreducible algebraic curve defined over* **Q**. *Suppose that C contains infinitely many points with integral coordinates. Then C is rational, and its smooth model has at most two points at infinity.*

This Theorem was proved by Siegel in his famous 1929 paper [25] (for an English translation, as well as comments on it, see [31]; an alternative shorter proof, making use of the advanced Subspace Theorem in Diophantine approximation, but avoiding any tool from the theory of Jacobians, was recently provided in [2]).

Remark: it can happen that the curve C is smooth, but its natural projective completion in $\mathbf{P}_2$ is singular at infinity; its desingularisation might then have more points at infinity than C itself. The conclusion of Siegel's theorem asserts that the number of points at infinity *after desingularising* is ≤ 2.

Recall that every (abstract) affine curve C (over the complex number field) can be parametrised by a smooth affine curve X, called its normalisation, which in turn can be viewed as the complement $\widetilde{X} \setminus S$ of a finite set S in a smooth projective curve $\widetilde{X}$. In other words, one can find such a curve X and a degree-one morphism $X \to C$ which is an isomorphism outside the singular locus of C. Siegel's theorem asserts that whenever C contains infinitely many integral points, the parametrising curve X is obtained from $\widetilde{X} = \mathbf{P}_1$ by removing one or two points, so $X \simeq \mathbb{A}^1$ or $X \simeq \mathbb{A}^1 \setminus \{0\}$. Hence our affine algebraic curve C is parametrised (at least over the complex number field) by a morphism $\mathbb{A}^1 \to C$ or a morphism $\mathbb{A}^1 \setminus \{0\} \to C$.

Moreover, if X is the complement of a single point, i.e. $X \simeq \mathbb{A}^1$, then this single point must be rational; we shall see that if X is obtained by removing two points, these points must be real quadratic irrational.

3 Characterisation of Rank-1 Diophantine Sets

This section offers a complete classification of Diophantine sets of rank one.

Notation. For an affine algebraic curve $X \subset \mathbb{A}^n$, defined by polynomial equations with integral coefficients, we denote by $X(\mathbf{Z})$ the set of its

---

[1]We will denote n-dimensional affine space by $\mathbb{A}^n$. For our purposes, we can identify it with the set of n-tuples in an algebraically closed field which we write $\mathbb{A}$. This field can be taken to be either the field of complex numbers, or the field of complex algebraic numbers.

integral points, i.e. the points on X whose coordinates are integers.

Let $\mathbb{K}$ be a real quadratic number field, $\epsilon > 1$ a fundamental unit of norm 1 in its ring of integers. Recall that after writing $\mathbb{K} = \mathbf{Q}(\sqrt{\delta})$ for some square-free integer $\delta > 1$, we obtain that ϵ can be written in the form $a + b\sqrt{\delta}$ if $\delta \not\equiv 1 \pmod 4$, and the pair (a, b) constitutes the minimal solution to Pell's equation $x^2 - \delta y^2 = 1$. If, on the other hand, $\delta \equiv 1 \pmod 4$, then ϵ can be written in the form $\frac{a + b\sqrt{\delta}}{2}$ for some odd numbers a, b.

We denote by $'$ the Galois conjugation in $\mathbb{K}$. Then $\epsilon' = \epsilon^{-1}$. For every algebraic integer $\alpha \in \mathbb{K}$, the sequence $n \mapsto \alpha \epsilon^n + \alpha' \epsilon^{-n} = \mathrm{Tr}_{\mathbb{K}/\mathbf{Q}}(\alpha \, \epsilon^n)$ is a sequence of rational integers. It satisfies a linear recurrence relation of order two.

Definition 3. A sequence $n \mapsto u_n$ of integers will be called of EXPONENTIAL TYPE with splitting field $\mathbb{K}$ if it can be defined as

$$u_n = p(\epsilon^n)$$

for some Laurent polynomial $p(T) \in \mathbb{K}[T, T^{-1}]$. Since the sequence u_n takes rational values, necessarily the Laurent polynomial $p(T)$ can be written in the form

$$p(T) = \alpha_0 + \alpha_1 T + \ldots + \alpha_d T^d + \alpha_1' T^{-1} + \ldots + \alpha_d' T^{-d},$$

where $\alpha_0 \in \mathbf{Q}$ and $\alpha_1, \ldots, \alpha_d \in \mathbb{K}$.

Note that such a sequence is either constant or tends to infinity exponentially.

Definition 4. We say that a sequence of integers $n \mapsto u_n$ is of POLYNOMIAL TYPE if there exists a polynomial $p(T) \in \mathbf{Q}[T]$ such that for all $n \in \mathbf{N}$

$$u_n = p(n) \, .$$

Statement of results. We can now formulate a key result:

Theorem 2. *A subset of $\mathbf{Z}$ is a Diophantine set of rank one if and only if it is a finite union of sequences of exponential or polynomial type.*

Rather than pausing here to prove this claim, we prefer to first exhibit some of its applications—the proof is postponed to Sec. 5.

A consequence of our theorem Thm. 2 is that the distribution of points of a Diophantine set of rank 1 enjoys a kind of regularity, which we will describe after one additional definition.

Definition 5. Let $A \subset \mathbf{N}$ be a set of natural numbers. We say that A has LOGARITHMIC DENSITY if the following limit exists, is finite and non-zero:

$$\lim_{N\to\infty} \frac{\#(A \cap \{1,\dots,N\})}{\log N}.$$

We say that A has POLYNOMIAL DENSITY if there exists a real number $0 < \alpha \leq 1$ such that the following limit is finite and non-zero:

$$\lim_{N\to\infty} \frac{\#(A \cap \{1,\dots,N\})}{N^\alpha}.$$

Clearly, sequences of polynomial type have a polynomial density; likewise, sequences of exponential type have logarithmic density. It is not true in general that a union of sequences of polynomial (resp. logarithmic) density still has polynomial (resp. logarithmic) density. However, as we shall see in connection with the proof of the following theorem, the union of sequences with polynomial or logarithmic density arising from Diophantine sets of rank one always has polynomial or logarithmic density:

Theorem 3. *Let* $P(X,Y) \in \mathbf{Z}[X,Y]$ *be a polynomial with integral coefficients. Then the set*

$$\left\{ \boldsymbol{x} \in \mathbf{N} \mid \text{there exists } \boldsymbol{y} \in \mathbf{N} \text{ such that } P(\boldsymbol{x},\boldsymbol{y}) = 0 \right\} \qquad (4)$$

either is finite, or has logarithmic density, or has polynomial density.

4 Estimating a Lower Rank Bound for $\mathcal{P}$

From the results stated in the preceding section, we get the lower rank bound

Theorem 4. rank $\mathcal{P} \geq 2$,

improving the estimate made at the end of Sec. 2.1.

The argument runs as follows. The set $\mathcal{P}$ of prime numbers satisfies

$$\lim_{N\to\infty} \frac{\#(\mathcal{P} \cap \{1,\dots,N\})}{N/\log N} = 1,$$

a fact known as the Prime Number Theorem, proved at the end of the nineteenth century by Hadamard and de la Vallée Poussin. Therefore $\mathcal{P}$ has neither polynomial nor logarithmic density. Hence, in the light of Thm. 3, we readily conclude that $\mathcal{P}$ is not Diophantine of rank 1.

Note that in order to derive the proposition just stated, we could rely on the easier estimates

$$\lim_{N\to\infty} \sup \frac{\#(\mathcal{P} \cap \{1,\dots,N\})}{N/\log N} < \infty, \quad \lim_{N\to\infty} \inf \frac{\#(\mathcal{P} \cap \{1,\dots,N\})}{N/\log N} > 0$$

obtained by Chebychev before the Prime Number Theorem was proved.

We end by noting that the fact that $\mathcal{P}$ cannot be Diophantine of rank one follows from Theorem 2, and thus can be seen to be true without any use use of density considerations. Actually, a polynomial sequence, if not constant, always takes infinitely many non-prime values, and the same is true for an exponential one. This fact can be proved by congruence considerations: for instance, in the case of a sequence $n \mapsto p(n)$, for a polynomial $p(x) \in \mathbf{Z}[x]$, if for some $n \in \mathbf{N}$ the value $p(n)$ is a prime q, then all the values $p(n + kq)$, for $k \in \mathbf{Z}$, are divisible by q, hence they cannot all be prime unless $p(x)$ is constant. The same kind of argument holds for exponential sequences.

5   Proof of Theorem 2

Let $P(X, Y) \in \mathbf{Z}[X, Y]$ be a polynomial with integral coefficients. For the moment, we study first the (Diophantine) set

$$A_P = \Big\{ x \in \mathbf{Z} \mid \exists y \in \mathbf{Z} \big(P(x, y) = 0 \big) \Big\}, \tag{5}$$

obtaining that it is given by a finite union of exponential or polynomial sequences. At the end of the proof we shall see that the conclusion does not change if we consider only points with non-negative coordinates.

Letting $C \subset \mathbb{A}^2$ be the algebraic curve of equation $P(X, Y) = 0$, we are interested in describing the possible values of the x-coordinates of the integral points on C.

The polynomial $P(X, Y)$, hence the curve C, can be reducible. We start by decomposing it as a product $P_1(X, Y) \cdots P_h(X, Y)$ of irreducible factors over $\mathbf{Q}$. The corresponding Diophantine sets A_{P_i} will decompose our set A_P into a finite union $\bigcup_i A_{P_i}$. Hence, in order to prove our Theorem 2, it suffices to prove it in the case of a polynomial irreducible in $\mathbf{Q}[X, Y]$.

We further notice that if it is irreducible in $\mathbf{Q}[X, Y]$ but reducible in $\mathbf{C}[X, Y]$, then the corresponding algebraic curve can only contain finitely many rational points. Hence we can restrict our attention to an absolutely irreducible algebraic curve C.

By Siegel's theorem, the desingularisation X of the curve C is a smooth rational curve with one or two points at infinity. More precisely, there exists a degree-one morphism $\varphi : X \to C$, where X is smooth, affine, of genus zero with one or two points at infinity.

Letting $\pi : C \to \mathbb{A}^1$ be the projection onto the first coordinate, we can view $\pi \circ \varphi$ as a regular function on the affine curve X. Our set A_P is then the set of values $\pi \circ \varphi(p)$, for $p \in X(\mathbf{Q})$ such that $\varphi(p)$ has integral coordinates.

The proof now splits into two cases, according to the number of points at infinity in X.

First case: $X \simeq \mathbb{A}^1$.

Now, the morphisms $\varphi : \mathbb{A}^1 \to C$, where C is a plane curve, are always expressed by a pair $(x(T), y(T))$ of polynomials; namely, the point $t \in \mathbb{A}^1$ is sent to the point $(x(t), y(t)) \in C$ and the condition that $(x(t), y(t))$ belongs to C amounts to saying that the pair $(x(t), y(t))$ satisfies the polynomial equation of the curve C. If C contains infinitely many rational points, then one can choose the polynomials $x(T), y(T)$ to have rational coefficients. Also, all but finitely many rational points of C are images of rational points of X, since φ has degree one as a morphism $\mathbb{A}^1 \to C$. Hence in particular all the integral points of C are of the form $(x(t), y(t))$ for $t \in \mathbb{Q}$, with finitely many exceptions.

Now, given two polynomials with rational coefficients $x(T), y(T) \in \mathbb{Q}[T]$, the set of rational values $t \in \mathbb{Q}$ such that $x(t), y(t)$ are both integer forms a finite union of arithmetic progressions of the form $t = a + nb$, $n \in \mathbb{Z}$, for suitable $a, b \in \mathbb{Q}$. Hence, after a suitable variable change we obtain finitely many polynomial parametrisations $(x_i(T), y_i(T))$, for $i = 1, \dots, k$, such that the x-coordinates of the integral points of C will be of the form $x_i(n), n \in \mathbb{Z}, i = 1, \dots, k$.

In this case we have obtained a sequence of polynomial type.

Second case. X has two points at infinity. It is in this case that Pell's equation, or Pell-type equations, come into play. The points at infinity must be real quadratic irrational; otherwise, by Runge's theorem, X would have only finitely many integral points, and also every non-constant regular function on X could take only finitely many integral values.

Let then $\mathbb{K}$ be the (real quadratic) field of definition of the points at infinity. Then X is isomorphic over $\mathbb{Q}$ to the complement on the projective line of a pair of points α, α' which are defined over $\mathbb{K}$ and conjugated over $\mathbb{Q}$ (up to change of coordinates, we can suppose that α, α' are algebraic integers). A plane model of X is the hyperbola of equation $(x - \alpha y)(x - \alpha' y) = 1$. Putting $a = \alpha + \alpha' = \mathrm{Tr}_{\mathbb{K}/\mathbb{Q}}(\alpha)$ and $b = \mathrm{N}_{\mathbb{K}/\mathbb{Q}}(\alpha)$, we obtain $x^2 - axy + by^2 = 1$. Every other integral model can be represented by the equation

$$X : x^2 - axy + by^2 = c \qquad (6)$$

for some non-zero integer c. The curve X is a homogenous space over $\mathbb{Q}$ for the orthogonal group $G = \mathrm{SO}(q)$, where $q(x, y) = x^2 - axy + by^2$. Over the integers, the action of $G(\mathbb{Z})$ on $X(\mathbb{Z})$ can be non-transitive, but in any case the number of orbits is finite (Lemma 4). We then obtain that there exist finitely many points (ξ_i, η_i), $i = 1, \dots, k$, on $X(\mathbb{Z})$ such that every integral point in $X(\mathbb{Z})$ is of the form $g(\xi_i, \eta_i)$ for some $g \in G(\mathbf{Z})$ and some $i = 1, \ldots, k$. Since $G = \{\pm g_0^n : n \in \mathbf{N}\}$ for a single matrix g_0 having eigenvalues ϵ, ϵ^{-1}, we deduce that the points $(\boldsymbol{x}, \boldsymbol{y}) \in X(\mathbf{Z})$, i.e. the integral solutions to (6), belong to finitely many sequences of the form $n \mapsto g_0^n(\xi, \eta)$, where (ξ, η) is fixed. Now, each of the four coefficients of the matrix g_0^n is of the form $\beta \epsilon^n + \beta' \epsilon^{-n}$, for a suitable $\beta \in \mathbb{K}$, so the coordinates $(\boldsymbol{x}, \boldsymbol{y})$ will be given by finitely many sequences of that type.

Now, let us come back to our original curve C. Recall that there is a degree-one morphism $\varphi : X \to C \subset \mathbb{A}^2$, defined over the rationals. So we are given two plane curves $X, C \subset \mathbb{A}^2$, one of which is a conic, and a morphism $\varphi : X \to C$ which is a bijection outside the complement of finite sets. We already have given a clear description of the integral points on X and want to describe the integral points on C.

Every morphism between plane curves is defined by a pair of polynomials $x(T_1, T_2), y(T_1, T_2) \in \mathbf{Q}[T_1, T_2]$ (this is a consequence of Hilbert's Nullstellensatz). As in the previous case, since φ has degree one as a morphism $X \to C$, up to finitely many exceptions every rational point of C must be the image of a rational point of X and after passing to a finite set of polynomial maps $(x_i(T_1, T_2), y_i(T_1, T_2))$, for $i = 1, \ldots, k$, we can also assert that each integral point of C comes from an integral point of X.

Hence, we obtain in particular that the x-coordinates of the integral points of C are of the form $p(\alpha \epsilon^n + \alpha' \epsilon^{-n})$ for a polynomial $p(T) \in \mathbf{Q}[T]$. Letting $\widetilde{p}(T) := p(\alpha T + \alpha' T^{-1})$, we obtain finitely many sequences of the form $\widetilde{p}(\epsilon^n)$, i.e. sequences of exponential type according to Definition 3. We have then characterised the set A_P defined in (5).

In order to finish the proof, we must study the subset of A_P formed by the *positive* integer values of x such that one of the solutions y of the equation $P(x, Y) = 0$ is also *positive*.

Now, as we have just proved, the pairs (x, y) with $P(x, y) = 0$ are parametrised as $(x(n), y(n))$, $n \in \mathbf{Z}$, by polynomial or by exponential sequences and we are interested in understanding when are they both positive: since each such sequence can change sign only finitely many times, if they take simultaneously positive values infinitely often, they do so for all but a finite number of positive n's, or for all but a finite number of negative n's.

6   Proof of Theorem 3

Here we want to deduce Theorem 3 from Theorem 2. This task is not related to the determination of a lower rank bound for the primes, as we have noted at the end of Sec. 4; on the other hand, it is relevant for the full characterisation of the Diophantine subsets of $\mathbf{Z}$ of rank 1.

Take a polynomial $P(X, Y) \in \mathbf{Z}[X, Y]$ such that for infinitely many $x \in \mathbf{N}$ there exists $y \in \mathbf{N}$ with $P(x, y) = 0$.

Then, by Theorem 2, the set (4) defined in Theorem 3 is a finite union of sequences of exponential or polynomial type. We have already remarked that each such sequence has a well defined density, either a polynomial or a logarithmic one. However, in principle the union of two sequences with comparable densities might fail to have a density (whenever their intersection has a density which is not neglible).

Clearly, adding an exponential sequence (which has logarithmic density) to a finite union of sequences with polynomial density does not change the density of the union.

Also, if two sequences $A = \{a_1 < a_2 < a_3 < \cdots\}$ and $B = \{b_1 < b_2 < \cdots\}$ satisfy

$$\lim_{N \to \infty} \# \left(A \cap \{1, \ldots, N\} \right) / N^{\alpha} \quad =: \quad a \, ,$$

$$\lim_{N \to \infty} \# \left(B \cap \{1, \ldots, N\} \right) / N^{\beta} \quad =: \quad b \, ,$$

where $0 < a < \infty, 0 < b < \infty$ and $\alpha < \beta$, their union has the same (polynomial) density as B.

So the only problem in proving that a finite union of polynomial and exponential sequences has a density arises whenever either all sequences are exponential or there are (at least) two dominant polynomial sequences A, B with the same asymptotic growth N^{α}.

In either case, we can prove that their intersection $A \cap B$ has infinitesimal density with respect to the densities of A and B, unless it has exactly the same growth as A and B.

Let us treat in detail the case of polynomial sequences: we are given two polynomials $p(X), q(Y)$, of the same degree d, and define the sets A, B to be

$$A \;=\; \{p(n) : n \in \mathbf{N}\} \quad \text{and} \quad B \;=\; \{q(m) : m \in \mathbf{N}\} \, .$$

Then $\# \left(A \cap \{1, \ldots, N\} \right) / N^{1/d}$ and $\# \left(B \cap \{1, \ldots, N\} \right) / N^{1/d}$ both have finite non-zero limit: in symbols,

$$\# \left(A \cap \{1, \ldots, N\} \right) \;\sim\; a N^{1/d} \quad \text{and} \quad \# \left(B \cap \{1, \ldots, N\} \right) \sim b N^{1/d} \, ,$$

for suitable constants a, b.

The set $A \cap B$ is hence defined to be the set of those natural numbers which are expressible both in the form $p(n)$ and in the form $q(m)$. Let us consider the plane algebraic curve C of equation

$$C : \quad p(x) = q(y) \, .$$

Then the intersection $A \cap B$ is the projection of the set of points with positive integral coordinates of C via the map $(x, y) \mapsto p(x)$ (or

$(x, y) \mapsto q(y))$. Now, if $C(\mathbf{Z})$ is finite, then $A \cap B$ is finite; otherwise the points of $C(\mathbf{Z})$ are parametrised, either exponentially or polynomially. In the former case the intersection $A \cap B$ has logarithmic growth and we are done; in the latter case the intersection $A \cap B$ can be described by finitely many polynomial parametrisations $n \mapsto p(x_i(n))$ for some polynomials $x_i(T) \in \mathbf{Z}[T]$. In that case, if all of the polynomials $x_i(T)$ have degree > 1, then again $A \cap B$ has infinitesimal growth (with respect to $N^{1/d}$).

The only remaining case is when one polynomial $x_i(T)$ has degree 1; in that case $\#(A \cap B \cap \{1, \ldots, N\}) \sim cN^{1/d}$ for some non-zero constant c. If $\#(A \cap \{1, \ldots, N\}) \sim aN^{1/d}$ and $\#(B \cap \{1, \ldots, N\}) \sim bN^{1/d}$, then

$$\#((A \cup B) \cap \{1, \ldots, N\}) \quad \sim \quad (a + b - c)N^{1/d},$$

which proves that also $A \cap B$ has a density.

The case of exponential sequences A, B is analogous: again, their intersection is contained in the image of the projection of the set of integral points of a suitable algebraic curve.

7 Historical Note on Prime-Representing Polynomials

In her seminal paper [22] on existentially definable relations among natural numbers, Julia Bowman Robinson proposed the hypothesis that a Diophantine relation of exponential rate of growth exists, and proved that it would imply that the ternary relation $a^b = c$ would in its turn be Diophantine. Accordingly, any n-ary relation specifiable in the form

$$E_1(a_1, \ldots, a_n, x_1, \ldots, x_m) \quad = \quad E_2(a_1, \ldots, a_n, x_1, \ldots, x_m),$$

where E_1 and E_2 are expressions constructed from parameters a_i, unknowns x_j, and particular natural numbers using addition, multiplication, and exponentiation, would turn out to be Diophantine. J. Robinson also showed that primality is existentially definable in terms of exponentiation (see Fig. 1); therefore, under her hypothesis, a polynomial representing primes would exist.

Wilson's theorem enables one to state that p is a prime number through the simple formula $\exists k \, \exists u \, (p = 2 + k \, \& \, p u = (k+1)! + 1)$. Hence, by resorting to a trick present in [4], one can specify primality by means of a *generator* of the form $2 + k \, 0^E$, where E is a nonnegative-

$$a = \binom{r}{j} \quad \leftrightarrow \quad a = \left\lfloor \frac{(u+1)^r}{u^j} \right\rfloor \%u \quad \& \quad u = 2^r + 1$$

$$j! = \left\lfloor \frac{r^j}{\binom{r}{j}} \right\rfloor \quad \text{for any } r > (2j)^{j+1}$$

$$\neg \exists x \, \exists y \, \bigl(p = (x+2)(y+2) \lor p = 0 \lor p = 1\bigr)$$
$$\leftrightarrow \exists k \, \exists u \, \exists v \, \bigl(p = 2 + k \, \& \, pu = (k+1)! \, v + 1\bigr)$$

FIGURE 1 Binomial coefficient, factorial, and "p is a prime" are existentially definable in terms of exponentiation, cf. [22] pp. 446–447.

Here, '%' designates the integer remainder operation.

valued expression.[2] For that purpose, it suffices to take

$$E(k, u) = \bigl((k+1)! - (2+k)\,u + 1\bigr)^2,$$

so that

$$\mathsf{Prime}(p) \quad \leftrightarrow \quad \exists k \, \exists u \qquad p = 2 + k \, 0^{E(k,u)}.$$

Yuri Matiyasevich's proof of J. Robinson's hypothesis, in [13], made it clear that exponentiation could be eliminated from similar specifications of primality, thus leading to various prime-representing (and corresponding prime-generating) polynomials. For a survey on these polynomials see [17], Commentary in Chap. 3; see also Fig. 2.[3]

Although methods have significantly evolved over time, the techniques for getting prime-representing polynomials usually results from the combination of ideas already present in [22] with a Diophantine polynomial specification of exponentiation, such as the masterpiece proposed in [15], which Maxim A. Vsemirnov refined somewhat in [29] (see below). A decisive enhancement in the formulation of a prime-representing polynomial would result if one could remove factorial from the pipeline and could avoid exploiting the binomial coefficient in its full strength.

---

[2]The reader should bear in mind that exponentiation obeys, over $\mathbf{N}$, the rule:

$$0^m = \begin{cases} 1 & \text{if } m = 0, \\ 0 & \text{if } m > 0. \end{cases}$$

[3]In 1977 [16], before the universal rank bound 9 appeared in press, Matiyasevich exhibited a prime-representing polynomial in one parameter and 9 unknowns. Unsurprisingly, that earlier polynomial better fits the goal than any polynomial that results from the construction of a universal Diophantine equation, which must rely on versatile coding techniques. Owing to the use of a specific characterisation of the set of primes, the degree of the earlier representing polynomial was kept lower by several orders of magnitude.

	year	nr. of unknowns aka *rank*	degree	source
Matiyasevich	1971	23 ⤳ 20	18 ⤳ 10	[14]
Robinson / Matiyasevich	1975	13	?	[15]
Jones–Sato–Wada–Wiens	1976	11	6848	[7]
Matiyasevich	1977	9	7952 ⤳ 5640	[16]
Vsemirnov	1994	9	5000	[29]

FIGURE 2 Quick view of Diophantine polynomial representations of the set of primes. Twice, as indicated by '⤳', the rank and/or degree of the original polynomial were lowered in the English translation of a Russian paper.

With the transformation of a Diophantine polynomial defining primality into a prime-generating polynomial, the number of unknowns increases by 1 and the degree d gets transformed into $2d + 1$.

8   Estimating an Upper Rank Bound for $\mathcal{P}$

In this section we will prove:

Theorem 5. *The set consisting of those integers $\kappa > 3$ which satisfy the congruence*

$$\binom{2\kappa}{\kappa} \equiv 2 \mod \kappa^3$$

is Diophantine, of rank not exceeding 7.

An improved upper estimate of rank $\mathcal{P}$ might result, as Matiyasevich suggests in [17] (Open Questions and Commentary on p. 53 and p. 56 of the English transl.), from a positive answer to a specific conjecture made by James P. Jones, which *"would enable us to provide a rather simple Diophantine representation for the set of prime numbers"*.[4] Here we produce such a Diophantine representation of the set of primes— novel, to the best of our knowledge.[5]

As it often happens with challenging Diophantine representations of sets of integers, we will need most of the techniques invented along the study of Hilbert's Tenth Problem; in particular our main proposition, Lemma 10, will rely on *ad hoc* modifications of [9], Lemma 2.25.

Throughout this section, we will designate positive integers by low-

---

[4]Matiyasevich suggests, therein (Exercise 4 p. 51), recourse to the identity

$$\frac{1}{\sqrt{1-4\lambda}} = \sum_{\ell=0}^{\infty} \binom{2\ell}{\ell}\lambda^\ell, \qquad \text{holding for} \quad |\lambda| < \frac{1}{4},$$

for an improved Diophantine representation of the central binomial coefficient, but the authors have not been able to take advantage of that clue and will in its stead use Lemma 7 below. For clarifications about this identity, see [10] and [11].

[5]This representation already appears in [27], but it is proved correct here for the first time.

ercase Latin letters, and integers and polynomials by uppercase Latin letters. When using the assignment symbol $\risingdotseq$, we will place on its right-hand side a polynomial which the letter occurring as left-hand side will stand for. We state that Q is a perfect square by the notation $Q = \square$, and designate by $\lfloor\cdot\rfloor$ the floor function. When the distinction between parameters and unknowns turns out to be irrelevant, we refer generically by the word 'variable' to either.

We will need a Diophantine representation of the sequence $\psi_a(B)$ associated with Pell's equation (3): the one recalled next is a variant of the one proposed by Matiyasevich and Robinson in [15], as refined by Vsemirnov [29], Lemma 8.[6]

Lemma 5. *Let* A, B, C, L *be integers with* $A > 1$, $B > 1$, $C > 1$, B *odd, and* $L > 0$. *The condition* $\psi_A(B) = C$ *is satisfied if and only if there exist positive integers* i, j *such that*

$$D \risingdotseq (A^2 - 1)\,C^2 + 1 \quad (P4)$$

$$
\begin{cases}
DFI = \square & (P1) \\
F \mid H - C & (P2) \\
B \leq C & (P3)
\end{cases}
\qquad
\begin{aligned}
E &\risingdotseq 2\,i\,C^2 D\,L & (P5) \\
F &\risingdotseq (A^2 - 1)\,E^2 + 1 & (P6) \\
G &\risingdotseq -A + F\,(A + 1) & (P7) \\
H &\risingdotseq B + 2\,j\,C & (P8) \\
I &\risingdotseq (G^2 - 1)\,H^2 + 1 & (P9)
\end{aligned}
$$

Crucial for a Diophantine representation of exponentiation, which we will also need, we recall here the following claim, first proved by Julia Robinson:

Lemma 6 ([8], Lemma 2.8). *The relationship* $Y = S^B$ *holds for integers* Y, S, B *with* $S > 0$ *if and only if there exist integers* A *and* C *such that*

(1) $\quad S < A$,

(2) $\quad Y^3 < A$,

(3) $\quad S^{3B} < A$,

(4) $\quad C = \psi_A(B)$,

(5) $\quad (S^2 - 1)\,Y\,C \equiv S\,(Y^2 - 1) \mod (2\,A\,S - S^2 - 1)$.

---

[6]The system proposed by [29] enforces that B be even; accordingly, the condition (P2) reads $F \mid H + C$. The variant system reported here works properly when B is odd, as noted in [29], Remark 2, and as evident with marginal changes to the correctness proof.

Generally (see, e.g., [14],[7],[16]), Diophantine representations of primality are based on Wilson's theorem which characterises primes as being those integers $p > 1$ which satisfy the congruence $(p - 1)! \equiv -1$ mod p. This calls into play factorial, a polynomial representation of which passes through the binomial coefficient and exponentiation. An alternative approach seems worth being pursued if we are to avoid the factorial and the binomial coefficient in its full strength.

Joseph Wolstenholme showed in 1862 that every prime number p greater than 3 satisfies the congruence $\binom{2p-1}{p-1} \equiv 1 \mod p^3$ — or equivalently $\binom{2p}{p} \equiv 2 \mod p^3$, taking into account that $\binom{2k}{k} = 2 \cdot \binom{2k-1}{k-1}$ holds in general. J. P. Jones conjectured that the converse also holds, so that the biimplication

$$\kappa \text{ is prime} \iff \binom{2\kappa}{\kappa} \equiv 2 \mod \kappa^3 \qquad (7)$$

holds for every integer $\kappa > 3$ (cf. [6], p. 47, [19], p. 381, and [21] p. 23).[7]

In order to represent the set of all positive integers which satisfy the congruence on the right-hand side of (7), we must express in a Diophantine way the *central* binomial coefficient $\binom{2k}{k}$; to do that, we need three characterising results.

Lemma 7 ([9], Lemma 2.24). *When $U > 2^{2k}$, the following holds:*

$$\left\lfloor \frac{(U+1)^{2k}}{U^k} \right\rfloor = \binom{2k}{k} + \binom{2k}{k+1} U + \cdots + \binom{2k}{2k} U^k, \qquad (8)$$

$$\frac{(U+1)^{2k}}{U^k} - \left\lfloor \frac{(U+1)^{2k}}{U^k} \right\rfloor < \frac{2^{2k}}{U}. \qquad (9)$$

Lemma 8. *Let α and n be a real number and a positive integer. Then:*

1) if $0 \le \alpha < 1$, then $(1 - \alpha)^n \ge 1 - n\alpha$;

2) if $0 \le \alpha \le \frac{1}{2}$, then $(1 - \alpha)^{-1} \le 1 + 2\alpha$.

Lemma 9. *When $k \ge 5$, $M > 0$, and $U > 0$, then*

$$\lim_{M \to \infty} \frac{\psi_{M(U+1)}(2k+1)}{\psi_{2M^2 U}(k+1)} = \frac{(U+1)^{2k}}{U^k};$$

more specifically,

$$\frac{(U+1)^{2k}}{U^k}\left(1 - \frac{k}{M(U+1)}\right) < \frac{\psi_{M(U+1)}(2k+1)}{\psi_{2M^2 U}(k+1)} < \frac{(U+1)^{2k}}{U^k}\left(1 + \frac{k}{2M^2 U}\right)$$

$$(10)$$

---

[7]For a comprehensive account of equivalent formulations, generalisations and extensions of Wolstenholme's theorem, see [20]. McIntosh [19] verified Jones's conjecture for every natural number up to 10^9.

Proof. Using Lemma 3 and the inequalities stated in Lemma 8:

$$\frac{\psi_{M(U+1)}(2k+1)}{\psi_{2M^2U}(k+1)} \leq \frac{(2\,M\,(U+1))^{2k}}{(4\,M^2\,U-1)^k}$$

$$= \frac{(U+1)^{2k}}{U^k}\,\frac{1}{\left(1-\frac{1}{4\,M^2\,U}\right)^k}$$

$$< \frac{(U+1)^{2k}}{U^k}\left(1+\frac{k}{2\,M^2\,U}\right)$$

and

$$\frac{\psi_{M(U+1)}(2k+1)}{\psi_{2M^2U}(k+1)} \geq \frac{(2\,M\,(U+1)-1)^{2k}}{(4\,M^2\,U)^k}$$

$$= \frac{(U+1)^{2k}}{U^k}\left(1-\frac{1}{2\,M\,(U+1)}\right)$$

$$> \frac{(U+1)^{2k}}{U^k}\left(1-\frac{k}{M\,(U+1)}\right).$$

The claim readily follows. ⊣

Lemma 10. *Let κ be an integer such that $\kappa \geq 5$. Then κ^3 divides $\binom{2\kappa}{\kappa} - 2$ if and only if there exist positive integers w, s, h, z such that*

$$
\begin{cases}
\left(\frac{C}{K}-Y\right)^2 < \frac{1}{4} & (C1) \\
(P^2-1)\,K^2 + 1 = \square & (C2) \\
(2^2-1)\,w\,C \equiv 2(w^2-1) \bmod Q & (C3) \\
C = \psi_A(B) & (C4)
\end{cases}
$$

$$
\begin{aligned}
A &\coloneqq M(U+1) & (A1) \\
B &\coloneqq 2\kappa + 1 & (A2) \\
M &\coloneqq \kappa Y & (A3) \\
Y &\coloneqq \kappa^3 s + 2 & (A4) \\
P &\coloneqq 2M^2 U & (A5) \\
Q &\coloneqq 4A - 5 & (A6) \\
U &\coloneqq \kappa^3 w & (A7) \\
K &\coloneqq \kappa + 1 + h\,(P-1) & (A8) \\
C &\coloneqq B + z & (A9)
\end{aligned}
$$

(In Diophantine format, condition $(C1)$ becomes
$$K^2 - 4\,(C - K\,Y)^2 > 0\,.)$$

Proof. **Sufficiency:**
Assuming that the conditions $\kappa \geq 5$ and $(C1)$–$(C4)$ hold under the assignments $(A1)$–$(A9)$, we must show that κ^3 divides $\binom{2\kappa}{\kappa} - 2$.

Notice that (in order of derivation): $U \geq 125$, $Y \geq 127$, $M \geq 635$, $P \geq 2\cdot 125^3$, $A \geq 125^2$, $Q \geq 125^2$, $B \geq 11$, $C \geq 12$, $K \geq 2\cdot 125^3$; hence $A, B, M, Y, P, Q, U, K, C$ are positive integers.

Thanks to Lemma 1, condition $(C2)$, and the assignments $(A5)$ e $(A8)$, we can write $K = \psi_{2M^2U}\big(\kappa + 1 + \bar{K}\,(2M^2U - 1)\big)$, for a suitable

$\bar{K}$. It readily follows from $\kappa + 1 < 2M^2U - 1$ that $\bar{K} \geq 0$, but we will show that $\bar{K} = 0$ by arguing by contradiction as follows. Assume that $\bar{K} > 0$; then, from $(A1)$, $(A2)$, Lemma 2, and Lemma 3, we get the inequalities

$$0 < \frac{C}{K} \leq \frac{\psi_{M(U+1)}(2\kappa + 1)}{\psi_{2M^2U}(\kappa + 1 + 2M^2U - 1)} \leq$$

$$\leq \frac{(2M(U+1))^{2\kappa}}{(4M^2U - 1)^{2M^2U+\kappa-1}} \leq$$

$$\leq \frac{(U+1)^{2\kappa}}{U^\kappa} \left(1 - \frac{\kappa}{M(U+1)}\right) \leq$$

$$\leq \left(\frac{2M(U+1)}{4M^2U - 1}\right)^{2\kappa} < \frac{1}{2}.$$

We hence get the desired contradiction because of $(C1)$, since $Y \geq 127$. Next, from $\bar{K} = 0$, we get $K = \psi_{2M^2U}(\kappa + 1)$. Now the inequality (10), combined with the assignment $(A3)$, gives us

$$\frac{C}{K} > \frac{(U+1)^{2\kappa}}{U^\kappa} \left(1 - \frac{\kappa}{M(U+1)}\right) = \qquad (11)$$

$$= \frac{(U+1)^{2\kappa}}{U^\kappa} \left(1 - \frac{1}{Y(U+1)}\right) >$$

$$> \frac{1}{2} \frac{(U+1)^{2\kappa}}{U^\kappa} >$$

$$> \frac{1}{2} U^\kappa >$$

$$> U^{\kappa-1} + \frac{1}{2},$$

whence, taking $(C1)$ into account, $U^{\kappa-1} < Y$.

Next we prove that $w = 2^B$; to see this, it suffices to show that the hypotheses of Lemma 6 are met by $S = 2$ and $Y = w$. Of those, only two are non-trivial in our case; specifically, we must prove that $2^{3B} < A$ and $w^3 < A$. In fact, it follows from $U^{\kappa-1} < Y$ (just seen) and from the assignments $(A1)$, $(A3)$, $(A7)$, $(A2)$ that

$$A = M(U+1) = \kappa Y(U+1) > \kappa YU > \kappa U^\kappa = \kappa (\kappa^3 w)^\kappa >$$
$$> \kappa^{3\kappa+1} \geq 5^{3\kappa+1} > 4^{3\kappa+2} > 2^{6\kappa+3} = 2^{3B}. \qquad (12)$$

Moreover, $A > \kappa (\kappa^3 w)^\kappa > w^\kappa > w^3$. So the hypotheses of Lemma 6 are met.

By the definition $(A7)$ of U, we have $U = \kappa^3 w = \kappa^3 2^{2\kappa+1}$. Let us prove:

$$\left| \frac{C}{K} - \frac{(U+1)^{2\kappa}}{U^\kappa} \right| < \frac{1}{4}. \tag{13}$$

From condition $(C1)$ and the inequalities (11) we get

$$\frac{1}{2} \frac{(U+1)^{2\kappa}}{U^\kappa} < 2Y + 1 \tag{14}$$

which, together with the assignment $(A3)$, gives us

$$\frac{(U+1)^{2\kappa}}{U^\kappa} \frac{\kappa}{M(U+1)} < 2(2Y+1) \frac{\kappa}{M(U+1)} < \frac{5Y}{Y(U+1)} < \frac{5}{U+1} < \frac{1}{4}; \tag{15}$$

therefore, since $M(U+1) < M(2U) < 2M^2U$, we have:

$$\frac{(U+1)^{2\kappa}}{U^\kappa} \frac{\kappa}{2M^2U} < \frac{(U+1)^{2\kappa}}{U^\kappa} \frac{\kappa}{M(U+1)} < \frac{1}{4}. \tag{16}$$

From condition $(C1)$ we know that, trivially, $\left| \frac{C}{K} - Y \right| < \frac{1}{2}$; then, by combining this inequality with (13), we are led to

$$\left| Y - \frac{(U+1)^{2\kappa}}{U^\kappa} \right| < \frac{3}{4}. \tag{17}$$

Now, as we already know that $U = \kappa^3 2^{2\kappa+1}$, we get $U > 2^{2\kappa}$; Lemma 7 thus gives us $\frac{(U+1)^{2\kappa}}{U^\kappa} - \left\lfloor \frac{(U+1)^{2\kappa}}{U^\kappa} \right\rfloor < \frac{1}{4}$. It follows from (17) that $\left| Y - \left\lfloor \frac{(U+1)^{2\kappa}}{U^\kappa} \right\rfloor \right| < 1$, and therefore $Y = \left\lfloor \frac{(U+1)^{2\kappa}}{U^\kappa} \right\rfloor$. From Lemma 7 we know that:

$$Y - 2 = \left\lfloor \frac{(U+1)^{2\kappa}}{U^\kappa} \right\rfloor - 2 = \binom{2\kappa}{2\kappa} U^\kappa + \cdots + \binom{2\kappa}{\kappa+1} U + \binom{2\kappa}{\kappa} - 2. \tag{18}$$

Since the assignment $(A7)$ implies that κ^3 divides U, and the assignment $(A4)$ implies that κ^3 divides $Y-2$, we conclude that $\kappa^3 \mid \binom{2\kappa}{\kappa} - 2$.

Necessity:
Consider an integer $\kappa \geq 5$ such that κ^3 divides $\binom{2\kappa}{\kappa} - 2$; we must show that there exist w, s, h, z satisfying the conditions $(C1)$–$(C4)$, $(A1)$–$(A9)$.

Put $w = 2^{2\kappa+1}$. According to $(A7)$, we get $U = \kappa^3 w = \kappa^3 2^{2\kappa+1} > 8 \cdot 2^{2\kappa}$. Lemma 7 then gives us a clue on how to instantiate s: for that purpose, it will suffice to impose that $Y = \left\lfloor \frac{(U+1)^{2\kappa}}{U^\kappa} \right\rfloor$, so that $\kappa^3 \mid Y-2$ by (8), whence we get s through $(A4)$. Continuing, notice also that (9)

yields the inequality

$$\frac{(U+1)^{2\kappa}}{U^\kappa} - \left\lfloor \frac{(U+1)^{2\kappa}}{U^\kappa} \right\rfloor = \frac{(U+1)^{2\kappa}}{U^\kappa} - Y < \frac{1}{8} . \tag{19}$$

After s we get, in succession: M, P, A, and Q on the basis of $(A3)$, $(A5)$, $(A1)$, and $(A6)$. Moreover, we put $C = \psi_A(B)$ and $K = \psi_P(\kappa+1)$, in agreement with conditions $(C4)$ and $(C2)$.

In order to complete the proof, we must show that $(C3)$ and $(C1)$ are satisfied; and that $P - 1 \mid K - \kappa - 1$ and $C > B$ also hold, so that h and z can be instantiated on the basis of $(A8)$ and $(A9)$.

Regarding $(C3)$, we refer the reader to Jones's [8], Lemmas 2.3 and 2.4.

Next we show that condition $(C1)$ is satisfied under the variable-instantiations made so far. From the inequality (10) it follows directly:

$$-\frac{(U+1)^{2\kappa}}{U^\kappa}\frac{\kappa}{M(U+1)} < \frac{C}{K} - \frac{(U+1)^{2\kappa}}{U^\kappa} < \frac{(U+1)^{2\kappa}}{U^\kappa}\frac{\kappa}{2M^2 U} \tag{20}$$

whence, by $(A3)$ and $M\,(U+1) < M^2\,U$:

$$\left| \frac{C}{K} - \frac{(U+1)^{2\kappa}}{U^\kappa} \right| < \frac{(U+1)^{2\kappa}}{U^\kappa}\frac{\kappa}{M(U+1)} =$$

$$= \frac{(U+1)^{2\kappa}}{U^\kappa}\frac{1}{Y(U+1)} <$$

$$< 2\left\lfloor \frac{(U+1)^{2\kappa}}{U^\kappa} \right\rfloor \frac{1}{Y(U+1)} =$$

$$= \frac{2Y}{Y(U+1)} = \frac{2}{U+1} \leq \frac{2}{126} < \frac{1}{8} . \tag{21}$$

By (19) and (21), we have:

$$\left| \frac{C}{K} - Y \right| \leq \left| \frac{C}{K} - \frac{(U+1)^{2\kappa}}{U^\kappa} \right| + \left| \frac{(U+1)^{2\kappa}}{U^\kappa} - Y \right| < \frac{1}{8} + \frac{1}{8} = \frac{1}{4} . \tag{22}$$

Hence $\left(\frac{C}{K} - Y \right)^2 < \frac{1}{16}$, and therefore $(C1)$ holds.

To conclude, observe that a positive h satisfying condition $(A8)$ exists since, by Lemma 1, $K \equiv \kappa + 1 \mod (P - 1)$ holds. Moreover, a value can be assigned to the variable z so as to satisfy $(A9)$: this follows directly, taking Lemma 3 into account, from the inequalities

$$C = \psi_A(B) = \psi_A(2\kappa + 1) >$$

$$> (2A - 1)^{2\kappa} = \left(2(\kappa Y)(\kappa^3 w + 1) - 1 \right)^{2\kappa} > 2\kappa + 1 = B . \tag{23}$$

⊣

The lemma just proved yields a system of Diophantine conditions which, once combined with the ψ-representing system of Lemma 5 (where the value of L will be suitably fixed), will represent the set of all integers greater than or equal to 5 that satisfy the congruence $\binom{2\kappa}{\kappa} \equiv 2 \mod \kappa^3$. Since we aim at defining this set by means of a single equation, we must join together our Diophantine conditions; to do this, we can use a classical result due to Matiyasevich and Robinson (see [15], Theorems 1, 2, 3), named *Relation Combining Theorem*[8]:

Theorem 6 (RCT75). *To each q in $\mathbf{N}$ there corresponds a polynomial M_q with integer coefficients such that, for all integers $X_1, \ldots, X_q$, J, R, V with $J \neq 0$, the conditions*

$$\begin{cases} X_1 = \square, \ldots, X_q = \square \\ J \mid R \\ V > 0 \end{cases} \tag{24}$$

are all satisfied if and only if $M_q(X_1, \ldots, X_q, J, R, V, n) = 0$ admits solutions for some value n in $\mathbf{N}$ of the variable n.

After eliminating assignments from the conditions of Lemma 10 and Lemma 5 through replacement of their left-hand sides by their right-hand sides, we are left with two perfect square conditions, one inequality, and two divisibility conditions which we must reduce to a single one in order to fulfill the hypothesis of RCT75. This reduction can be guided by the following easy lemma:

Lemma 11. *If d_1, d_2, z_1, z_2 are positive integers and d_1, d_2 are relatively prime, then*

$$(d_1 \mid z_1 \wedge d_2 \mid z_2) \Longleftrightarrow d_1 d_2 \mid z_1 d_2 + d_1 z_2 .$$

Our divisibility conditions are:

$$(P2) \qquad F \mid H - C ,$$

$$(C3) \qquad Q \mid 2 (w^2 - 1) - 3 w C .$$

Note that F and L are relatively prime, because the assignments $(P5)$ e $(P6)$ yield $F = (A^2 - 1)(2 i C^2 D L)^2 + 1 \equiv 1 \mod L$; by fixing $L \leftrightharpoons Q$ we can hence apply Lemma 11 and, consequently, RCT75.[9]

---

[8]Although lowering the degree of the representing polynomial was not a primary goal of this work, we tried to improve that degree by resorting to a recent variant of the relation-combining theorem: [26], Lemmas 10 and 11; unfortunately, our Diophantine system does not seem to be amenable to the hypothesis of that proposition.

[9]The polynomial equation thus resulting from conditions $(A1)$–$(A9)$, $(C1)$–$(C3)$, $(P4)$–$(P9)$, $(P1)$, and $(P2)$ has degree 5488, as explained in detail in [27], p. 10.

To conclude, we must bring the requirement $\kappa \geq 5$ inside our conditions. For that purpose, it suffices to instantiate V as

$$V = (\kappa - 1)\,(\kappa - 2)\,(\kappa - 3)\,(\kappa - 4)\,\left(K^2 - 4\,(C - K\,Y)^2\right)$$

(instead of, simply, as $V = K^2 - 4\,(C - K\,Y)^2$) in the inequality condition $V > 0$ when applying RCT75.

We thus get from M_2, through replacement of X_1, X_2, J, R, V by specific polynomials, a polynomial $\hat{M}_2$ representing the set of all integers $\kappa > 4$ which satisfy the congruence $\binom{2\kappa}{\kappa} \equiv 2 \mod \kappa^3$. This new polynomial involves one parameter, κ, and 7 unknowns: i, j, w, s, h, z, and n.

Summing up: If Jones's conjecture is true then the polynomial

$$(\kappa - 2)\,(\kappa - 3)\,\hat{M}_2(\,\kappa,\ i, j,\ w, s, h, z,\ n\,)$$

with parameter κ represents the set $\mathcal{P}$ of all prime numbers, so that

$$\mathsf{rank}\,\mathcal{P} \leq 7\,.$$

Conclusions and Future Work

We have drawn a non-trivial estimate of the lower rank bound of $\mathcal{P}$ from Siegel's theorem on integral points on curves, of which Paul Vojta [28] envisages a generalisation to integral points on surfaces (and beyond). It is reasonable to expect that on the basis of Vojta's conjecture our estimate of the lower rank bound of $\mathcal{P}$ can be increased from 2 to 3.

As regards finding an upper rank bound for $\mathcal{P}$, the contribution given in this paper somehow resembles one result discussed in [18]: here, much as there, a superset of $\mathcal{P}$ has been represented by means of a Diophantine polynomial in 7 variables. The set represented in [18] comprises, in addition to primes, the infinitely many Carmichael numbers (i.e. composite odd numbers which behave like primes according to a classical criterion based on Fermat's little theorem); on the other hand, it is not known whether composite numbers exist that belong to the set which we have represented: if Jones's conjecture is true, there are none, and the estimated upper rank bound of $\mathcal{P}$ diminishes from 9 to 7.

Acknowledgements

We are grateful to Martin Davis, who kindly suggested many stylistic improvements to this paper.

This work has been partially supported by the project FRA-UniTS (2014) *"Learning specifications and robustness in signal analysis."*

References

[1] John William Scott Cassels. *Rational Quadratic Forms*. Dover Publications, New York, 2008.

[2] Pietro Corvaja and Umberto Zannier. A subspace theorem approach to integral points on curves. *C. R. Acad. Sci. Paris, Ser. I*, 334(4):267–271, 2002.

[3] Martin Davis. *Computability and Unsolvability*. McGraw-Hill, New York, 1958. Reprinted with an additional appendix, Dover 1983.

[4] Martin Davis, Hilary Putnam, and Julia Robinson. The decision problem for exponential Diophantine equations. *Annals of Mathematics, Second Series*, 74(3):425–436, 1961.

[5] Martin Davis. Hilbert's tenth problem is unsolvable. *Amer. Math. Monthly*, 80(3):233–269, 1973. Reprinted with corrections in the Dover edition of *Computability and Unsolvability* [3], pp.199–235.

[6] Richard K. Guy. *Unsolved problems in Number Theory*. Springer-Verlag, New York, 1st edition, 1981.

[7] James P. Jones, Daihachiro Sato, Hideo Wada, and Wiens, Douglas. Diophantine representation of the set of prime numbers. *Amer. Math. Monthly*, 83(6):449–464, 1976.

[8] James P. Jones. Diophantine representation of Mersenne and Fermat primes. *Acta Arithmetica*, XXXV(3):209–221, 1979.

[9] James P. Jones. Universal Diophantine equation. *The Journal of Symbolic Logic*, 47(3):549–571, 1982.

[10] J. P. Jones and Ju. V. Matijasevič. A new representation for the symmetric binomial coefficient and its applications. *Ann. Sc. math. Québec*, VI(1):81–97, 1982. (Errata, ibid., VI(2):223, 1982).

[11] James P. Jones and P. Kiss. Exponential Diophantine representation of binomial coefficients, factorials and Lucas sequences. *Discussiones Mathematicae*, 12:53–65, 1992.

[12] Yuri Ivanovich Manin. *A course in mathematical logic*. Graduate texts in Mathematics. Springer-Verlag, 1977.

[13] Yu. V. Matiyasevich. Diofantovost' perechislimykh mnozhestv. *Doklady Akademii Nauk SSSR*, 191(2):279–282, 1970. (Russian. English translation, improved, by A. Doohovskoy as Ju. V. Matijasevič, Enumerable sets are Diophantine, *Soviet Mathematics. Doklady*, 11(3):354–358, 1970; see Errata, ibid., 11(6):vi (for 1970, pub. 1971). Translation reprinted in [24], pp. 269–273).

[14] Yu. V. Matiyasevich. Diofantovo predvstavlenie mnozhestva prostykh chisel. *Doklady Akademii Nauk SSSR*, 196(4):770–773, 1971.

(Russian. English translation by R. N. Goss as Ju. V. Matijasevič, Diophantine representation of the set of prime numbers, *Soviet Mathematics. Doklady*, 12(1):249–254, 1971).

[15] Yuri Matijasevič and Julia Robinson. Reduction of an arbitrary diophantine equation to one in 13 unknowns. *Acta Arithmetica*, XXVII:521–553, 1975. Reprinted in [23], p. 235ff.

[16] Yu. V. Matiyasevich. Prostye chisla perechislyayutsya polinomom ot 10 peremennykh. *Zap. Nauchn. Seminarov Leningradskogo Otdeleniya Matematicheskogo Instituta im. V. A. Steklova AN SSSR (LOMI)*, 68:62–82, 1977. (Russian. Translated into English as Yu. V. Matijasevič, Primes are nonnegative values of a polynomial in 10 variables, *J. of Soviet Mathematics*, 15(1):33–44, 1981).

[17] Yuri Vladimirovich Matiyasevich. *Desyataya Problema Gilberta*. Fizmatlit, Moscow, 1993. English translation: *Hilbert's Tenth problem*. The MIT Press, Cambridge (MA) and London, 1993. French translation: *Le dixième Problème de Hilbert: son indécidabilité*, Masson, Paris Milan Barcelone, 1995. URL: http://logic.pdmi.ras.ru/~yumat/H10Pbook/

[18] Yu. V. Matiyasevich. Diofantovo predstavleniye chisel Bernulli i yego prilozheniya. *Trudy Mat. Inst. Steklova*, 242:98–102, 2003. (Russian. Translated into English as *A Diophantine representation of Bernoulli numbers and its applications*, in *Mathematical Logic and Algebra – Collected papers dedicated to the 100th birthday of academician Petr Sergeevich Novikov, Proc. Steklov Inst. Math.*, 242(3):86–91, 2003).

[19] Richard J. McIntosh. On the converse of Wolstenholme's theorem. *Acta Arithmetica*, LXXI(4):381–389, 1995.

[20] Romeo Mestrovic. Wolstenholme's theorem: Its generalizations and extensions in the last hundred and fifty years (1862–2012). *arXiv preprint*, arXiv:1111.3057, 2012.

[21] Paulo Ribenboim. *The little book of bigger primes*. Springer, 2nd edition, 2004.

[22] Julia Robinson. Existential definability in arithmetic. *Trans. Amer. Math. Soc.*, 72(3):437–449, 1952. Reprinted in [23], p. 47ff.

[23] Julia Robinson. *The collected works of Julia Robinson*, volume 6 of *Collected Works*. American Mathematical Society, Providence, RI, 1996. ISBN 0-8218-0575-4. With an introduction by Constance Reid. Edited and with a foreword by Solomon Feferman. xliv | 338 pp.

[24] Gerald E. Sacks, editor. *Mathematical Logic in the 20th Century.* Singapore University Press, Singapore; World Scientific Publishing Co., Inc., River Edge, NJ, 2003.

[25] Carl Ludwig Siegel. Über einige Anwendungen diophantischer Approximationen. *Abhandlungen der Preussischer Akademie der Wissenschaften*, 1, 1929. An English translation by Clemens Fuchs is available in [31].

[26] Zhi-Wei Sun. A new relation-combining theorem and its application. *Zeitschrift für mathematische Logik und Grundlagen der Mathematik*, 38(1):209–212, 1992.

[27] Luca Vallata and Eugenio G. Omodeo. A Diophantine representation of Wolstenholme's pseudoprimality. In Davide Ancona, Marco Maratea, and Viviana Mascardi, eds, *Proc. of the 30th Italian Conference on Computational Logic, Genova, July 1-3, 2015*, volume 1459 of *CEUR Workshop Proc.*, pp. 2–12. CEUR-WS.org, 2015.

[28] Paul Vojta. *Diophantine approximations and value distribution theory*, volume 1239 of *Lect. Notes Math.* Springer-Verlag, Berlin, 1987. x+132 pp.

[29] Maxim Aleksandrovich Vsemirnov. Beskonechnye mnozhestva prostykh chisel, dopuskayuschie diofantovy predstavleniya s vosemyu peremennymi. *Zap. Nauchn. Sem. S.-Peterburg. Otdel. Mat. Inst. Steklov (POMI)*, 220:36–48, 1995. (Russian. Translated into English as M. A. Vsemirnov, Infinite sets of primes, admitting Diophantine representations in eight variables, *Journal of Mathematical Sciences*, 87(1):3200–3208, 1997).

[30] Joseph Wolstenholme. On certain properties of prime numbers. *The Quarterly Journal of Pure and Applied Mathematics*, 5:35–39, 1862.

[31] U. Zannier, editor. *On some applications of Diophantine approximations.* Edizioni della Normale, Scuola Normale Superiore, 2014.

Δ_0–Definability of Dedekind Sums

Henri-Alex Esbelin

Abstract: It is known that various generalizations of Dedekind sums are computable in polynomial time. We prove that the Dedekind sums and the more general Rademacher Dedekind sums have a Δ_0-definable graph. The proofs are based on the coding of a sequence as the continued fraction development of a rational number: we recently prove that such a coding is available in the framework of Δ_0-definability.

1 Introduction and Results

1.1 Dedekind Sums and Generalization

Let d and c be integers, $c \neq 0$. The famous Dedekind sum is the rational number defined by

$$s(d,c) = \sum_{k=1}^{k=|c|} \left(\left(\frac{k}{c}\right)\right)\left(\left(\frac{kd}{c}\right)\right)$$

where $((x)) = 0$ if $x \in \mathbb{Z}$, and $((x)) = x-[x]-\frac{1}{2}$ otherwise. Let us recall that $12c \times s(d,c)$ is a rational number (most of the classical knowledge on Dedekind sums may be found in [10]). It is known to be polynomial-time computable: this fact follows easily from the reciprocity law for the Dedekind sums (see [2], some details are given further).

A well known identity is $s(d,c) = \frac{1}{4c}\sum_{k=1}^{k=c-1} \cot\left(\frac{\pi kd}{c}\right)\cot\left(\frac{\pi k}{c}\right)$. Building on this identity, Zagier introduced a d-dimensional Dedekind

Studies in Weak Arithmetics, Volume 3.
Patrick Cégielski, Ali Enayat,
Roman Kossak.
Copyright © 2016, CSLI Publications.

sum defined by

$$s(a_0; a_1, ..., a_d) = \frac{(-1)^{d/2}}{a_0} \sum_{k=1}^{k=a_0-1} \cot\left(\frac{\pi k a_1}{a_0}\right) ... \cot\left(\frac{\pi k a_d}{a_0}\right)$$

so that $s(c; d, 1) = s(d, c)$. The polynomial-time complexity of these sums have been proved in [2]. Here we study the complexity of some of these sums in the framework of weak arithmetic definability.

1.2 Δ_0–Definability

This paper is not self-contained: the reader is supposed to be familiar with the elementary results and classical methods in Δ_0–definability (for an extended overview, see [5]).

Definition 1. *Let us denote by $\Delta_0^{\mathbb{N}}$ the class of relations over non negative integers definable by Δ_0 formulae. Relations in $\Delta_0^{\mathbb{N}}$ are often called rudimentary. A set A is said to be rudimentary whenever the relation $z \in A$ is rudimentary.*

It is well known that the Δ_0-definable relations have *LINSPACE* complexity and that the relations having *LOGSPACE* complexity are Δ_0-definable. Δ_0-definability is a natural question about relations that have been proved to have *POLYTIME* and *LINSPACE* complexity before tackling their *LOGSPACE* complexity. Let us denote $\mathbf{x} = (x_1, ..., x_k)$.

Definition 2. *A function f with domain a power of $\mathbb{N}$ and codomain $\mathbb{N}$ is rudimentary if (a) it is polynomially bounded (i.e. there exists a polynomial function ϕ with positive integer coefficients, such that $f(i, \mathbf{x}) \leq \phi(i, \mathbf{x})$) and (b) its graph is Δ_0–definable.*

Definition 3. *A function f with domain a power of $\mathbb{N}$ and codomain $\mathbb{Z}$ is rudimentary if its absolute value $|f|$ is rudimentary and the relation $f(\vec{x}) \geq 0$ is Δ_0–definable.*

Definition 4. *A function f with domain $\mathbb{Z}^k$ and codomain $\mathbb{Z}$ is rudimentary if for all $(\varepsilon_1, ..., \varepsilon_k) \in \{-1; +1\}^k$, the 2^k functions with domain $\mathbb{N}^k$ and codomain $\mathbb{Z}$ defined by $f(\varepsilon_1 x_1, ..., \varepsilon_k x_k)$ are rudimentary.*

Definition 5. *A relation on $\mathbb{Z}^k$ is rudimentary if its characteristic function is rudimentary.*

One of the main problems in Δ_0-definability is to prove that recursively defined functions have a Δ_0–definable graph, even in the case of an iterated summation such as $\sum_{i=1}^{i=y} f(i, \mathbf{x})$, where f is rudimentary. E.g., it is not known wether the binary relation *z is the x-th prime number* or even the unary relation *z is a prime number of odd index* are Δ_0-definable.

1.3   Results

In this paper, we prove the two following theorems:

Theorem 1. *The two-variables function defined by* $12c \times s(d, c)$ *is rudimentary.*

It is worth noticing that, as a corollary of this theorem, and from the formula

$$s(d, c) = \frac{c-1}{12c}(4cd - 2d - 3c) - \frac{1}{c}\sum_{n=1}^{c-1} n \left\lfloor \frac{dn}{c} \right\rfloor$$

(see [13]) the function defined by $\sum_{n=1}^{c-1} n \left\lfloor \frac{dn}{c} \right\rfloor$ of variables c and d turns out to be rudimentary, which is not obvious.

Let us consider now the following generalization known as the Rademacher-Dedekind sum (see [7]):

$$r(d, c, u) = \sum_{k=1}^{k=|c|} \left(\!\left(\frac{k}{c}\right)\!\right) \left(\!\left(\frac{kd+u}{c}\right)\!\right)$$

where $u \in \mathbb{Z}$. Here again $r(d, c, u)$ is a rational number, but $12c \times r(d, c, u)$ is an integer. Using the reciprocity law for these sums, Knuth provides an algorithm for computing them. Building on this algorithm, we prove the following result:

Theorem 2. *The three-variables function defined by* $12c \times r(d, c, u)$ *is* Δ_0–*definable.*

2   Proof of the Main Results

The first theorem will be proved using a formula due to Barkan, the second one using a formula due to Knuth. The first subsection provides logical material for both proofs; the second subsection provides mathematical material for the first one; then follows the proof of the first theorem; the fourth subsection provides additional mathematical material for the second theorem and the last subsection provides its proof.

2.1   Logical Material

The following easy proposition is of frequent use:

Proposition 1. *If R is a rudimentary relation and f is a rudimentary function then the relation* $R(x_1, ..., x_{i-1}, f(x_1, ..., x_n), x_{i+1}, ..., x_n)$ *is rudimentary.*

The following theorem is fundamental in the proof of the main theorem (see [11] in [12]).

Theorem 3. *(Woods) Let $f(i, \mathbf{x})$ and $g(\mathbf{x})$ be two positive functions with domains a power of $\mathbb{N}$, the graphs of which are Δ_0-definable. Suppose that f is polynomially bounded and g is polylogarithmicaly bounded (i.e. there exists a polynomial function ψ with positive integer coefficients, such that $g(\mathbf{x}) \leq \psi(\lfloor log_2(1 + x_1) \rfloor, ..., \lfloor log_2(1 + x_k) \rfloor))$. Then the relation $z = \sum_{i=0}^{i=g(\mathbf{x})} f(i, \mathbf{x})$ is Δ_0-definable.*

2.2 Mathematical Material

First we fix notations around the Euclidean Algorithm.

Definition 6. *Let c and d be positive integers. Let ρ be the following recursively defined function:*

$$\begin{cases} \rho(d, c, 0) = d \\ \rho(d, c, 1) = c \\ \rho(d, c, i + 2) = \begin{cases} \rho(d, c, i) \bmod \rho(d, c, i+1) \text{ if } \rho(d, c, i+1) \neq 0 \\ 0 \text{ otherwise} \end{cases} \end{cases}$$

Let λ be defined by $\lambda(d, c) = Max\{i; \rho(d, c, i) \neq 0\}$ and α be defined by $\alpha(d, c, i) = \left\lfloor \frac{\rho(d,c,i)}{\rho(d,c,i+1)} \right\rfloor$ for $0 \leq i < \lambda(d, c)$ and 0 otherwise.

Although it is quite easy, the following result plays a fundamental role in the proofs, because it allows to use Theorem 3.

Theorem 4. *(Lamé) $\lambda(d, c) = O(\log(Max\{c, d\}))$*

The proof of theorem 1 comes from the following Barkan's formula.

Theorem 5. *(Barkan) $s(d, c)$ is equal to*

$$\frac{1}{12} \left(-3 + \frac{d + d^{-1}\text{mod}c}{c} - \sum_{i=1}^{i=\lambda(d,c)} (-1)^i \alpha(d, c, i) \right).$$

It is a consequence of the reciprocity law for Dedekind sums :

$$12hk\left(s(h, k) + s(k, h) \right) = h^2 + k^2 - 3hk + 1$$

combined with $s(a, b) = s(a^{-1} \bmod b, b)$, where $a^{-1} \bmod b$ is the least positive integer u such that $au \equiv 1 \bmod b$. For a proof, see [1].

2.3 Proof of Theorem 1

We use the formula in theorem 5. The proof of theorem 1 needs the Δ_0-definability of the summation (which comes from Theorem 3 and 4) and of the graph of the function α which comes from the following theorem, which has been proved in [4].

Theorem 6. *The functions ρ, λ and α have Δ_0-definable graphs.*

Now the proof of theorem 1 goes easily to its end since the functions defined by $d^{-1} \bmod c$, $\displaystyle\sum_{j=1}^{\lambda(d,c)/2} \alpha(d,c,2j)$ and $\displaystyle\sum_{j=1}^{\lambda(d,c)/2} \alpha(d,c,2j+1)$ are rudimentary ones. Let us notice that from its definition, α is obviously bounded by the variable $Max\{c,d\}$.

2.4 More Mathematical Material for the Proof of Theorem 2

The Knuth's formula for the Rademacher-Dedekind is based on the following reciprocity law of this sum (see[13, 3]):

$$r(d,c,u) + r(c,d,u) = \frac{1}{12}\left(\frac{d}{c} + \frac{c}{d} + \frac{1 + 6\lfloor u\rfloor\lceil u\rceil}{cd} - 6\lfloor\frac{u}{c}\rfloor - 3e(d,u)\right)$$

for $d > u > 0$ and $d \geq c > 0$, where $e(d,u) = 0$ if $u > 0$ and $u \equiv 0 \bmod c$, and $e(c,u) = 1$ otherwise. It needs some additional functions.

Definition 7. *Let* $\mathbf{a} = (a_i)_{0 \leq i \leq l}$ *be a finite sequence of positive integers. Let* $p_\mathbf{a}$ *and* $q_\mathbf{a}$ *be the following recursively defined sequences:*

$$\begin{cases} p_\mathbf{a}(0) = a_0 \\ p_\mathbf{a}(1) = a_0 a_1 + 1 \\ p_\mathbf{a}(i+2) = a_{i+2}p_\mathbf{a}(i+1) + p_\mathbf{a}(i) \\ q_\mathbf{a}(0) = 1 \\ q_\mathbf{a}(1) = a_1 \\ q_\mathbf{a}(i+2) = a_{i+2}q_\mathbf{a}(i+1) + q_\mathbf{a}(i) \end{cases}$$

Hence we have:

$$\frac{p_\mathbf{a}(i)}{q_\mathbf{a}(i)} = a_0 + \cfrac{1}{a_1 + \cfrac{1}{\dots + \cfrac{1}{a_i}}}$$

It will be convenient to make use of $p_\mathbf{a}(-1) = 1$. A classical fact of continued fractions theory is that, for $j \leq i$, we have $\alpha(p_\mathbf{a}(i), q_\mathbf{a}(i), j) = a_j$. In the main part of the proof, we need the following property of continued fractions:

Lemma 1. *Let* $\mathbf{a} = (a_i)_{0 \leq i \leq l}$ *and* $\mathbf{b} = (b_i)_{0 \leq i \leq l}$ *be two finite sequences of positive integers. Suppose that for all* $i \geq 0$, *we have* $1 \leq b_i \leq 2a_i$. *Then for all* $i \geq 0$, *we have* $p_\mathbf{b}(i) \leq 2p_\mathbf{a}(i)^3$.

Proof. First step. Let us prove by induction on i that $p_\mathbf{b}(i) \leq 2^{i+1}p_\mathbf{a}(i)$: the basis case results of $p_\mathbf{b}(0) = b_0 \leq 2a_0 = 2p_\mathbf{a}(0)$ and $p_\mathbf{b}(1) =$

$b_0 b_1 + 1 \leq 4(a_0 a_1 + 1) = 4p_{\mathbf{a}}(1)$; the induction step is the following:

$$p_{\mathbf{b}}(i+2) = b_{i+2} p_{\mathbf{b}}(i+1) + p_{\mathbf{b}}(i) \leq 2a_{i+2} 2^{i+2} p_{\mathbf{a}}(i+1) + 2^{i+1} p_{\mathbf{a}}(i)$$

$$p_{\mathbf{b}}(i+2) \leq 2^{i+3}(a_{i+2} p_{\mathbf{a}}(i+1) + p_{\mathbf{a}}(i)) = 2^{i+3} p_{\mathbf{a}}(i+2).$$

Second step. Let us prove that $2^i \leq p_{\mathbf{a}}(i)^2$ for all i: let us notice first that $p_{\mathbf{a}}(i+1) \geq p_{\mathbf{a}}(i)$; hence from

$$p_{\mathbf{a}}(i+2) = a_{i+2} p_{\mathbf{a}}(i+1) + p_{\mathbf{a}}(i) \geq 2p_{\mathbf{a}}(i)$$

follows $p_{\mathbf{a}}(2k) \geq 2^k$ and $p_{\mathbf{a}}(2k+1) \geq 2^{k+1}$. $\dashv$

Definition 8. *Let φ and β be defined by*

$$\begin{cases} \varphi(d,c,u,0) = u \\ \varphi(d,c,u,i+1) = \begin{cases} 0 \text{ if } \rho(d,c,i+1) = 0 \\ \varphi(d,c,u,i) \bmod \rho(d,c,i+1) \text{ otherwise} \end{cases} \end{cases}$$

and $\beta(d,c,u,i) = \begin{cases} \left\lfloor \frac{\varphi(d,c,u,i)}{\rho(d,c,i+1)} \right\rfloor & \text{for } 0 \leq i < \lambda(d,c) \\ 0 & \text{otherwise} \end{cases}$.

The following formula was proved by Knuth (see [7]). For simplicity, we omit the variables d, c and u of the functions α, β, ρ, λ.

Theorem 7. *(Knuth)* $12c \times r(d,c,u)$ *is equal to*

$$c\left(\sum_{j=0}^{j=\lambda-1} (-1)^j \left[\alpha(j) - 6\beta(j) - 3e(\rho(j+1), \varphi(j))\right] \right) +$$

$$(-1)^{\lambda-1} p_{(\alpha(0),\ldots,\alpha(\lambda))}(\lambda-1) + d +$$

$$6\sum_{j=0}^{j=\lambda-1} (-1)^j \beta(j)(\varphi(j) + \varphi(j+1)) p_{(\alpha(0),\ldots,\alpha(\lambda))}(j-1)$$

2.5 Proof of Theorem 2

We use the formula in theorem 7. The proof of theorem 2 needs the Δ_0-definability of the summation (which comes from Theorem 3 and 4), of the graphs of the functions α, ρ, λ (which comes from Theorem 6), and of the graphs of the functions β, ϕ and $p_{(\alpha(0),\ldots,\alpha(\lambda))}$ which is proved now.

The following theorem has been proved in [4].

Theorem 8. *The function of variables d, c, i defined by*

$$p_{(\alpha(d,c,j))_{0 \leq j \leq \lambda(d,c)}}(i)$$

has a Δ_0-definable graph.

Lemma 2. *The functions φ and β have Δ_0-definable graphs.*

Proof. The idea is to code a sequence of positive integers using the fraction, the coefficients of the continued fractions development (CFD for short) of which are the successive terms of this sequence (or equivalently by the pair of integers which provide the successive terms of the coded sequence as the successive quotients in the Euclidean algorithm). Unfortunately, the sequence to be coded here has non negative terms, but coefficients of the CFD of a positive real number are positive. A useful trick will be to add 1 to each terms of the sequence before coding it.

Let us introduce the functions Γ_1 and Γ_2 defined by

$$\Gamma_1(d, c, u) = p_{(1+\beta(d,c,0),\ldots,1+\beta(d,c,\lambda(d,c)))}(\lambda(d, c))$$

and

$$\Gamma_2(d, c, u) = q_{(1+\beta(d,c,0),\ldots,1+\beta(d,c,\lambda(d,c)))}(\lambda(d, c))$$

These functions provide the code we intend to use.

Let us first notice that Γ_1 and Γ_2 are polynomially bounded: indeed the definition of φ provides $0 \le \varphi(d, c, u, i) < \rho(d, c, i)$, hence from the definitions of α and β, we have $1 + \beta(d, c, u, i) \le 1 + \alpha(d, c, u, i) \le 2\alpha(d, c, u, i)$ hence from Lemma 1 we have $\Gamma_1(d, c, u) < 2d^3$; moreover $\Gamma_1(d, c, u) \ge \Gamma_2(d, c, u)$.

We want now to prove that $z = \beta(d, c, u, j)$ is equivalent to

$$\exists (\gamma_1)_{<2d^3} \exists (\gamma_2)_{<2d^3} (z = \alpha(\gamma_1, \gamma_2, j) - 1) \wedge$$
$$\left((\forall i)_{\le j} (\exists v)_{<\rho(d,c,i+1)} \left(v + \sum_{k=0}^{k=i} (1 + \alpha(\gamma_1, \gamma_2, k)) \, \rho(d, c, k + 1) = u \right) \right)$$

Let us denote this last relation $R(d, c, u, j)$. *First step.* From the definition of the functions Γ_1 and Γ_2 , we have

$$\beta(d, c, u, j) = \alpha(\Gamma_1(d, c, u), \Gamma_2(d, c, u), j) - 1$$

Moreover a straightforward recursion on i proves that

$$\varphi(d, c, u, i) = u - \sum_{k=0}^{k=i-1} \beta(d, c, u, k) \rho(d, c, k + 1)$$

Suppose now that the relation $z = \beta(d, c, u, j)$ is true. Then the relation $R(d, c, u, j)$ is satisfied: just consider $\gamma_1 = \Gamma_1(d, c, u)$, $\gamma_2 = \Gamma_2(d, c, u)$ and $v = \varphi(d, c, u, i)$.

Conversely, suppose that $R(d, c, u, j)$ is true. It asserts the existence of some γ_1 and some γ_2 and for all $i \le j$ some integer v_i such that $v_0 = u$ and

$$v_i = u - \sum_{k=0}^{k=i-1} (1 + \alpha(\gamma_1, \gamma_2, k)) \, \rho(d, c, k + 1)$$

for $1 \leq i \leq j$. Then, we easily get $v_{i-1} = v_i \bmod \rho(d, c, i+1)$ and therefore $v_i = \varphi(d, c, u, i)$ and $\beta(d, c, u, i) = \alpha(\gamma_1, \gamma_2, j) - 1$. ⊣

3 Conclusion and Open Questions

3.1 Reciprocity Laws and Recursivity

Let us introduce the following recursive schema, which we denote by SRL:

$$
\begin{cases}
f(a, b, \vec{x}) = f(a - b, b, \vec{x}) \text{ if } a > b \\
f(a, 0, \vec{x}) = g(a, \vec{x}) \\
f(a, b, \vec{x}) + f(b, a, \vec{x}) = h(a, b, \vec{x}) \\
f(a, b, \vec{x}) \leq poly(a, b, \vec{x})
\end{cases}
$$

where *poly* is a polynomial with non negative integer coefficients. Using the previous notations, we easily get the equality

$$
f(a, b, \vec{x}) = \sum_{j=1}^{j=\lambda(a,b)} (-1)^{j+1} h(\rho(a, b, j+1), \rho(a, b, j), \vec{x})
$$

$$
+ (-1)^\lambda g(\rho(a, b, \lambda(a, b)), \vec{x})
$$

Hence the rudimentarity of g and h implies the rudimentarity of f.

More generally, let us introduce the folowing recursive schema, which we denote by RL:

$$
\begin{cases}
f(a, b, \vec{x}) = f(a - b, b, \vec{x}) \text{ if } a > b \\
f(a, 0, \vec{x}) = g(a, \vec{x}) \\
f(a, b, \vec{x}) = h(f(b, a, \vec{x}), a, b, \vec{x}) \\
f(a, b, \vec{x}) \leq poly(a, b, \vec{x})
\end{cases}
$$

In this case, we get

$$
f(\rho(a, b, i), \rho(a, b, i+1), \vec{x}) =
$$

$$
h(f(\rho(a, b, i+1), \rho(a, b, i+2), \vec{x}), \rho(a, b, i), \rho(a, b, i+1), \vec{x})
$$

hence the closure under RL reduces to the closure under recursions with a polynomial bound and a logarithmic length. It is an open question whether the set of rudimentary functions is closed under such recursions, and even whether such recursion may be reduced to summation (or equivalently counting).

3.2 More Generalizations of Dedekind Sums

Let us consider the following generalization of Dedekind sums:

$$
s(a, b, c) = \sum_{k=1}^{k=c} \left(\left(\frac{ak}{c} \right) \right) \left(\left(\frac{bk}{c} \right) \right)
$$

The reciprocity law for these sums is due to Rademacher [9]

$$s(a,b,c) + s(b,c,a) + s(c,a,b) = \frac{1}{4} - \frac{1}{12} \left(\frac{c}{ab} + \frac{a}{bc} + \frac{b}{ac} \right)$$

An interesting question is whether its graph is $\Delta_0^\sharp$ definable, where $\Delta_0^\sharp$ is roughly speaking Δ_0 whith the additonal *counting schema*, which allows to introduce in formulae the number of integers least than a variable satisfying a given relation (see [5] for details).

The author thanks the anonymous referee for his / her careful reading and for valuable suggestions.

References

[1] Barkan, Ph., Sur les sommes de Dedekind et les fractions continues finies, *C.R. Acad. Paris Sér. A-B*, vol. 16, 1977, pp. A923–A926.

[2] Beck, M. and Robins, S., Dedekind sums: a combinatorial-geometric viewpoint, *DIMACS Series in Discrete Mathematics and Theoretical computer Science*, vol. 10 (64), 2004, pp. 25–36.

[3] Dieter, U., Beziehungen zwischen Dedekindschen Summen, *Abh. math. Sere. Univ, Hamburg*, vol. 1/2 (21), 1957, pp. 109–125.

[4] Esbelin, H.-A., Δ_0 definability of the denumerant with one plus three variables, *New Studies in Weak Arithmetics. Edited by Patrick Cégielski, Charalampos Cornaros, and Costas Dimitracopoulos*, Lecture Notes vol. 211, CSLI Pubications, Stanford, 2013, pp 103–114.

[5] Esbelin, H.A. and More, M., Rudimentary relations and primitive recursion : a toolbox, *Theor. Comput. Sci.*, vol. 193, 1998, pp. 129–148.

[6] Khinchin, A. Ya., Continued Fractions, 1997, Dover.

[7] Knuth, D., Note on generalized Dedekind sums, *Acta Arithmetica*, vol. 33, 1977, pp. 129–148.

[8] Paris, J. and Wilkie, A., Counting Δ_0-sets, *Fund. Math.*, vol. 127, 1987, pp. 67–76.

[9] Rademacher, H., Generalization of the Reciprocity Formula for Dedekind Sums, *Duke Math. J.*, vol 21, 1954, pp. 391–398.

[10] Rademacher, H. and Grosswald, E., Dedekind Sums, *The Carus Mathematical Monographs Washington*, DC: Math. Assoc. Amer., 1972.

[11] Woods, A., Some Problems in Logic and Number Theory and their Connections, *Ph.D., University of Manchester*, 1981, University of Manchester.

[12] Woods, A. , Some Problems in Logic and Number Theory and their Connections, Thesis (1981) of Alan Robert Woods, *New Studies in*

Weak Arithmetics. Edited by Patrick Cégielski, Charalampos Cornaros, and Costas Dimitracopoulos, Lecture Notes vol. 211, CSLI Pubications, Stanford, 2013, pp. 273–390.

[13] Zagier, D., Higher dimensional Dedekind sums, *Math. Ann.*, vol. 202, 1973, pp. 149–172.

4

Characterizations of Interpretability in Bounded Arithmetic

JOOST J. JOOSTEN

Abstract: This paper deals with three tools to compare proof-theoretic strength of formal arithmetical theories: interpretability, Π^0_1-conservativity and proving restricted consistency. It is well known that under certain conditions these three notions are equivalent and this equivalence is often referred to as the Orey-Hájek characterization of interpretability.

In this paper we look with detail at the Orey-Hájek characterization and study what conditions are needed and in what meta-theory the characterizations can be formalized.

1 Introduction

Interpretations are everywhere used in mathematics and mathematical logic. Basically, a theory U interprets a theory V –we write $U \rhd V$– whenever there is some translation from the symbols of the language of V to formulas of the language of U so that under a natural extension of this translation the axioms of V are mapped to theorems of U.

The corresponding intuition should be that U is at least as strong or expressible as V. And indeed, interpretations are used for example to give relative consistency proofs or to establish undecidability of theories. As such, interpretations are considered an important meta-mathematical notion. Probably, the first time that interpretations re-

Studies in Weak Arithmetics, Volume 3.
Patrick Cégielski, Ali Enayat,
Roman Kossak.
Copyright © 2016, CSLI Publications.

ceived a formal and systematic treatment has been in the book by A. Tarski, A. Mostowski and R. Robinson (1, [17]). In the current paper we will study that notion of interpretability and also some related notions. Sometimes we speak of *relative* interpretability as to indicate that quantifications become relativized to some domain specifier as we shall define precisely later on.

We will relate the notion of relative interpretability to two other basic metamathematical notions. The first such notion is the notion of consistency. The notion of consistency is central to mathematical logic and considered key and fundamental.

A second notion is that of Π_1^0 conservativity. Below we will exactly define what Π_1^0 formulas are, but basically, those are formulas in the language of arithmetic which are of the form $\forall x \psi(x)$ where ψ is some decidable predicate. On the other hand, Σ_1^0 formulas are those of the form $\exists x \psi(x)$ for decidable ψ. Since all true theories prove exactly the same set of Σ_1^0 sentences, the first natural and interesting class of formulas to distinguish theories is on the Π_1^0 level. Therefore, the notion of Π_1^0 conservativity has been very central in mathematical logic and foundational discussions. We say that a theory U is Π_1^0 conservative over V whenever any Π_1^0 sentence provable by V is also provable by U.

The main purpose of this paper is to discuss how these three different notions are related to each other in certain circumstances. This relation is known as the *Orey-Hájek* characterization of relative interpretability.

As such, the paper contains many well-known results and various formulations are taken from [8]. However, we think that it is instructive that all these results are put together and moreover that a clear focus is on the requirements needed so that various implications are formalizable in weak theories.

Apart from the main focus –which is bringing together facts of the Orey-Hájek characterization of relative interpretability and formalizations thereof– the paper contains a collection of new observations that might come in handy. For example, our simple generalization of Pudlák's lemma as formulated in Lemma 8 has been a main tool in proving arithmetical correctness of a new series of interpretability principles in [9].

2   Preliminaries

As mentioned before, a central notion in this paper is that of consistency. Consistency is a notion that concerns syntax: no sequence of symbols that constitute a proof will yield the conclusion that $0 = 1$. It shall be an important criterion whether or not a theory proves the consistency of another. As such we want that theories can talk about syntax.

The standard choice to represent syntax is by Gödel numbering, assigning natural numbers to syntax. Thus, our theories should contain a modicum of arithmetic. In this section we shall make some basic observations on coding and then fix what minimal arithmetic we should have in our base theory. We shall formulate some fundamental properties of this base theory and refer to the literature for further background. Further, we shall fix the notation that is used in the remainder of this paper.

2.1 A Short Word on Coding

Formalization calls for coding of syntax. At some places in this paper we shall need estimates of codes of syntactical objects. Therefore it is good to discuss the nature of the coding process we will employ. However we shall not consider the implementation details of our coding.

We shall code strings over some finite alphabet A with cardinality a. A typical coding protocol could be the following. First we define an alphabetic order on A. Next we enumerate all finite strings over A in the following way (pseudo-lexicographic order).

To start, we enumerate all strings of length 0, then of length 1, etcetera. For every n, we enumerate the strings of length n in alphabetic order. The coding of a finite string over A will just be its ordinal number in this enumeration. We shall now see some easy arithmetical properties of this coding. We shall often refrain from distinguishing syntactical objects and their codes.

1. There are a^n many strings of length n.
2. There are $a^n + a^{n-1} \cdots + 1 = \frac{a^{n+1}-1}{a-1}$ many strings of length $\leq n$.
3. From (2) it follows that the code of a syntactical object of length n, is $\mathcal{O}(\frac{a^{n+1}-1}{a-1}) = \mathcal{O}(a^n)$ big.
4. Conversely, the length of a syntactical object that has code φ is $\mathcal{O}(|\varphi|)$ (logarithm/length of φ) big.
5. If φ and ψ are codes of syntactical objects, the concatenation $\varphi \star \psi$ of φ and ψ is $\mathcal{O}(\varphi \cdot \psi)$ big. For, $|\varphi \star \psi| = |\varphi| + |\psi|$, whence by (3), $\varphi \star \psi \approx a^{|\varphi|+|\psi|} = a^{|\varphi|} \cdot a^{|\psi|} = \varphi \cdot \psi$.
6. If φ and t are (codes of) syntactical objects, then $\varphi_x(t)$ is $\mathcal{O}(\varphi^{|t|})$ big. Here $\varphi_x(t)$ denotes the syntactical object that results from φ by replacing every (unbounded) occurrence of x by t. The length of φ is about $|\varphi|$. In the worst case, these are all x-symbols. In this case, the length of $\varphi_x(t)$ is $|\varphi| \cdot |t|$ and thus $\varphi_x(t)$ is $\mathcal{O}(a^{|\varphi| \cdot |t|}) = \mathcal{O}(t^{|\varphi|}) = \mathcal{O}(\varphi^{|t|}) = \mathcal{O}(2^{|\varphi| \cdot |t|})$ big.

As mentioned, we shall refrain from the technical characteristics of our coding and refer to the literature for examples. Rather, we shall keep in mind restrictions on the sizes and bounds as mentioned above. Also, we shall assume that we work with a natural poly-time coding with poly-time decoding functions so that the code of substrings is always smaller than the code of the entire string.

2.2 Arithmetical Theories

Since substitution is key to manipulating syntax we need, by our observations above, a function whose growth-rate can capture substitution. Thus, we choose to work with the smash function $\sharp$ defined by $x \sharp y := 2^{|x| \cdot |y|}$ where $|x| := \lceil \log_2(x+1) \rceil$ is the length of the number x in binary. We shall often also employ the function ω_1 which is of similar growth-rate and defined by $\omega_1(x) := 2^{|x|^2}$.

Next, we need a certain amount of induction. For a formula φ, the regular induction formula I_φ is given by

$$\varphi(0) \ \wedge \ \forall x \ (\varphi(x) \to \varphi(x+1)) \ \to \ \forall x \ \varphi(x).$$

However, it turns out that we can work with a weaker version of induction called polynomial induction denoted by PIND:

$$\varphi(0) \wedge \forall x \ (\varphi(\lfloor \tfrac{1}{2}x \rfloor) \to \varphi(x)) \ \to \ \forall x \ \varphi(x)$$

or equivalently

$$\varphi(0) \wedge \forall x \ (\varphi(x) \to \varphi(2x)) \wedge \forall x \ (\varphi(x) \to \varphi(2x+1)) \ \to \ \forall x \ \varphi(x).$$

The idea is that one can conclude $\varphi(x)$ by only logarithmically many calls upon the induction hypothesis with this PIND principle. For example to conclude $\varphi(18)$ we'ld go $\varphi(0) \to \varphi(1) \to \varphi(2) \to \varphi(4) \to \varphi(9) \to \varphi(18)$.

Typically, induction on syntax is of this nature and in order to conclude a property of (the Gödel number of) some formula ψ we need to apply the induction hypothesis to the number of subformulas of ψ which is linear in the length of ψ. Thus, most inductions over syntax can be established by PIND rather than the regular induction schema.

Moreover, we shall restrict the formulas on which we allow ourselves to apply PIND to so to end up with a weak base theory. As we shall see, most of our arguments can be formalized within Buss' theory[1] S_2^1.

---

[1]As mentioned, the substitution operation on codes of syntactical objects asks for a function of growth rate $x^{|x|}$. In Buss's S_2^1 this is the smash function $\sharp$. In the theory $I\Delta_0 + \Omega_1$ this is the function $\omega_1(x)$. However, contrary to S_2^1, the theory $I\Delta_0 + \Omega_1$ –aka S_2– is not known to be finitely axiomatizable.

The theory S_2^1 is formulated in the language of arithmetic

$$\{0, S, +, \cdot, \sharp, |x|, \lfloor \tfrac{1}{2}x \rfloor, \leq\}.$$

Apart from some basic axioms that define the symbols in the language, S_2^1 is axiomatized by PIND induction for Σ_1^b formulas. The Σ_1^b formulas are those formed from atomic formulas via the boolean operators, sharply bounded quantification and bounded existential quantification. Sharply bounded quantification is quantification of the from $\mathcal{Q}\, x<|t|$ for $\mathcal{Q} \in \{\forall, \exists\}$. Bounded existential quantification in contrast, is of the form $\exists x<t$. We refer the reader for [1] or [6] for further details and for the definitions of the related Σ_n^b and Π_n^b hierarchies.

Equivalent to the PIND principle (see [11, Lemma 5.2.5]) is the length induction principle LIND:

$$\varphi(0) \wedge \forall x\, (\varphi(x) \to \varphi(x+1)) \; \to \forall x\; \varphi(|x|).$$

So, from the progressiveness of φ, we can conclude $\varphi(x)$ for any x for which the exponentiation is defined. We shall later see that if we are working with definable cuts (definable initial segments of the natural numbers with some natural closure properties) we can without loss of generality assume that exponentiation is defined for elements of this cut.

Although most of our reasoning can be performed in S_2^1, we sometimes mention stronger theories. As always Peano Arithmetic (PA) contains open axioms that define the symbols $0, S, +$ and $\cdot$ and induction axioms I_φ for any arithmetical formula φ. Similarly, IΣ_n is as PA where instead we only have induction axioms I_φ for $\varphi \in \Sigma_n$. Here, Σ_n refers to the usual arithmetical hierarchy (see e.g. [6]) in that such formulas are written as a decidable formula preceded by a string of n alternating quantifiers with an existential quantifier up front. In case no free variables are allowed in the induction formulas, we flag this by a superscript "$-$" as in IΣ_n^-.

Another important arithmetical principle that we will encounter frequently is collection. For example BΣ_n is the so-called collection scheme for Σ_n-formulae. Roughly, BΣ_n says that the range of a Σ_n-definable function on a finite interval is again finite. A mathematical formulation is $\forall x{\leq}u\, \exists y\; \sigma(x,y) \to \exists z\, \forall x{\leq}u\, \exists y{\leq}z\; \sigma(x,y) \in \Sigma_n$ may contain other variables too.

The least number principle LΓ for a class of formulas is the collection $\exists x\; \varphi(x) \to \exists x\; (\varphi(x) \wedge \forall y{<}x\; \neg\varphi(y))$ for $\varphi \in \Gamma$.

2.3 Numberized Theories

The notion of interpretability applies to any pair of theories and not necessarily need they contain any arithmetic. However, in this paper we will prove that $U \rhd V$ can in various occasions be equivalent to other properties that are stated in terms of numbers. For example, in certain situations we have that $U \rhd V$ is equivalent to U proving all the Π_1^0 formulas that V does. Clearly, in this situation we should understand that U and V come with a natural interpretation of numbers.

Definition 1. *We will call a pair* $\langle U, k \rangle$ *a numberized theory if* k : $U \rhd S_2^1$. *A theory* U *is* numberizable *or* arithmetical *if for some* j, $\langle U, j \rangle$ *is a numberized theory.*

From now on, we shall only consider numberizable or numberized theories. Often however, we will fix a numberization j and reason about the theory $\langle U, j \rangle$ as if it were formulated in the language of arithmetic.

A disadvantage of doing so is clearly that our statements may be somehow misleading; when we think of, e.g., ZFC we do not like to think of it as coming with a fixed numberization. However, for the kind of characterizations treated in this paper, it is really needed to have numbers around.

In this paper, we shall always assume that our theories are sequential. Basically, sequentiality means that any finite sequence of objects can be coded. We refer the reader to [22] for a proper treatment on sequentiality. All known examples of first order numberized theories that are not sequential however are rather artificial and as such our restriction to sequential theories does not exclude any of the well known numberized theories.

2.4 Metamathematics in Numberized Theories

On many occasions, we want to represent numbers by terms (numerals) and then consider the code of that term. It is not a good idea to represent a number n by

$$\overbrace{S \ldots S}^{n \text{ times}} 0.$$

For, the length of this object is $n + 1$ whence its code is about 2^{n+1} and we would like to avoid the use of exponentiation. In the setting of weaker arithmetics it is common practice to use so-called *efficient numerals*. These numerals are defined by recursion as follows. $\bar{0} = 0$; $\overline{2 \cdot n} = (SS0) \cdot \bar{n}$ and $\overline{2 \cdot n + 1} = S((SS0) \cdot \bar{n})$. Clearly, these numerals implement the system of dyadic notation which perfectly ties up with the PIND principle. Often we shall refrain between distinguishing n from its numeral $\bar{n}$ or even the Gödel number $\ulcorner \bar{n} \urcorner$ of its numeral.

As we want to do arithmetization of syntax, our theories should be coded in a simple way. We will assume that all our theories U have an axiom set that is decidable in polynomial time. In particular, we require that there is some formula $\mathsf{Axiom}_U(x)$ which is Δ_1^b (both the formula and its negation are provably equivalent to a Σ_1^b formula) in S_2^1, with

$$\mathsf{S}_2^1 \vdash \mathsf{Axiom}_U(\varphi) \text{ iff } \varphi \text{ is an axiom of } U.$$

The choice of Δ_1^b-axiomatizations is also motivated by Lemma 1 below. Most natural theories like ZFC or PA indeed have Δ_1^b-axiomatizations. Moreover, by a sharpening of Craig's trick, any recursive theory is deductively equivalent to one with a Δ_1^b-axiomatization.

We shall employ the standard techniques and concepts necessary for the arithmetization of syntax. Thus, we shall work with provability predicates $\Box_U$ corresponding uniformly to arithmetical theories U. We shall adhere to the standard dot notation so that, for example, $\Box_U\varphi(\dot{x})$ denotes a formula with one free variable x so that for each value of x, $\Box_U\varphi(\dot{x})$ is provably equivalent to $\Box_U\varphi(\overline{x})$.

We shall always write the formalized version of a concept in sans-serif style. For example, $\mathsf{Proof}_U(p,\varphi)$ stands for the formalization of "p is a U-proof of φ", $\mathsf{Con}(U)$ stands for the formalization of "U is a consistent theory" and so forth. It is known that for theories U with a poly-time axiom set, the formula $\mathsf{Proof}_U(p,\varphi)$ can be taken to be in Δ_1^b being a poly-time decidable predicate. Again, [1] and [6] are adequate references.

For already really weak theories T we have Σ_1-completeness in the sense that T proves any true Σ_1 sentence. However, proofs of Σ_1-sentences σ are multi-exponentially big, that is, 2_n^σ for some n depending on σ. (See e.g., [6].) As such, we cannot expect that we can formalize the Σ_1 completeness theorem in theories where exponentiation is not necessarily total.

However, for $\exists\Sigma_1^b$-formulas we do have a completeness theorem (see [1]) in bounded arithmetic. From now on, we shall often write a supindex to a quantifier to specify the domain of quantification.

Lemma 1. *If* $\alpha(x) \in \exists\Sigma_1^b$, *then*

$$\mathsf{S}_2^1 \vdash \forall x \, [\alpha(x) \rightarrow \exists p \, \mathsf{Proof}_U(p,\alpha(\dot{x}))].$$

This holds for any reasonable arithmetical theory U. Moreover, we have also a formalized version of this statement.

$$\mathsf{S}_2^1 \vdash \forall^{\exists\Sigma_1^b}\alpha \, \Box_{\mathsf{S}_2^1}(\forall x \, [\dot{\alpha}(x) \rightarrow \exists p \, \mathsf{Proof}_U(p,\dot{\alpha}(\dot{x}))]).$$

2.5 Consistency and Reflexive Theories

Since Gödel's second incompleteness theorem, we know that no recursive theory that is consistent can prove its own consistency. For a large class of natural theories we do have a good approximation of proving consistency though. A theory is *reflexive* if it proves the consistency of all of its finite subtheories. Reflexivity is a natural notion and most natural non-finitely axiomatized theories are reflexive like, for example, primitive recursive arithmetic and PA.

Many meta-mathematical statements involve the notion of reflexivity. There exist various ways in which reflexivity can be formalized, and throughout the literature we can find many different formalizations. For stronger theories, all these formalizations coincide. But for weaker theories, the differences are essential. We give some formalizations of reflexivity.

1. $\forall n \; U \vdash \mathsf{Con}(U[n])$ where $U[n]$ denotes the conjunction of the first n axioms of U.
2. $\forall n \; U \vdash \mathsf{Con}(U \restriction n)$ where $\mathsf{Con}(U \restriction n)$ denotes that there is no proof of falsity using only axioms of U with Gödel numbers $\leq n$.
3. $\forall n \; U \vdash \mathsf{Con}_n(U)$ where $\mathsf{Con}_n(U)$ denotes that there is no proof of falsity with a proof p where p has the following properties. All non-logical axioms of U that occur in p have Gödel numbers $\leq n$. All formulas φ that occur in p have a logical complexity $\rho(\varphi) \leq n$. Here ρ is some complexity measure that basically counts the number of quantifier alternations in φ. Important features of this ρ are that for every n, there are truth predicates for formulas with complexity n. Moreover, the ρ-measure of a formula should be more or less (modulo some poly-time difference, see Remark 2) preserved under translations. Examples of such a ρ are given in [19] and [25]. An more recent version will appear in [24].

Note that for the first notion it is not even clear that $U[n]$ would make much sense for weak arithmetics, since the code of $U[n]$ is exponential in the size of n. This can be mitigated only in case $U[n]$ occurs under the scope of a provability predicate and the n is quantified outside this scope as we shall see in Lemma 4.

It is clear that $(2) \Rightarrow (3)$ can be proven in a reasonably weak base theory. For the corresponding provability notions, the implication reverses. In this paper, our notion of reflexivity shall be the third one.

We shall write $\square_{U,n}\varphi$ for $\neg\mathsf{Con}_n(U + \neg\varphi)$ or, equivalently,

$$\exists p \; \mathsf{Proof}_{U,n}(p, \varphi).$$

Here, $\mathsf{Proof}_{U,n}(p, \varphi)$ denotes that p is a U-proof of φ with all axioms in p are $\leq n$ and for all formulas ψ that occur in p, we have $\rho(\psi) \leq n$.

Remark 1. *An inspection of the proof of provable Σ_1-completeness (Lemma 1) gives us some more information. The proof p that witnesses the provability in U of some $\exists \Sigma_1^b$-sentence α, can easily be taken so that all axioms occurring in p are about as big and complex as α. Thus, from α, we get for some n (depending linearly on α) that $\mathsf{Proof}_{U,n}(p, \alpha)$.*

If we wish to emphasize the fact that our theories are not necessarily in the language of arithmetic, but just can be numberized, our formulations of reflexivity should be slightly changed. For example, (3) will for some $\langle U, j \rangle$ look like $j : U \rhd \mathsf{S}_2^1 + \{\mathsf{Con}_n(U) \mid n \in \omega\}$. This route is followed, for example in [23, 21].

We know that regular consistency is provably equivalent to Π_1 reflection. That is, provably $\mathsf{Con}(V) \equiv \forall^{\Pi_1^0} \pi \, (\square_V \pi \to \pi)$. However, for restricted consistency we do not have a natural corresponding principle and in particular, if U is a reflexive theory, we do not necessarily have any reflection principles. That is, we do not have $U \vdash \square_V \varphi \to \varphi$ for some natural $V \subset U$ and for some natural class of formulae φ. We do have, however, a weak form of $\forall \Pi_1^b$-reflection. This is expressed in the following lemma.

Lemma 2. *Let U be a theory of which S_2^1 can prove that it is reflexive. Then*

$$\mathsf{S}_2^1 \vdash \forall^{\forall \Pi_1^b} \pi \, \forall n \, \square_U \forall x \, (\square_{U,n} \pi(\dot{x}) \to \pi(x)).$$

Proof. Reason in S_2^1 and fix π and n. Let m be such that we have (see Lemma 1 and Remark 1)

$$\square_U \forall x \, (\neg \pi(x) \to \square_{U,m} \neg \pi(\dot{x})).$$

Furthermore, let $k := \max\{n, m\}$. Now, reason in U, fix some x and assume $\square_{U,n} \pi(x)$. Thus, clearly also $\square_{U,k} \pi(x)$. If now $\neg \pi(x)$, then also $\square_{U,k} \neg \pi(x)$, whence $\square_{U,k} \bot$. This contradicts the reflexivity, whence $\pi(x)$. As x was arbitrary we get $\forall x \, (\square_{U,n} \pi(x) \to \pi(x))$. $\dashv$

We note that this lemma also holds for the other notions of restricted provability we introduced in this subsection.

3 Formalized Interpretability

As we already mentioned, our notion of interpretability is the one studied by Tarski et al in [17]. In that notion, any axiom needs to be provable after translation. Under some fairly weak conditions this implies that also theorems are translated to theorems. However, in the domain of bounded arithmetics we do not generally have this. In the realm of formalized interpretation therefore, there has been a tendency to consider a small adaptation of the original notion Tarski. This adaptation as introduced by Visser is called *smooth* interpretability. In this subsection we shall exactly define this notion and see how it relates to other notions of formalized interpretability. In various ways, one can hold that theorems interpretability as discussed below is actually the more natural formalized version of interpretability.

The theories that we study in this paper are theories formulated in first order predicate logic. All theories have a finite signature that contains identity. For simplicity we shall assume that all our theories are formulated in a purely relational way. Here is the formal definition of a relative interpretation.

Definition 2. *A translation k of the language of a theory S into the language of a theory T is a pair $\langle \delta, F \rangle$ for which the following holds.*

The first component δ, is called the domain specifier *and is a formula in the language of T with a single free variable. This formula is used to specify the domain of our interpretation.*

The second component, F, is a finite map that sends relation symbols R (including identity) from the language of S, to formulas $F(R)$ in the language of T. We demand for all R that the number of free variables of $F(R)$ equals the arity of R.[2] Recursively we define the translation φ^k of a formula φ in the language of S as follows.

- $(R(\vec{x}))^k = F(R)(\vec{x})$;

- $(\varphi \wedge \psi)^k = \varphi^k \wedge \psi^k$ *and likewise for other boolean connectives; (in particular, this implies $\perp^k = \perp$);*

- $(\forall x \, \varphi(x))^k = \forall x \, (\delta(x) \to \varphi^k)$ *and analogously for the existential quantifier.*

A relative interpretation k of a theory S into a theory T is a translation $\langle \delta, F \rangle$ so that $T \vdash \varphi^k$ for all axioms φ of S.

To formalize insights about interpretability in weak meta-theories like S_2^1 we need to be very careful. Definitions of interpretability that are unproblematically equivalent in a strong theory like, say, $I\Sigma_1$ diverge in weak theories. As we shall see, the major source of problems is the absence of $B\Sigma_1$.

In this subsection, we study various divergent definitions of interpretability. We start by making an elementary observation on interpretations. Basically, the next definition and lemma say that translations transform proofs into translated proofs.

---

[2] Formally, we should be more precise and specify our variables.

Definition 3. *Let k be a translation. By recursion on a proof p in natural deduction we define the translation of p under k, we write p^k. For this purpose, we first define $k(\varphi)$ for formulae φ to be[3]*

$$\bigwedge_{x_i \in \mathsf{FV}(\varphi)} \delta(x_i) \to \varphi^k.$$

Here $\mathsf{FV}(\varphi)$ denotes the set of free variables of φ. Clearly, this set cannot contain more than $|\varphi|$ elements, whence $k(\varphi)$ will not be too big. Obviously, for sentences φ, we have $k(\varphi) = \varphi^k$.

If p is just a single assumption φ, then p^k is $k(\varphi)$. The translation of the proof constructions are defined precisely in such a way that we can prove Lemma 3 below. For example, the translation of

$$\frac{\varphi \quad \psi}{\varphi \wedge \psi}$$

will be

$$\frac{[\bigwedge_{x_i \in \mathsf{FV}(\varphi \wedge \psi)} \delta(x_i)]_1}{\dfrac{\bigwedge_{x_i \in \mathsf{FV}(\varphi)} \delta(x_i) \quad \bigwedge_{x_i \in \mathsf{FV}(\varphi)} \delta(x_i) \to \varphi^k \quad \mathcal{D}}{\dfrac{\dfrac{\varphi^k}{\varphi^k \wedge \psi^k} \qquad \dfrac{}{\psi^k}}{\bigwedge_{x_i \in \mathsf{FV}(\varphi \wedge \psi)} \delta(x_i) \to \varphi^k \wedge \psi^k}}} \to I,1$$

where $\mathcal{D}$ is just a symmetric copy of the part above φ^k. We note that the translation of the proof constructions is available[4] in S^1_2, as the number of free variables in $\varphi \wedge \psi$ is bounded by $|\varphi \wedge \psi|$.

Lemma 3. *If p is a proof of a sentence φ with assumptions in some set of sentences Γ, then for any translation k, p^k is a proof of φ^k with assumptions in Γ^k.*

Proof. Note that the restriction on sentences is needed. For example

$$\frac{\forall x\ \varphi(x) \quad \forall x\ (\varphi(x) \to \psi(x))}{\psi(x)}$$

but

$$\frac{(\forall x\ \varphi(x))^k \quad (\forall x\ (\varphi(x) \to \psi(x)))^k}{\delta(x) \to \psi^k(x)}$$

and in general $\nvdash (\delta(x) \to \psi^k) \leftrightarrow \psi^k$. The lemma is proved by induction on p. To account for formulas in the induction, we use the notion $k(\varphi)$

---

[3]To be really precise we should say that, for example, we let smaller x_i come first in $\bigwedge_{x_i \in \mathsf{FV}(\varphi)} \delta(x_i)$.

[4]More efficient translations on proofs are also available. However they are less uniform.

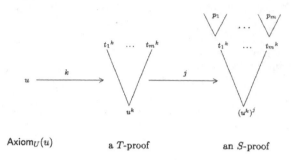

FIGURE 1 Transitivity of interpretability

from Definition 3, which is tailored precisely to let the induction go through. ⊣

Remark 2. *The proof translation leaves all the structure invariant. Thus, there is a provably total (in S_2^1) function f such that, if p is a U, n-proof of φ, then p^k is a proof of φ^k, where p^k has the following properties. All axioms in p^k are $\leq f(n, k)$ and all formulas ψ in p^k have $\rho(\psi) \leq f(n, k)$.*

There are various reasons to give, why we want the notion of interpretability to be provably transitive, that is, provably $S \rhd U$ whenever both $S \rhd T$ and $T \rhd U$. The obvious way of proving this would be by composing (doing the one after the other) two interpretations. Thus, if we have $j : S \rhd T$ and $k : T \rhd U$ we would like to have $j \circ k : S \rhd U$ where $j \circ k$ denotes a natural composition of translations.

If we try to perform a proof as depicted in Figure 1, at a certain point we would like to collect the S-proofs $p_1, \cdots, p_m$ of the j-translated T-axioms used in a proof of a k-translation of an axiom u of U, and take the maximum of all such proofs. But to see that such a maximum exists, we precisely need Σ_1-collection.

However, it is desirable to also reason about interpretability in the absence of BΣ_1. A trick is needed to circumvent the problem of the unprovability of transitivity (and many other elementary desiderata).

One way to solve the problem is by switching to a notion of interpretability where the needed collection has been built in. This is the notion of smooth (axioms) interpretability as in Definition 4. In this

In S_2^1:

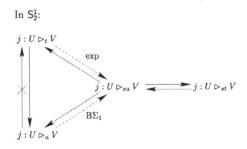

FIGURE 2 Versions of relative interpretability. The dotted arrows indicate that an additional condition is needed in our proof; the condition written next to it. The arrow with a cross through it, indicates that we know that the implication fails in S_2^1.

paper we shall mean by interpretability, unless mentioned otherwise, always smooth interpretability. In the presence of $B\Sigma_1$ this notion will coincide with the earlier defined notion of interpretability, as Theorem 1 tells us.

Definition 4. *We define the notions of axioms interpretability $\rhd_a$, theorems interpretability $\rhd_t$, smooth axioms interpretability $\rhd_{sa}$ and smooth theorems interpretability $\rhd_{st}$.*

$$
\begin{aligned}
j : U \rhd_a V &:= \forall v\, \exists p\ (\mathsf{Axiom}_V(v) \to \mathsf{Proof}_U(p, v^j)) \\
j : U \rhd_t V &:= \forall \varphi\, \forall p\, \exists p'\ (\mathsf{Proof}_V(p, \varphi) \to \mathsf{Proof}_U(p', \varphi^j)) \\
j : U \rhd_{sa} V &:= \forall x\, \exists y\, \forall v{\leq}x\, \exists p{\leq}y\ (\mathsf{Axiom}_V(v) \to \mathsf{Proof}_U(p, v^j)) \\
j : U \rhd_{st} V &:= \forall x\, \exists y\, \forall \varphi{\leq}x\, \forall p{\leq}x\, \exists p'{\leq}y \\
&\qquad (\mathsf{Proof}_V(p, \varphi) \to \mathsf{Proof}_U(p', \varphi^j))
\end{aligned}
$$

It is now easy to see that $\rhd_a$ is indeed provably transitive over very weak base theories. For $\rhd_t$ this follows almost directly from the definition.

Theorem 1. *In S_2^1 we have all the arrows as depicted in Figure 2.*

Proof. We shall only comment on the arrows that are not completely trivial.

- $T \vdash j : U \rhd_a V \to j : U \rhd_{sa} V$, if $T \vdash B\Sigma_1$. So, reason in T and suppose

$$\forall v\, \exists p\ (\mathsf{Axiom}_V(v) \to \mathsf{Proof}_U(p, v^j)).$$

If we fix some x, we get

$$\forall v \leq x \, \exists p \, (\mathsf{Axiom}_V(v) \to \mathsf{Proof}_U(p, v^j)).$$

By $B\Sigma_1$ we get the required

$$\exists y \, \forall v \leq x \, \exists p \leq y \, (\mathsf{Axiom}_V(v) \to \mathsf{Proof}_U(v^j)).$$

It is not clear if $T \vdash B\Sigma_1^-$, parameter-free collection, is a necessary condition.

- $\mathsf{S}_2^1 \nvdash j : U \rhd_a V \to j : U \rhd_t V$. A counter-example is given in [18].
- $T \vdash j : U \rhd_t V \to j : U \rhd_{sa} V$, if $T \vdash \mathsf{exp}$. If V is reflexive, we get by Corollary 3 that $\vdash U \rhd_t V \leftrightarrow U \rhd_{sa} V$. However, different interpretations are used to witness the different notions of interpretability in this case. If $T \vdash \mathsf{exp}$, we reason as follows. We reason in T and suppose that $\forall \varphi \, \forall p \, \exists p' \, (\mathsf{Proof}_V(p, \varphi) \to \mathsf{Proof}_U(p', \varphi^j))$. We wish to see

$$\forall x \, \exists y \, \forall v \leq x \, \exists p \leq y \, (\mathsf{Axiom}_V(v) \to \mathsf{Proof}_U(v^j)). \tag{1}$$

So, we pick x arbitrarily and consider[5] $\nu := \bigwedge\limits_{\mathsf{Axiom}_V(v_i) \wedge v_i \leq x} v_i$. Notice that in the worst case, for all $y \leq x$, we have $\mathsf{Axiom}_V(y)$, whence the length of ν can be bounded by $x \cdot |x|$. Thus, ν itself can be bounded by x^x, which exists whenever $T \vdash \mathsf{exp}$. Clearly, $\exists p \, \mathsf{Proof}_V(p, \nu)$ whence by our assumption $\exists p' \, \mathsf{Proof}_U(p', \nu^j)$. In a uniform way, with just a slightly larger proof p'', every $v_i{}^j$ can be extracted from the proof p' of ν^j. We may take this $p'' \approx y$ to obtain (1). Note that $T \vdash \mathsf{exp}$ is not a necessary condition since $\rhd_t$ implies $\rhd_a$ and if we have $B\Sigma_1$ the latter implies $\rhd_{sa}$.

- $\mathsf{S}_2^1 \vdash j : U \rhd_{sa} V \to j : U \rhd_{st} V$. So, we wish to see that

$$\forall x \, \exists y \, \forall \varphi \leq x \, \forall p \leq x \, \exists p' \leq y \, (\mathsf{Proof}_V(p, \varphi) \to \mathsf{Proof}_U(p', \varphi^j))$$

from the assumption that $j : U \rhd_{sa} V$. So, we pick x arbitrarily. If now for some $p \leq x$ we have $\mathsf{Proof}_V(p, \varphi)$, then clearly $\varphi \leq x$ and all axioms v_i of V that occur in p are $\leq x$. By our assumption $j : U \rhd_{sa} V$, we can find a y_0 such that we can find proofs $p_i \leq y_0$ for all the $v_i{}^j$. Now, with some sloppy notation, let $p^j[v_i^j/p_i]$ denote the j-translation of p where each j-translated axiom v_i^j is replaced by p_i.

Clearly, $p^j[v_i^j/p_i]$ is a proof for φ^j. The size of this proof can be estimated (again with sloppy notations):

$$p^j[v_i^j/p_i] \leq p^j[v_i^j/y_0] \leq (p^j)^{|y_0|} \leq (x^j)^{|y_0|}.$$

The latter bound is clearly present in S_2^1. ⊣

---

[5] To see that ν exists, we seem to also use some collection; we collect all the $v_i \leq x$ for which $\mathsf{Axiom}_V(v_i)$. However, it is not hard to see that we can consider ν also without collection since we use a natural coding.

We note that we have many admissible rules from one notion of interpretability to another. For example, by Buss's theorem on the provably total recursive functions of S_2^1, it is not hard to see that

$$S_2^1 \vdash j : U \rhd_a V \Rightarrow S_2^1 \vdash j : U \rhd_t V.$$

In the rest of this paper, we shall at most places no longer write subscripts to the $\rhd$'s. Our reading convention is then that we take that notion of interpretability that is best to perform the argument. Often this is just smooth interpretability $\rhd_s$, which from now on is the notation for $\rhd_{sa}$.

Moreover, in [18] some sort of conservation result concerning $\rhd_a$ and $\rhd_s$ is proved. For a considerable class of formulas φ and theories T, and for a considerable class of arguments we have that $T \vdash \varphi_a \Rightarrow T \vdash \varphi_s$. Here φ_a denotes the formula φ using the notion $\rhd_a$ and likewise for φ_s. Thus indeed, in many cases a sharp distinction between the notions involved is not needed.

We could also consider the following notion of interpretability.

$$j : U \rhd_{st_1} V := \forall x \, \exists y \, \forall \, \varphi {\leq} x \, \exists \, p' {\leq} y \, (\Box_V \varphi \to \mathsf{Proof}_U(p', \varphi^j))$$

Clearly, $j : U \rhd_{st_1} V \to U \rhd_{st} V$. However, for the reverse implication one seems to need $B\Pi_1^-$. Also, a straightforward proof of $U \vdash id : U \rhd_{st_1} U$ seems to need $B\Pi_1^-$. Thus, the notion $\rhd_{st_1}$ seems to say more on the nature of a theory than on the nature of interpretability.

4 Cuts and Induction

Inductive reasoning is a central feature of everyday mathematical practice. We are so used to it, that it enters a proof almost unnoticed. It is when one works with weak theories and in the absence of sufficient induction, that its all pervading nature is best felt.

A main tool to compensate for the lack of induction are the so-called definable cuts. They are definable initial segments of the natural numbers of a possibly non-standard model that possess some desirable properties that we could not infer for all numbers to hold by means of induction.

The idea is really simple. So, if we can derive $\varphi(0) \wedge \forall x \, (\varphi(x) \to \varphi(x + 1))$ and do not have access to an induction axiom for φ, we just consider $J(x) : \forall y {\leq} x \, \varphi(y)$. Clearly J now defines an initial segment on which φ holds. As we shall see, for a lot of reasoning we can restrict ourselves to initial segments rather than quantifying over all numbers.

4.1 Basic Properties of Cuts

Throughout the literature one can find some variations on the definition of a cut. At some places, a cut is only supposed to be an initial segment of the natural numbers. At other places some additional closure properties are demanded. By a well known technique due to Solovay (see for example [6]) any definable initial segment can be shortened in a definable way, so that it has a lot of desirable closure properties. Therefore, and as we almost always need the closure properties, we include them in our definition.

Definition 5. *A definable U-cut is a formula $J(x)$ with only x free, for which we have the following.*

1. $U \vdash J(0) \wedge \forall x\ (J(x) \to J(x+1))$
2. $U \vdash J(x) \wedge y \leq x \to J(y)$
3. $U \vdash J(x) \wedge J(y) \to J(x+y) \wedge J(x \cdot y)$
4. $U \vdash J(x) \to J(\omega_1(x))$

We shall sometimes also write $x \in J$ instead of $J(x)$. A first fundamental insight about cuts is the principle of *outside big, inside small*. Although not every number x is in J, we can find for every x a proof p_x that witnesses $x \in J$.

Lemma 4. *Let T and U be reasonable arithmetical theories and let J be a U-cut. We have that*

$$T \vdash \forall x\ \Box_U J(x).$$

Actually, we can have the quantifier over all cuts within the theory T, that is

$$T \vdash \forall^{U\text{-Cut}} J \forall x\ \Box_U J(x).$$

Proof. Let us start by making the quantifier $\forall^{U\text{-Cut}} J$ a bit more precise. By $\forall^{U\text{-Cut}} J$ we shall mean $\forall J\ (\Box_U \mathsf{Cut}(J) \to \ldots)$. Here $\mathsf{Cut}(J)$ is the definable function that sends the code of a formula χ with one free variable to the code of the formula that expresses that χ defines a cut.

For a number a, we start with the standard proof of $J(0)$. This proof is combined with $a-1$ many instantiations of the standard proof of $\forall x\ (J(x) \to J(x+1))$. In the case of weaker theories, we have to switch to efficient numerals to keep the bound of the proof within range. ⊣

Remark 3. *The proof sketch actually tells us that (provably in S_2^1) for every U-cut J, there is an $n \in \omega$ such that $\forall x\ \Box_{U,n} J(x)$.*

Lemma 5. *Cuts are provably closed under terms, that is*

$$T \vdash \forall^{U\text{-Cut}} J \forall^{\mathsf{Term}} t\ \Box_U \forall \vec{x} {\in} J\ t(\vec{x}) \in J.$$

Proof. By an easy induction on terms, fixing some U-cut J. Prima facie this looks like a Σ_1-induction but it is easy to see that the proofs have poly-time (in t) bounds, whence the induction is $\Delta_0(\omega_1)$. ⊣

As all U-cuts are closed under $\omega_1(x)$ and the smash function $\sharp$, simply relativizing all quantors to a cut is an example of an interpretation of S_2^1 in U. We shall always denote both the cut and the interpretation that it defines by the same letter.

4.2 Cuts and the Henkin Construction

It is well known that we can perform the Henkin construction in a rather weak meta-theory. As the Henkin model has a uniform description, we can link it to interpretations. The following theorem makes this precise.

Theorem 2. *If* $U \vdash \mathsf{Con}(V)$, *then* $U \rhd V$.

Early treatments of this theorem were given in [26] and [7]. A first fully formalized version was given in [2]. A proof of Theorem 2 would closely follow the Henkin construction.

Thus, first the language of V is extended so that it contains a witness $c_{\exists x \varphi(x)}$ for every existential sentence $\exists x \; \varphi(x)$. Then we can extend V to a maximal consistent V' in the enriched language, containing all sentences of the form $\exists x \varphi(x) \to \varphi(c_{\exists x \varphi(x)})$. This V' can be seen as a term model with a corresponding truth predicate. Clearly, if $V \vdash \varphi$ then $\varphi \in V'$. It is not hard to see that V' is representable (close inspection yields a Δ_2-representation) in U.

At first sight the argument uses quite some induction in extending V to V'. Miraculously enough, the whole argument can be adapted to S_2^1. The trick consists in replacing the use of induction by employing definable cuts as is explained above. We get the following theorem.

Theorem 3. *For any numberizable theories* U *and* V, *we have that*

$$S_2^1 \vdash \Box_U \mathsf{Con}(V) \to \exists k \; (k : U \rhd V \; \& \; \forall \varphi \; \Box_U(\Box_V \varphi \to \varphi^k)).$$

Proof. A proof can be found in [18]. Actually something stronger is proved there. Namely, that for some standard number m we have

$$\forall \varphi \; \exists p {\leq} \omega_1^m(\varphi) \; \mathsf{Proof}_U(p, \Box_V \varphi \to \varphi^k).$$

$\dashv$

As cuts have nice closure properties, many arguments can be performed within that cut. The numbers in the cut will, so to say, play the role of the normal numbers. It turns out that the whole Henkin argument can be carried out using only the consistency on a cut.

We shall write $\Box_T^J \varphi$ for $\exists p {\in} J \; \mathsf{Proof}_T(p, \varphi)$. Thus, it is also clear what $\Diamond_T^J \varphi$ and $\mathsf{Con}^J(V)$ mean.

Theorem 4. *We have Theorem 3 also in the following form.*

$$T \vdash \forall^{U\text{-Cut}} I \; \left[\Box_U \mathsf{Con}^I(V) \to \exists k \; (k : U \rhd V \; \& \; \forall \varphi \; \Box_U(\Box_V \varphi \to \varphi^k)) \right]$$

Proof. By close inspection of the proof of Theorem 3. All operations on hypothetical proofs p can be bounded by some $\omega_1^k(p)$, for some standard k. As I is closed under $\omega_1(x)$, all the bounds remain within I. ⊣

We conclude this subsection with two asides, closely related to the Henkin construction.

Lemma 6. *Let U contain S_2^1. We have that $U \vdash \mathsf{Con}(\mathsf{Pred})$. Here, $\mathsf{Con}(\mathsf{Pred})$ is a natural formalization of the statement that predicate logic is consistent.*

Proof. By defining a simple (one-point) model within S_2^1. ⊣

Remark 4. *If U proves $L\Delta_2^0$, then it holds that $U \rhd V$ iff V is interpretable in U by some interpretation that maps identity to identity.*

Proof. Suppose $j : U \rhd V$ with $j = \langle \delta, F \rangle$. We can define $j' := \langle \delta', F' \rangle$ with $\delta'(x) := \delta(x) \wedge \forall y{<}x \ (\delta(y) \rightarrow y{\neq}^j x)$. F' agrees with F on all symbols except that it maps identity to identity. By the minimal number principle we can prove $\forall x \ (\delta(x) \rightarrow \exists x' \ (x'{=}^j x) \wedge \delta'(x))$, and thus $\forall \vec{x} \ (\delta'(\vec{x}) \rightarrow (\varphi^j(\vec{x}) \leftrightarrow \varphi^{j'}(\vec{x})))$ for all formulae φ. ⊣

5 Pudlák's Lemma

In this section we will state and prove what is known as Pudlák's lemma. Moreover, we shall prove a very useful consequence of this lemma. Roughly speaking, Pudlák's lemma tells us how interpretations bear on the models that they induce. Therefore, let us first see how interpretations and models are related.

5.1 Interpretations and Models

We can view interpretations $j : U \rhd V$ as a way of defining uniformly a model $\mathcal{N}$ of V inside a model $\mathcal{M}$ of U. Interpretations in foundational papers mostly bear the guise of a uniform model construction.

Definition 6. *Let $j : U \rhd V$ with $j = \langle \delta, F \rangle$. If $\mathcal{M} \models U$, we denote by $\mathcal{M}^j$ the following model.*

- $|\mathcal{M}^j| = \{x \in |\mathcal{M}| \mid \mathcal{M} \models \delta(x)\}/ \equiv$, *where $a \equiv b$ iff $\mathcal{M} \models a =^j b$.*
- $\mathcal{M}^j \models R(\alpha_1, \ldots, \alpha_n)$ *iff $\mathcal{M} \models F(R)(a_1, \ldots, a_n)$, for some $a_1 \in \alpha_1$, $\ldots, a_n \in \alpha_n$.*

The fact that $j : U \rhd V$ is now reflected in the observation that, whenever $\mathcal{M} \models U$, then $\mathcal{M}^j \models V$.

On many occasions viewing interpretations as uniform model constructions provides the right heuristics.

5.2 Pudlák's Isomorphic Cut

Pudlák's lemma is central to many arguments in the field of interpretability logics. It provides a means to compare a model $\mathcal{M}$ of U and its internally defined model $\mathcal{M}^j$ of V if $j : U \rhd V$. If U has full induction, this comparison is fairly easy.

Theorem 5. *Suppose $j : U \rhd V$ and U has full induction. Let $\mathcal{M}$ be a model of U. We have that $\mathcal{M} \preceq_{\mathsf{end}} \mathcal{M}^j$ via a definable embedding.*

Proof. If U has full induction and $j : U \rhd V$, we may by Remark 4 actually assume that j maps identity in V to identity in U. Thus, we can define the following function.

$$f := \left\{ \begin{array}{l} 0 \mapsto 0^j \\ x + 1 \mapsto f(x) +^j 1^j \end{array} \right.$$

Now, by induction, f can be proved to be total. Note that full induction is needed here, as we have a-priori no bound on the complexity of 0^j and $+^j$. Moreover, it can be proved that $f(a + b) = f(a) +^j f(b)$, $f(a \cdot b) = f(a) \cdot^j f(b)$ and that $y \leq^j f(b) \to \exists a < b \ f(a) = y$. In other words, that f is an isomorphism between its domain and its co-domain and the co-domain is an initial segment of $\mathcal{M}^j$. $\dashv$

If U does not have full induction, a comparison between $\mathcal{M}$ and $\mathcal{M}^j$ is given by Pudlák's lemma, first explicitly mentioned in [15]. Roughly, Pudlák's lemma says that in the general case, we can find a definable U-cut I of $\mathcal{M}$ and a definable embedding $f : I \longrightarrow \mathcal{M}^j$ such that $f[I] \preceq_{\mathsf{end}} \mathcal{M}^j$.

In formulating the statement we have to be careful as we can no longer assume that identity is mapped to identity. A precise formulation of Pudlák's lemma in terms of an isomorphism between two initial segments can for example be found in [10]. We have chosen here to formulate and prove the most general syntactic consequence of Pudlák's lemma, namely that I and $f[I]$, as substructures of $\mathcal{M}$ and $\mathcal{M}^j$ respectively, make true the same Δ_0-formulas.

We shall make the quantifier $\exists^{j,J\text{-function}} h$ explicit in the proof of Pudlák's lemma. It basically means that h defines a function from a cut J to the $=^j$-equivalence classes of the numbers defined by the interpretation j.

Lemma 7 (Pudlák's Lemma).

$$S_2^1 \vdash j : U \rhd V \to$$

$$\exists^{U\text{-Cut}} J \ \exists^{j,J\text{-function}} h \ \forall^{\Delta_0} \varphi \ \Box_U \forall \vec{x} \in J \ (\varphi^j(h(\vec{x})) \leftrightarrow \varphi(\vec{x}))$$

Moreover, the h and J can be obtained uniformly from j by a function that is provably total in S_2^1.

Proof. Again, by $\exists^{U\text{-}\mathrm{Cut}} J \, \psi$ we shall mean $\exists J \, (\Box_U \mathsf{Cut}(J) \wedge \psi)$, where $\mathsf{Cut}(J)$ is the definable function that sends the code of a formula χ to the code of a formula that expresses that χ defines a cut. We apply a similar strategy for quantifying over j, J-functions. Given a translation j, the defining property for a relation H to be a j, J-function is

$$\forall \vec{x}, y, y' \in J \, (H(\vec{x}, y) \;\&\; H(\vec{x}, y') \to y =^j y').$$

We will often consider H as a function h and write for example $\psi(h(\vec{x}))$ instead of

$$\forall y \, (H(\vec{x}, y) \to \psi(y)).$$

The idea of the proof is very easy. Just map the numbers of U via h to the numbers of V so that 0 goes to 0^j and the mapping commutes with the successor relation. If we want to prove a property of this mapping, we might run into problems as the intuitive proof appeals to induction. And sufficient induction is precisely what we lack in weaker theories.

The way out here is to just put all the properties that we need our function h to possess into its definition. Of course, then the work is in checking that we still have a good definition. Being good means here that the set of numbers on which h is defined induces a definable U-cut.

In a sense, we want an (definable) initial part of the numbers of U to be isomorphic under h to an initial part of the numbers of V. Thus, h should definitely commute with successor, addition and multiplication. Moreover, the image of h should define an initial segment, that is, be closed under the smaller than relation. All these requirements are reflected in the definition of Goodsequence. Let δ denote the domain specifier of the translatio j. We define

$$
\begin{aligned}
\mathsf{Goodsequence}(\sigma, x, y) \quad := \quad & \mathsf{lh}(\sigma) = x + 1 \wedge \sigma_0 =^j 0^j \wedge \sigma_x =^j y \\
& \wedge \; \forall i \leq x \; \delta(\sigma_i) \\
& \wedge \; \forall i < x \; (\sigma_{i+1} =^j \sigma_i +^j 1^j) \\
& \wedge \; \forall k + l \leq x \; (\sigma_k +^j \sigma_l =^j \sigma_{k+l}) \\
& \wedge \; \forall k \cdot l \leq x \; (\sigma_k \cdot^j \sigma_l =^j \sigma_{k \cdot l}) \\
& \wedge \; \forall a \; (a \leq^j y \to \exists i \leq x \; \sigma_i =^j a).
\end{aligned}
$$

Subsequently, we define

$$
\begin{aligned}
H(x, y) \quad := \quad & \exists \sigma \; \mathsf{Goodsequence}(\sigma, x, y) \\
& \wedge \; \forall \sigma' \, \forall y' \, (\mathsf{Goodsequence}(\sigma', x, y') \to y =^j y'),
\end{aligned}
$$

and

$$J'(x) := \forall x' \leq x \, \exists y \; H(x', y).$$

Finally, we define J to be the closure of J' under $+$, $\cdot$ and $\omega_1(x)$. Now that we have defined all the machinery we can start the real proof. The reader is encouraged to see at what place which defining property is used in the proof.

We first note that $J'(x)$ indeed defines a U-cut. For $\square_U J'(0)$ you basically need sequentiality of U, and the translations of the identity axioms and properties of 0.

To see $\square_U \forall x\ (J'(x) \rightarrow J'(x+1))$ is also not hard. It follows from the translation of basic properties provable in V, like $x = y \rightarrow x+1 = y+1$ and $x + (y+1) = (x+y)+1$, etc. The other properties of Definition 5 go similarly.

We should now see that h is a j, J-function. This is quite easy, as we have all the necessary conditions present in our definition. Actually, we have

$$\square_U \forall\, x, y {\in} J\ (h(x){=}^j h(y) \leftrightarrow x = y) \tag{2}$$

The $\leftarrow$ direction reflects that h is a j, J-function. The $\rightarrow$ direction follows from elementary reasoning in U using the translation of basic arithmetical facts provable in V. So, if $x \neq y$, say $x < y$, then $x + (z + 1) = y$ whence $h(x){+}^j h(z + 1){=}^j h(y)$ which implies $h(x){\neq}^j h(y)$.

We are now to see that for our U-cut J and for our j, J-function h we indeed have that[6]

$$\forall^{\Delta_0} \varphi\ \square_U \forall\, \vec{x} {\in} J\ (\varphi^j(h(\vec{x})) \leftrightarrow \varphi(\vec{x})).$$

First we shall proof this using a seemingly Σ_1-induction. A closer inspection of the proof shall show that we can provide at all places sufficiently small bounds, so that actually an $\omega_1(x)$-induction suffices. We first proof the following claim.

Claim 1. $\forall^{\mathsf{Term}} t\ \square_U \forall\, \vec{x}, y \in J\ (t^j(h(\vec{x})){=}^j h(y) \leftrightarrow t(\vec{x}) = y)$

Proof. The proof is by induction on t. The basis is trivial. To see for example

$$\square_U \forall\, y {\in} J\ (0^j {=}^j h(y) \leftrightarrow 0 = y)$$

we reason in U as follows. By the definition of h, we have that $h(0){=}^j 0^j$, and by (2) we moreover see that $0^j {=}^j h(y) \leftrightarrow 0 = y$. The other base case, that is, when t is an atom, is precisely (2).

For the induction step, we shall only do $+$, as $\cdot$ goes almost completely the same. Thus, we assume that $t(\vec{x}) = t_1(\vec{x}) + t_2(\vec{x})$ and set out to prove

$$\square_U \forall\, \vec{x}, y {\in} J\ (t_1{}^j(h(\vec{x})){+}^j t_2{}^j(h(\vec{x})){=}^j h(y) \leftrightarrow t_1(\vec{x}) + t_2(\vec{x}) = y).$$

Within U:

---

[6]We use $h(\vec{x})$ as short for $h(x_0), \cdots, h(x_n)$.

← If $t_1(\vec{x}) + t_2(\vec{x}) = y$, then by Lemma 5, we can find y_1 and y_2 with $t_1(\vec{x}) = y_1$ and $t_2(\vec{x}) = y_2$. The induction hypothesis tells us that $t_1{}^j(h(\vec{x}))=^jh(y_1)$ and $t_2{}^j(h(\vec{x}))=^jh(y_2)$. Now by (2), $h(y_1 + y_2)=^jh(y)$ and by the definition of h we get that

$$
\begin{aligned}
h(y_1 + y_2) \quad &=^j \quad h(y_1)+^jh(y_2) \\
&=^j{}_{\text{i.h.}} \quad t_1{}^j(h(\vec{x}))+^jt_2{}^j(h(\vec{x})) \\
&=^j \quad (t_1(h(\vec{x})) + t_2(h(\vec{x})))^j.
\end{aligned}
$$

→ Suppose now

$$
t_1{}^j(h(\vec{x}))+^jt_2{}^j(h(\vec{x}))=^jh(y).
$$

Then clearly $t_1{}^j(h(\vec{x}))\leq^jh(y)$ whence by the definition of h we can find some $y_1 \leq y$ such that $t_1{}^j(h(\vec{x}))=^j h(y_1)$ and likewise for t_2 (using the translation of the commutativity of addition). The induction hypothesis now yields $t_1(\vec{x}) = y_1$ and $t_2(\vec{x}) = y_2$. By the definition of h, we get $h(y)=^jh(y_1)+^jh(y_2)=^jh(y_1 + y_2)$, whence by (2), $y_1 + y_2 = y$, that is, $t_1(\vec{x}) + t_2(\vec{x}) = y$.

⊣

We now prove by induction on $\varphi \in \Delta_0$ that

$$
\Box_U\forall\,\vec{x}{\in}J\ (\varphi^j(h(\vec{x})) \leftrightarrow \varphi(\vec{x})). \tag{3}
$$

For the base case, we consider that $\varphi \equiv t_1(\vec{x}) + t_2(\vec{x})$. We can now use Lemma 5 to note that

$$
\Box_U\forall\,\vec{x}{\in}J\ (t_1(\vec{x}) = t_2(\vec{x}) \leftrightarrow \exists y{\in}J\ (t_1(\vec{x}) = y \wedge t_2(\vec{x}) = y))
$$

and then use Claim 1, the transitivity of $=$ and its translation to obtain the result.

The boolean connectives are really trivial, so we only need to consider bounded quantification. We show (still within U) that

$$
\forall\,y, \vec{z}{\in}J\ (\forall\,x\leq^jh(y)\ \varphi^j(x, h(\vec{z})) \leftrightarrow \forall\,x\leq y\ \varphi(x, \vec{z})).
$$

← Assume $\forall\,x\leq y\ \varphi(x, \vec{z})$ for some $y, \vec{z} \in J$. We are to show $\forall\,x\leq^jh(y)\ \varphi^j(x, h(\vec{z}))$. Now, pick some $x\leq^jh(y)$ (the translation of the universal quantifier actually gives us an additional $\delta(x)$ which we shall omit for the sake of readability). Now by the definition of h we find some $y' \leq y$ such that $h(y') = x$. As $y' \leq y$, by our assumption, $\varphi(y', \vec{z})$ whence by the induction hypothesis $\varphi^j(h(y'), h(\vec{z}))$, that is $\varphi^j(x, h(\vec{z}))$. As x was arbitrarily $\leq^jh(y)$, we are done.

→ Suppose $\forall\,x\leq^jh(y)\ \varphi^j(x, h(\vec{z}))$. We are to see that $\forall\,x\leq y\ \varphi(x, \vec{z}))$. So, pick $x \leq y$ arbitrarily. Clearly $h(x)\leq^jh(y)$, whence, by our assumption $\varphi^j(h(x), h(\vec{z}))$ and by the induction hypothesis, $\varphi(x, \vec{z})$.

Note that in our proof we have used twice a Σ_1-induction; In Claim 1 and in proving (3). Let us now see that we can dispense with the Σ_1 induction.

In both cases, at every induction step, a constant piece p' of proof is added to the total proof. This piece looks every time the same. Only some parameters in it have to be replaced by subterms of t. So, the addition to the total proof can be estimated by $p'_a(t)$ which is about $\mathcal{O}(t^k)$ for some standard k and indeed, our induction was really but a bounded one. Both our inductions went over syntax and whence are available in S^1_2.

Note that in proving (3) we dealt with the bounded quantification by appealing to the induction hypothesis only once, followed by a generalization. So, fortunately we did not need to apply the induction hypothesis to all $x \leq y$, which would have yielded an exponential blow-up. ⊣

Remark 5. *Pudlák's lemma is valid already if we employ the notion of theorems interpretability rather than smooth interpretability. If we work with theories in the language of arithmetic, we can do even better. In this case, axioms interpretability can suffice. In order to get this, all arithmetical facts whose translations were used in the proof of Lemma 7 have to be promoted to the status of axiom. However, a close inspection of the proof shows that these facts are very basic and that there are not so many of them.*

If j is an interpretation with $j : \alpha \rhd \beta$, we shall sometimes call the corresponding isomorphic cut that is given by Lemma 7, the *Pudlák cut* of j and denote it by the corresponding upper case letter J.

5.3 A Consequence of Pudlák's Lemma

The following consequence of Pudlák's Lemma is simple, yet can be very useful. For simplicity we state the consequence for sentential extensions of some base theory T extending S^1_2. Thus, $\alpha \rhd \beta$ will be short for $(T + \alpha) \rhd (T + \beta)$.

Lemma 8. *(In S^1_2:) If $j : \alpha \rhd \beta$ then, for every $T + \beta$ cut I there exists a $T + \alpha$ cut J such that for every γ we have that*

$$j : \alpha \wedge \square^J \gamma \rhd \beta \wedge \square^I \gamma.$$

Proof. By a minor adaptation of the standard argument. First, we define Goodsequence.

$$\text{Goodsequence}(\sigma, x, y) \quad := \quad \text{lh}(\sigma) = x + 1 \wedge \sigma_0 =^j 0^j \wedge \sigma_x =^j y$$
$$\wedge \, \forall i {<} x \, (\sigma_{i+1} =^j \sigma_i +^j 1^j)$$
$$\wedge \, \forall k {+} l {\leq} x \, (\sigma_k +^j \sigma_l =^j \sigma_{k+l})$$
$$\wedge \, \forall k {\cdot} l {\leq} x \, (\sigma_k {\cdot}^j \sigma_l =^j \sigma_{k \cdot l})$$
$$\wedge \, \forall a \, (a {\leq}^j y \rightarrow \exists i {\leq} x \, \sigma_i =^j a)$$
$$\wedge \, \forall i {<} x \, I^j(\sigma_i)$$

Next, we define

$$H(x, y) \quad := \quad \exists \sigma \, \text{Goodsequence}(\sigma, x, y)$$
$$\wedge \, \forall \sigma' \, \forall y' \, (\text{Goodsequence}(\sigma', x, y') \rightarrow y =^j y'),$$

and

$$J'(x) := \forall \, x' {\leq} x \, \exists y \, H(x', y).$$

Finally, we define J to be the closure of J' under $+$, $\cdot$ and $\omega_1(x)$.

As before, one can see $H(x, y)$ as defining a function (modulo $=^j$), call it h, that defines an isomorphism between J and the image of J. Moreover, in the definition of Goodsequence we demanded that the image of h is a subset of I in the clause $\forall i {<} x \, I^j(\sigma_i)$.

It is easy to see that J' is closed under successor, that is, $J'(x) \rightarrow J'(x + 1)$. We only comment on the new ingredient of the image of h being a subset of I. However, $T + \beta \vdash I(x) \rightarrow I(x + 1)$, as I is a definable cut. As $j : T + \alpha \rhd T + \beta$, clearly $T + \alpha \vdash I^j(x) \rightarrow I^j(x +^j 1^j)$ and indeed J' is closed under successor.

$\dashv$

In the literature, Lemma 8 was known only for I to be the trivial cut of all numbers defined by $x = x$.

6 The Orey-Hájek Characterizations

This final section contains the most substantial part of the paper. We consider the diagram from Figure 3. It is well known that all the implications hold when both U and V are reflexive. This fact is referred to as the Orey-Hájek characterizations ([2], [14], [4], [5], [18], [20]) for interpretability. However, for the Π_1-conservativity part, we should also mention work by Guaspari, Lindström and Pudlák ([3], [12], [13], [15]).

In this section we shall comment on all the implications in Figure 3, and study the conditions on U, V and the meta-theory, that are necessary or sufficient.

Lemma 9. *In S_2^1 we can prove* $\forall n \, \Box_U \text{Con}_n(V) \rightarrow U \rhd V$.

Proof. The only requirement for this implication to hold, is that $U \vdash \text{Con}(\text{Pred})$. But, by our assumptions on U and by Lemma 6 this is

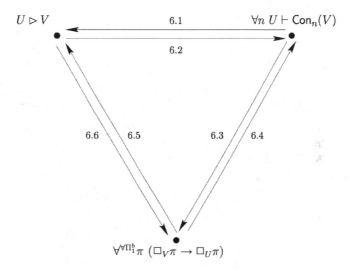

FIGURE 3 Characterizations of interpretability. The labels at the arrows are references to where in the paper this arrow is proven and what the conditions are for the arrow to hold. Moreover, we will discuss which conditions should hold for the base theory so that the implications become formalizable.

automatically satisfied.

Let us first give the informal proof. Thus, let $\mathsf{Axiom}_V(x)$ be the formula that defines the axiom set of V.

We now apply a trick due to Feferman and consider the theory V' that consists of those axioms of V up to which we have evidence for their consistency. Thus, $\mathsf{Axiom}_{V'}(x) := \mathsf{Axiom}_V(x) \wedge \mathsf{Con}_x(V)$.

We shall now prove that $U \rhd V$ in two steps. First, we will see that

$$U \vdash \mathsf{Con}(V').\tag{4}$$

Thus, by Theorem 2 we get that $U \rhd V'$. Second, we shall see that

$$V = V'.\tag{5}$$

To see (4), we reason in U, and assume for a contradiction that $\mathsf{Proof}_{V'}(p, \bot)$ for some proof p. We consider the largest axiom v that occurs in p. By assumption we have (in U) that $\mathsf{Axiom}_{V'}(v)$ whence $\mathsf{Con}_v(V)$. But, as clearly $V' \subseteq V$, we see that p is also a V-proof. We can now obtain a cut-free proof p' of $\bot$. Clearly $\mathsf{Proof}_{V,v}(p', \bot)$ and we have our contradiction.

If V' is empty, we cannot consider v. But in this case, $\mathsf{Con}(V') \leftrightarrow \mathsf{Con}(\mathsf{Pred})$, and by assumption, $U \vdash \mathsf{Con}(\mathsf{Pred})$.

We shall now see (5). Clearly $\mathbb{N} \models \mathsf{Axiom}_{V'}(v) \to \mathsf{Axiom}_V(v)$ for any $v \in \mathbb{N}$. To see that the converse also holds, we reason as follows.

Suppose $\mathbb{N} \models \mathsf{Axiom}_V(v)$. By assumption $U \vdash \mathsf{Con}_v(V)$, whence $\mathsf{Con}_v(V)$ holds on any model $\mathcal{M}$ of U. We now observe that $\mathbb{N}$ is an initial segment of (the numbers of) any model $\mathcal{M}$ of U, that is,

$$\mathbb{N} \preceq_{\mathsf{end}} \mathcal{M}.\tag{6}$$

As $\mathcal{M} \models \mathsf{Con}_v(V)$ and as $\mathsf{Con}_v(V)$ is a Π_1-sentence, we see that also $\mathbb{N} \models \mathsf{Con}_v(V)$. By assumption we had $\mathbb{N} \models \mathsf{Axiom}_V(v)$, thus we get that $\mathbb{N} \models \mathsf{Axiom}_{V'}(v)$. We conclude that

$$\mathbb{N} \models \mathsf{Axiom}_V(x) \leftrightarrow \mathsf{Axiom}_{V'}(x)\tag{7}$$

whence, that $V = V'$. As $U \vdash \mathsf{Con}(V')$, we get by Theorem 2 that $U \rhd V'$. We may thus infer the required $U \rhd V$.

It is not possible to directly formalize the informal proof. At (7) we concluded that $V = V'$. This actually uses some form of Π_1-reflection which is manifested in (6). The lack of reflection in the formal environment will be compensated by another sort of reflection, as formulated in Theorem 3.

Moreover, to see (4), we had to use a cut elimination. To avoid this, we shall need a sharper version of Feferman's trick.

Let us now start with the formal proof sketch and refer to [18] for more details. We shall reason in U. Without any induction we conclude

$\forall x\ (\mathsf{Con}_x(V) \to \mathsf{Con}_{x+1}(V))$ or $\exists x\ (\mathsf{Con}_x(V) \wedge \square_{V,x+1}\bot)$. In both cases we shall sketch a Henkin construction.

If $\forall x\ (\mathsf{Con}_x(V) \to \mathsf{Con}_{x+1}(V))$ and also $\mathsf{Con}_0(V)$, we can find a cut $J(x)$ with $J(x) \to \mathsf{Con}_x(V)$. We now consider the following non-standard proof predicate.

$$\square_W^* \varphi := \exists\, x {\in} J\ \square_{W,x}\varphi$$

We note that we have $\mathsf{Con}^*(V)$, where $\mathsf{Con}^*(V)$ of course denotes

$$\neg(\exists\, x {\in} J\ \square_{V,x}\bot).$$

As always, we extend the language on J by adding witnesses and define a series of theories in the usual way. That is, by adding more and more sentences (in J) to our theories while staying consistent (in our non-standard sense).

$$V = V_0 \subseteq V_1 \subseteq V_2 \subseteq \cdots \text{ with } \mathsf{Con}^*(V_i) \tag{8}$$

We note that $\square_{V_i}^*\varphi$ and $\square_{V_i}^*\neg\varphi$ is not possible, and that for $\varphi \in J$ we can not have $\mathsf{Con}^*(\varphi \wedge \neg\varphi)$. These observations seem to be too trivial to make, but actually many a non-standard proof predicate encountered in the literature does prove the consistency of inconsistent theories.

As always, the sequence (8) defines a cut $I \subseteq J$, that induces a Henkin set W and we can relate our required interpretation k to this Henkin set as was, for example, done in [18].

We now consider the case that for some fixed b we have $\mathsf{Con}_b(V) \wedge \square_{V,b+1}\bot$. We note that we can see the uniqueness of this b without using any substantial induction. Basically, we shall now do the same construction as before only that we now possibly stop at b.

For example the cut $J(x)$ will now be replaced by $x \leq b$. Thus, we may end up with a truncated Henkin set W. But this set is complete with respect to relatively small formulas. Moreover, W is certainly closed under subformulas and substitution of witnesses. Thus, W is sufficiently large to define the required interpretation k.

In both cases we can perform the following reasoning.

$$
\begin{aligned}
\square_V \varphi &\to \exists x\ \square_{V,x}\varphi \\
&\to \exists x\ \square_U(\mathsf{Con}_x(V) \wedge \square_{V,x}\varphi) \\
&\to \square_U \square_V^* \varphi \\
&\to \square_U \varphi^k \qquad\qquad\qquad\text{by Theorem 3.}
\end{aligned}
$$

The remarks from [18] on the bounds of our proofs are still applicable and we thus obtain a smooth interpretation. $\quad\dashv$

Lemma 10. *In the presence of* exp, *we can prove that for reflexive* U, $U \rhd V \to \forall x\ \square_U \mathsf{Con}_x(V)$.

In U:

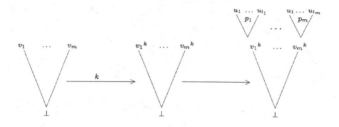

FIGURE 4 Transformations on proofs

Proof. The informal argument is conceptually very clear and we have depicted it in Figure 4. The accompanying reasoning is as follows.

We assume $U \triangleright V$, whence for some k we have $k : U \triangleright V$. Thus, for axioms interpretability we find that $\forall u \, \exists p \, (\mathsf{Axiom}_V(u) \rightarrow \mathsf{Proof}_U(p, u^k))$. We are now to see that $\forall x \; U \vdash \mathsf{Con}_x(V)$. So, we fix some x. By our assumption we get that for some l, that

$$\forall u{\leq}x \, \exists p \, (\mathsf{Axiom}_V(u) \rightarrow \mathsf{Proof}_{U,l}(p, u^k)). \tag{9}$$

This formula is actually equivalent to the Σ_1-formula

$$\exists n \, \forall u{\leq}x \, \exists p{\leq}n \, (\mathsf{Axiom}_V(u) \rightarrow \mathsf{Proof}_{U,l}(p, u^k)) \tag{10}$$

from which we may conclude by provable Σ_1-completeness,

$$U \vdash \exists n \, \forall u{\leq}x \, \exists p{\leq}n \, (\mathsf{Axiom}_V(u) \rightarrow \mathsf{Proof}_{U,l}(p, u^k)). \tag{11}$$

We now reason in U and suppose that there is some V, x-proof p of $\perp$. The assumptions in p are axioms $v_1 \ldots v_m$ of V, with each $v_i \leq x$. Moreover, all the formulas ψ in p have $\rho(\psi) \leq x$. By Lemma 3, this p transforms to a proof p^k of $\perp^k$ which is again $\perp$.

The assumptions in p^k are now among the $v_1{}^k \ldots v_m{}^k$. By Remark 2 we get that for some n' depending on x and k, we have that all the axioms in p^k are $\leq n'$ and all the ψ occurring in p^k have $\rho(\psi) \leq n'$.

Now by (11), we have U, l-proofs $p_i \leq n$ of $v_i{}^k$. The assumptions in the p_i are axioms of U. Clearly all of these axioms are $\leq l$. We can now form a $U, l{+}n'$-proof p' of $\perp$ by substituting all the p_i for the $(v_i)^k$. Thus we have shown $\mathsf{Proof}_{U,l+n'}(p', \perp)$. But this clearly contradicts the reflexivity of U.

The informal argument is readily formalized to obtain $T \vdash U \triangleright V \rightarrow$

$\forall x \ \Box_U \mathsf{Con}(V, x)$. However there are some subtleties.

First of all, to conclude that (9) is equivalent to (10), a genuine application of $\mathsf{B}\Sigma_1$ is needed. If U lacks $\mathsf{B}\Sigma_1$, we have to switch to smooth interpretability to still have the implication valid. Smoothness then automatically also provides the l that we used in 9.

In addition we need that T proves the totality of exponentiation. For weaker theories, we only have provable $\exists\Sigma_1^b$-completeness. But if $\mathsf{Axiom}_V(u)$ is Δ_1^b, we can only guarantee that

$$\forall u \leq m \ \exists p \leq n \ (\mathsf{Axiom}_V(u) \to \mathsf{Proof}_U(p, u^k))$$

is Π_2^b. As far as we know, exponentiation is needed to prove $\exists\Pi_2^b$-completeness.

All other transformations of objects in our proof only require the totality of $\omega_1(x)$. ⊣

The assumption that U is reflexive can in a sense not be dispensed with. That is, if

$$\forall V \ (U \rhd V \to \forall x \ \Box_U \mathsf{Con}_x(V)), \tag{12}$$

then U is reflexive, as clearly $U \rhd U$. In a similar way we see that if

$$\forall U \ (U \rhd V \to \forall x \ \Box_U \mathsf{Con}_x(V)), \tag{13}$$

then V is reflexive. However, V being reflexive could never be a sufficient condition for (13) to hold, as we know from [16] that interpreting reflexive theories in finitely many axioms is complete Σ_3.

Lemma 11. *In* S_2^1 *we can prove* $\forall x \ \Box_U \mathsf{Con}_x(V) \to \forall^{\forall\Pi_1^b} \pi \ (\Box_V \pi \to \Box_U \pi)$.

Proof. There are no conditions on U and V for this implication to hold. We shall directly give the formal proof as the informal proof does not give a clearer picture.

Thus, we reason in S_2^1 and assume $\forall x \ \Box_U \mathsf{Con}_x(V)$. Now we consider any $\pi \in \forall\Pi_1^b$ such that $\Box_V \pi$. Thus, for some x we have $\Box_{V,x} \pi$. We choose x large enough, so that we also have (see Remark 1)

$$\Box_U(\neg\pi \to \Box_{V,x} \neg\pi). \tag{14}$$

As $\Box_{V,x} \pi \to \Box_U \Box_{V,x} \pi$, we also have that

$$\Box_U \Box_{V,x} \pi. \tag{15}$$

Combining (14), (15) and the assumption that $\forall x \ \Box_U \mathsf{Con}_x(V)$, we see that indeed $\Box_U \pi$. |

Lemma 12. *In* S_2^1 *we can prove that for reflexive* V *we have*

$$\forall^{\forall\Pi_1^b} \pi \ (\Box_V \pi \to \Box_U \pi) \to \forall x \ \Box_U \mathsf{Con}_x(V).$$

Proof. If V is reflexive and $\forall^{\forall \Pi_1^b} \pi \, (\Box_V \pi \to \Box_U \pi)$ then, as for every x, $\mathsf{Con}_{\overline{x}}(V)$ is a $\forall \Pi_1^b$-formula, also $\forall x \, \Box_U \mathsf{Con}_x(V)$. ⊣

It is obvious that

$$\forall U \, [\forall^{\forall \Pi_1^b} \pi \, (\Box_V \pi \to \Box_U \pi) \to \forall x \, \Box_U \mathsf{Con}_x(V)] \qquad (16)$$

implies that V is reflexive. Likewise,

$$\forall V \, [\forall^{\forall \Pi_1^b} \pi \, (\Box_V \pi \to \Box_U \pi) \to \forall x \, \Box_U \mathsf{Con}_x(V)] \qquad (17)$$

implies that U is reflexive. However, U being reflexive can never be a sufficient condition for (17) to hold. An easy counterexample is obtained by taking U to be PRA and V to be $I\Sigma_1$ as it is well-known that $I\Sigma_1$ is provably Π_2 conservative over PRA and that $I\Sigma_1$ is finitely axiomatized.

Lemma 13. *(In S_2^1:) For reflexive V we have* $\forall^{\forall \Pi_1^b} \pi \, (\Box_V \pi \to \Box_U \pi) \to U \rhd V$.

Proof. We know of no direct proof of this implication. Also, all proofs in the literature go via Lemmata 12 and 9, and hence use reflexivity of V. ⊣

In our context, the reflexivity of V is not necessary, as $\forall U \; U \rhd \mathsf{S}_2^1$ and S_2^1 is not reflexive.

Lemma 14. *Let U be a reflexive theory. We have in S_2^1 that $U \rhd V \to \forall^{\forall \Pi_1^b} \pi \, (\Box_V \pi \to \Box_U \pi)$.*

If moreover $U \vdash \mathsf{exp}$ we also get $U \rhd V \to \forall^{\Pi_1} \pi \, (\Box_V \pi \to \Box_U \pi)$. If U is not reflexive, we still have that $U \rhd V \to \exists^{U\text{-Cut}} J \forall^{\Pi_1} \pi \, (\Box_V \pi \to \Box_U \pi^J)$.

For these implications, it is actually sufficient to work with the notion of theorems interpretability.

Proof. The intuition for the formal proof comes from Pudlák's lemma, which in turn is tailored to compensate a lack of induction. We shall first give an informal proof sketch if U has full induction. Then we shall give the formal proof using Pudlák's lemma.

If U has full induction and $j : U \rhd V$, we may assume by Remark 4 assume that j maps identity to identity. Let $\mathcal{M}$ be an arbitrary model of U. By Theorem 5 we now see that $\mathcal{M} \preceq_{\mathsf{end}} \mathcal{M}^j$. If for some $\pi \in \Pi_1$, $\Box_V \pi$ then by soundness $\mathcal{M}^j \models \pi$, whence $\mathcal{M} \models \pi$. As $\mathcal{M}$ was an arbitrary model of U, we get by the completeness theorem that $\Box_U \pi$.

To transform this argument into a formal one, valid for weak theories, there are two major adaptations to be made. First, the use of the soundness and completeness theorem have to be avoided . This can be done by simply staying in the realm of provability. Secondly, we should get rid of the use of full induction. This is done by switching to a cut in Pudlák's lemma.

Thus, the formal argument runs as follows. Reason in S_2^1 and assume $U \rhd V$.

We fix some $j : U \rhd V$. By Pudlák's lemma, Lemma 7, we now find[7] a definable U-cut J and a j, J-function h such that

$$\forall^{\Delta_0} \varphi \; \Box_U \forall \vec{x} {\in} J \; (\varphi^j(h(\vec{x})) \leftrightarrow \varphi(\vec{x})).$$

We shall see that for this cut J we have that

$$\forall^{\Pi_1} \pi \; (\Box_V \pi \to \Box_U \pi^J). \tag{18}$$

Therefore, we fix some $\pi \in \Pi_1$ and assume $\Box_V \pi$. Let $\varphi(x) \in \Delta_0$ be such that $\pi = \forall x \; \varphi(x)$. Thus we have $\Box_V \forall x \; \varphi(x)$, hence by theorems interpretability

$$\Box_U \forall x \; (\delta(x) \to \varphi^j(x)). \tag{19}$$

We are to see

$$\Box_U \forall x \; (J(x) \to \varphi(x)). \tag{20}$$

To see this, we reason in U and fix x such that $J(x)$. By definition of J, $h(x)$ is defined. By the definition of h, we have $\delta(h(x))$, whence by (19), $\varphi^j(h(x))$. Pudlák's lemma now yields the desired $\varphi(x)$. As x was arbitrary, we have proved (20).

So far, we have not used the reflexivity of U. We shall now see that

$$\forall^{\forall \Pi_1^b} \pi \; (\Box_U \pi^J \to \Box_U \pi)$$

holds for any U-cut J whenever U is reflexive. For this purpose, we fix some $\pi \in \forall \Pi_1^b$, some U-cut J and assume $\Box_U \pi^J$. Thus, $\exists n \; \Box_{U,n} \pi^J$ and also $\exists n \; \Box_U \Box_{U,n} \pi^J$. If $\pi = \forall x \; \varphi(x)$ with $\varphi(x) \in \Pi_1^b$, we get $\exists n \; \Box_U \Box_{U,n} \forall x \; (x \in J \to \varphi(x))$, whence also

$$\exists n \; \Box_U \forall x \; \Box_{U,n}(x \in J \to \varphi(x)).$$

By Lemma 4 and Remark 3, for large enough n, this implies

$$\exists n \; \Box_U \forall x \; \Box_{U,n} \varphi(x)$$

and by Lemma 2 (only here we use that $\pi \in \forall \Pi_1^b$) we obtain the required $\Box_U \forall x \; \varphi(x)$. ⊣

Again, by [16] we note that V being reflexive can never be a sufficient condition for $\forall U \; [U \rhd V \to \forall^{\forall \Pi_1^b} \pi \; (\Box_V \pi \to \Box_U \pi)]$. It is known that for Lemma 14 the sequentiality of U is essential (see [8]).

---

[7]Remark 5 ensures us that we can find them also in the case of theorems interpretability.

The main work on the Orey-Hájek characterization has now been done. We can easily extract some useful, mostly well-known corollaries.

Corollary 1. *If U is a reflexive theory, then*

$$T \vdash U \rhd V \leftrightarrow \forall x \; \Box_U \mathsf{Con}_x(V).$$

Here T contains exp *and $\rhd$ denotes smooth interpretability.*

Corollary 2. *(In S^1_2:) If V is a reflexive theory, then the following are equivalent.*

1. $U \rhd V$
2. $\exists^{U\text{-}\mathsf{Cut}} J \, \forall^{\Pi_1} \pi \; (\Box_V \pi \to \Box_U \pi^J)$
3. $\exists^{U\text{-}\mathsf{Cut}} J \, \forall x \; \Box_U \mathsf{Con}_x^J(V)$

Proof. This is part of Theorem 2.3 from [16]. (1) $\Rightarrow$ (2) is already proved in Lemma 14, (2) $\Rightarrow$ (3) follows from the transitivity of V and (3) $\Rightarrow$ (1) is a sharpening of Lemma 9. which closely follows Theorem 4. Note that $\rhd$ may denote denote smooth or theorems interpretability.
⊣

Corollary 3. *If V is reflexive, then*

$$\mathsf{S}^1_2 \vdash U \rhd_t V \leftrightarrow U \rhd_s V.$$

Proof. By Remark 5 and Corollary 2. ⊣

Corollary 4. *If U and V are both reflexive theories we have that the following are provably equivalent in S^1_2.*

1. $U \rhd V$
2. $\forall^{\forall \Pi_1^b} \pi \; (\Box_V \pi \to \Box_U \pi)$
3. $\forall x \; \Box_U \mathsf{Con}_x(V)$

Proof. If we go (1) $\Rightarrow$ (2) $\Rightarrow$ (3) $\Rightarrow$ (1) we do not need the totality of exp that was needed for (1) $\Rightarrow$ (3). ⊣

As an application we can, for example, see that $\mathsf{PA} \rhd \mathsf{PA} + \mathsf{InCon}(\mathsf{PA})$. It is well known that PA is essentially reflexive which means that any finite extension of it is reflexive. So, we use Corollary 4 and, it is sufficient to show that $\mathsf{PA} + \mathsf{InCon}(\mathsf{PA})$ is Π_1-conservative over PA.

So, suppose that $\mathsf{PA} + \mathsf{InCon}(\mathsf{PA}) \vdash \pi$ for some Π_1-sentence π. In other words $\mathsf{PA} \vdash \Box\bot \to \pi$. We shall now see that $\mathsf{PA} \vdash \Box\pi \to \pi$, which by Löb's Theorem gives us $\mathsf{PA} \vdash \pi$.

Thus, in PA, assume $\Box\pi$. Suppose for a contradiction that $\neg\pi$. By Σ_1-completeness we also get $\Box\neg\pi$, which yields $\Box\bot$ with the assumption $\Box\pi$. But we have $\Box\bot \to \pi$ and we conclude π. A contradiction, so that indeed $\mathsf{PA} \rhd \mathsf{PA} + \mathsf{InCon}(\mathsf{PA})$.

Acknowledgements

I am grateful to Lev Beklemishev, Félix Lara and Albert Visser for pointers to the literature and helpful discussions. Also, many thanks is due to an anonymous referee who pointed out various inaccuracies and helped to make some important corrections.

This research has been funded by Grant 2014 SGR 437 from the Catalan government and by Grant MTM2014-59178-P from the Spanish government.

References

[1] S.R. Buss. First-order proof theory of arithmetic. In S.R. Buss, editor, *Handbook of Proof Theory*, pages 79–148, Amsterdam, 1998. Elsevier, North-Holland.

[2] S. Feferman. Arithmetization of metamathematics in a general setting. *Fundamenta Mathematicae*, 49:35–92, 1960.

[3] D. Guaspari. Partially conservative sentences and interpretability. *Transactions of AMS*, 254:47–68, 1979.

[4] P. Hájek. On interpretability in set theories I. *Comm. Math. Univ. Carolinae*, 12:73–79, 1971.

[5] P. Hájek. On interpretability in set theories II. *Comm. Math. Univ. Carolinae*, 13:445–455, 1972.

[6] P. Hájek and P. Pudlák. *Metamathematics of First Order Arithmetic*. Springer-Verlag, Berlin, Heidelberg, New York, 1993.

[7] D. Hilbert and P. Bernays. *Grundlagen der Mathematik, Vols. I and II, 2d ed.* Springer-Verlag, Berlin, 1968.

[8] J. J. Joosten. *Interpretability Formalized*. PhD thesis, Utrecht University, 2004.

[9] J. J. Joosten. Two series of formalized interpretability principles for weak systems of arithmetic. *arXiv:1503.09130 [math.LO]*, 2015.

[10] J.J. Joosten and A. Visser. The interpretability logic of *all* reasonable arithmetical theories. *Erkenntnis*, 53(1–2):3–26, 2000.

[11] Jan Krajíek. *Bounded Arithmetic, Propositional Logic, and Complexity Theory*. Cambridge University Press, 1995.

[12] P. Lindström. Some results on interpretability. In *Proceedings of the 5th Scandinavian Logic Symposium*, pages 329–361. Aalborg University press, 1979.

[13] P. Lindström. On partially conservative sentences and interpretability. *Proceedings of the AMS*, 91(3):436–443, 1984.

[14] S. Orey. Relative interpretations. *Zeitschrift f. math. Logik und Grundlagen d. Math.*, 7:146–153, 1961.

[15] P. Pudlák. Cuts, consistency statements and interpretations. *Journal of Symbolic Logic*, 50:423–441, 1985.

[16] V.Yu. Shavrukov. Interpreting reflexive theories in finitely many axioms. *Fundamenta Mathematicae*, 152:99–116, 1997.

[17] A. Tarski, A. Mostowski, and R. Robinson. *Undecidable theories*. North–Holland, Amsterdam, 1953.

[18] A. Visser. The formalization of interpretability. *Studia Logica*, 50(1):81–106, 1991.

[19] A. Visser. The unprovability of small inconsistency. *Archive for Mathematical Logic*, 32:275–298, 1993.

[20] A. Visser. An Overview of Interpretability Logic. In M. Kracht, M. de Rijke, H. Wansing, and M. Zakharyaschev, editors, *Advances in Modal Logic*, volume 1, 87 of *CSLI Lecture Notes*, pages 307–359. Center for the Study of Language and Information, Stanford, 1998.

[21] A. Visser. Can we make the Second Incompleteness Theorem coordinate free. *Journal of Logic and Computation*, 2011, 21(4): 543–560. First published online August 12, 2009, doi: 10.1093/logcom/exp048.

[22] A. Visser. What is sequentiality? In P. Cégielski, Ch. Cornaros, and C. Dimitracopoulos, editors, *New Studies in Weak Arithmetics*, volume 211 of *CSLI Lecture Notes*, pages 229–269. CSLI Publications and Presses Universitaires du Pôle de Recherche et d'Enseingement Supérieur Paris-est, Stanford, 2013.

[23] A. Visser. Interpretability degrees of finitely axiomatized sequential theories. *Archive for Mathematical Logic*, 53(1-2):23–42, 2014.

[24] A. Visser. The interpretation existence lemma. To appear, 2016.

[25] A. Visser. Transductions in arithmetic. *Annals of Pure and Applied Logic*, 167(3):211–234, 2016.

[26] H Wang. Arithmetical models of formal systems. *Methodos 3*, pages 217–232, 1951.

5

An Imperative Language Characterizing PTIME Algorithms

YOANN MARQUER AND PIERRE VALARCHER

Abstract: Abstract State Machines of Y. Gurevich capture sequential algorithms, so we define the set of PTIME algorithms as the set of ASM programs computed in polynomial time (using timer-step principle). Then, we construct an imperative programming language PLoopC using bounded loop with break, and running with any data structure. Finally, we prove that PLoopC computes exactly PTIME algorithms in lock step (one step of the ASM is simulated by using a constant number of steps).

Context

The paper is devoted to answer the question:

> Is it possible to construct a programming language complete for the set of polynomial time algorithms, and no more?

The common definition of PTIME is the set of functions computable by a Turing Machine in polynomial time. This definition uses a step-timer principle (the polynomial bounding the running time), which can be criticized in two ways:

1. According to this definition, one-tape Turing Machines and two-tape Turing Machines should be equivalent. But the palindrome recognition can be done in $O(n)$ steps with a two-tape Turing machine while requiring (see [24]) at least $O(n^2/log(n))$ steps with a one-tape Turing machine. So, the concrete implementation

Studies in Weak Arithmetics, Volume 3.
Patrick Cégielski, Ali Enayat,
Roman Kossak.
Copyright © 2016, CSLI Publications.

and its associated complexity (more specifically, the degree of the time complexity) depend on the considered model. Therefore, we need a more precise definition of polynomial time algorithms.

2. The second issue comes from the step-timer principle itself. It is not convenient for programmers to use an external function attached to the algorithm. The answer comes from *implicit complexity computation* frameworks. Since a long time, there are many attempts (like [2, 22] and [4]) to capture polynomial time algorithms. The main motivations are usually to find "pleasant" syntactical (and semantical) languages, to capture more algorithms than the previous models, or to give simpler method to decide if a program is computable in polynomial time.

We answer both criticisms by defining a convincing set of algorithms running in polynomial time (using the step-timer principle), and by constructing a programming language computing those algorithms:

Algorithms with Step-Timer Principle

From now half a century, there is a growing interest in defining formally the notion of sequential algorithms [20, 10], and some of these definitions allow to specify classes of algorithms[1] (in [11, 10] and [8]).

An axiomatic definition of the sequential algorithms is mapped to the notion of Abstract State Machine with a *strict lock-step* simulation[2]. The model of ASM is a kind of super Turing Machine that works not on simple tapes (with finite alphabets) but on multi–sorted algebras (see the point of view in [13]). A program is a finite set of rules that updates terms. It is shown in [17], that the expressive power of ASM lies not in control structures but in data structures, which are modeled within a first order structure. According to Gurevich's Thesis (see [12]), every algorithm can be computed step-by-step by an ASM. So, we will define p. 96 the set $ASM_{\mathcal{P}}$ as the set of every ASM with polynomial running time.

The PLoopC language

Many attempts tries to restrict Loop imperative language (see [18]) in order to obtain relevant classes of algorithms.

In [1, 23] a class APRA of primitive recursive algorithms is defined for the basic data structure of (unary) integer coming from the *abstract state machine* theory. From there, two imperative

---

[1]Even if there are parallel, distributed, real-time, bio-inspired or quantum algorithms, this paper focuses only on sequential algorithms.

[2]See [9] for a definition of simulation and strict lock-step

programming languages and one functional programming languages are proved to be **complete** for this set of algorithms. In other words, every algorithm defined in APRA can be written in those languages without lost of time complexity (the simulation is in $O(1)$).

For polynomial time, a `Loop` programming language is presented in [22], where some properties are found to capture PTIME on data structure such as stack, trees or graphs, and Neergaard introduced in [21] his polynomial time `PLoop` programming language `PLoop`, which uses stacks.

Following [1], we will define p. 108 an imperative programming language `PLoopC`, and prove that `PLoopC` captures exactly the set $\text{ASM}_\mathcal{P}$ of polynomial time algorithms. We do that by following the Bellantoni and Cook's approach (see [2]) separating safe and normal variables. By restricting the use of iteration bounds to be inputs, we prove that `PLoopC` has polynomial time for every data structure.

The paper is organized as follows.

In the first section we briefly introduce Gurevich's framework of the sequential algorithms, and define p. 100 the notion of *fair simulation*. In the second section we introduce the Gurevich's ASM model of computation, and our programming language `LoopC`. Moreover, we will prove p. 109 that the sublanguage `PLoopC` has polynomial time.

The third section is devoted to prove p. 125 that `ASM` fairly simulates `LoopC`, by using the notion of *graph of execution* and a translation of the ASM program into an imperative program. Therefore, because `PLoopC` is a sublanguage of `LoopC` with polynomial time, $\text{ASM}_\mathcal{P}$ fairly simulates `PLoopC`. Reciprocally, in the fourth section, we will prove p. 115 that `PLoopC` fairly simulates $\text{ASM}_\mathcal{P}$, by using an imperative program translating one step of the ASM, and repeat it a sufficient number of times by using a translation of the complexity and a formula detecting the end of the execution.

Therefore, we will prove that `PLoopC` characterizes polynomial time algorithms.

Keywords. ASM, computability, imperative language, implicit complexity, polynomial time, sequential algorithms, simulation.

1 Sequential Algorithms

In [10] Gurevich introduced an axiomatic presentation of the sequential algorithms, by giving the three postulates of Sequential Time, Abstract States and Bounded Exploration. In our paper the set of "objects" satisfying these postulates is denoted by `Algo`.

We will introduce them briefly in the first subsection, as well as other notions from Gurevich's framework such as execution, time, structure and update. In the second subsection we will introduce our definition of a fair simulation between computation models.

1.1 Three Postulates

Postulate 1 (Sequential Time). A sequential algorithm A is given by:

1. A set of states $S(A)$
2. A set of initial states $I(A) \subseteq S(A)$
3. A transition function $\tau_A : S(A) \to S(A)$

Remark 1. According to this postulate, two sequential algorithms A and B are the same (see [3]) if they have the same set of states $S(A) = S(B)$, the same set of initial states $I(A) = I(B)$, and the same transition function $\tau_A = \tau_B$.

An **execution** of A is a sequence of states $\vec{X} = X_0, X_1, X_2, \ldots$ such that:

1. X_0 is an initial state
2. For every $t \in \mathbb{N}$, $X_{t+1} = \tau_A(X_t)$

A state X_t of an execution is final if $\tau_A(X_t) = X_t$. An execution is **terminal** if it contains a final state. The duration of an execution is defined by the number of steps[3] done before reaching a final state:

$$time(A, X_0) =_{def} min\{t \in \mathbb{N} \mid \tau_A^t(X_0) = \tau_A^{t+1}(X_0)\}$$

Notice that if the execution $\vec{X}$ is not terminal then $time(A, X_0) = \infty$.

Remark 2. Two algorithms A and B have the same set of executions if they have the same set of initial states $I(A) = I(B)$ and the same transition function $\tau_A = \tau_B$. In that case, they can only be different on the states which cannot be reached by an execution.

To state the second postulate, we need to introduce the notion of structure. Gurevich formalized the states of a sequential algorithm with first-order structures. A (first-order) **structure** X is given by:

1. An infinite[4] set $\mathcal{U}(X)$ called the **universe** (or domain) of X
2. A finite set of function symbols $\mathcal{L}(X)$ called the **signature** (or language) of X

---

[3]In the definition of *time*, f^i is the iteration of f defined by $f^0 = id$ and $f^{i+1} = f(f^i)$.

[4]Usually the universe is only required to be non-empty, we need the universe to be at least countable in order to define unary integers .

3. For every symbol $s \in \mathcal{L}(X)$ an **interpretation** $\overline{s}^X$ such that:
   (a) If c has arity 0 then $\overline{c}^X$ is an element of $\mathcal{U}(X)$
   (b) If f has an arity $\alpha > 0$ then $\overline{f}^X$ is an application: $\mathcal{U}(X)^\alpha \to \mathcal{U}(X)$

In order to have a uniform presentation, Gurevich considered constant symbols of the signature as 0-ary function symbols, and relation symbols R as their indicator function χ_R. Therefore, every symbol in $\mathcal{L}(X)$ is a function. Moreover, partial functions can be implemented with a special value *undef*.

This formalization can be seen as a representation of a computer data storage. For example, the interpretation $\overline{s}^X$ of the symbol s in the structure X represents the value in the register s for the state X.

The second postulate can be seen as a claim assuming that every data structure can be formalized as a first-order structure[5]. Moreover, since the representation of states should be independent from their concrete implementation (for example the name of objects), isomorphic states will be considered as equivalent:

Postulate 2 (Abstract States). For every algorithm A,

1. The states of A are (first-order) structures with the same signature $\mathcal{L}(A)$
2. $S(A)$ and $I(A)$ are closed by isomorphism
3. The transition function τ_A preserves the universes and commutes with the isomorphisms

The symbols of $\mathcal{L}(A)$ are distinguished between:

1. $Dyn(A)$ the set of **dynamic symbols**, whose interpretation can change during an execution (as an example, a variable x)
2. $Stat(A)$ the set of **static symbols**, which have a fixed interpretation during an execution. They are also distinguished between:
   (a) $Init(A)$, the set of **parameters**, whose interpretation depends only on the initial state (as an example, two given integers m and n).

   The symbols depending on the initial state are the dynamic symbols and the parameters, so we call them the **inputs**.

   The other symbols have a uniform interpretation in every state (up to isomorphism), and they are also distinguished between:

---

[5]We tried to characterize common data types (such as integers, words, lists, arrays, and graphs) in [16]. But we will not go into details, because this is not the point of this article.

(b) $Cons(A)$ the set of **constructors** (*true* and *false* for the booleans, 0 and S for the unary integers, ...)

(c) $Oper(A)$ the set of **operations** ($\neg$ and $\wedge$ for the booleans, $+$ and $\times$ for the unary integers, ...)

The **size** of an element of the universe is the length of its representation (see [16] for more details), in other words the number of constructors necessary to write it. As an example, $|\overline{\neg\neg true}^X| = |\overline{true}^X| = 1$, $|\overline{1+2}^X| = |\overline{S(S(S(0)))}^X| = 4$ in the unary numeral system[6], and $|\overline{1+2}^X| = |\overline{11}^X| = 2$ in the binary numeral system.

The size of a state is the maximum (or the sum, which is equivalent because the signature is finite) of the size of its inputs:

$$|X| =_{def} \max_{f \in Dyn(A) \sqcup Init(A)} \{|f|_X\}, \text{ where } |f|_X =_{def} \sup_{a_i \in \mathcal{U}(A)} |\overline{f}^X(\vec{a})|$$

Definition 1 (Time Complexity).

The algorithm A is C-time if there exists $\varphi_A \in C$ such that for all $X \in I(A)$:

$$time(A, X) \leq \varphi_A(|X|)$$

Let Algo_C *be the set of C-time algorithms. In particular, we denote by* $\text{Algo}_{\mathcal{P}}$ *the set of polynomial-time algorithms.*

The logical variables are not used in this paper: every term and every formula is closed, and every formula is without quantifier. In this framework the **variables** are the 0-ary dynamic function symbols.

For a sequential algorithm A, let X be a state of A, $f \in \mathcal{L}(A)$ be a dynamic α-ary function symbol, and $a_1, \ldots, a_\alpha, b \in \mathcal{U}(X)$. $(f, a_1, \ldots, a_\alpha)$ denotes a location of X and $(f, a_1, \ldots, a_\alpha, b)$ denotes an **update** on X at the location $(f, a_1, \ldots, a_\alpha)$.

If u is an update then[7] $X \oplus u$ is a new structure of signature $\mathcal{L}(A)$ and universe $\mathcal{U}(X)$ such that the interpretation of a function symbol $f \in \mathcal{L}(A)$ is:

$$\overline{f}^{X \oplus u}(\vec{a}) =_{def} \begin{cases} b & \text{if } u = (f, \vec{a}, b) \\ \overline{f}^X(\vec{a}) & \text{else} \end{cases}$$

If $\overline{f}^X(\vec{a}) = b$ then the update $(f, \vec{a}, b)$ is trivial in X, because nothing has changed. Indeed, if $(f, \vec{a}, b)$ is trivial in X then $X \oplus (f, \vec{a}, b) = X$.

---

[6]More generally, in the unary numeral system $|\overline{n}^X| = n + 1$.

[7]The update is denoted $\oplus$, not $+$ like in [12] or [17], because the associativity has no meaning here, and because we don't have the commutativity:

$$(X \oplus (x, 0)) \oplus (x, 1) \neq (X \oplus (x, 1)) \oplus (x, 0)$$

If Δ is a set of updates then Δ is **consistent** on X if it does not contain two distinct updates with the same location. If Δ is inconsistent, there exists $(f, \vec{a}, b), (f, \vec{a}, b') \in \Delta$ with $b \neq b'$, so the entire set of updates clashes:

$$\overline{f}^{X \oplus \Delta}(\vec{a}) =_{def} \begin{cases} b & \text{if } (f, \vec{a}, b) \in \Delta \text{ and } \Delta \text{ is consistent on } X \\ \overline{f}^X(\vec{a}) & \text{else} \end{cases}$$

If X and Y are two states of the same algorithm A then there exists a unique consistent set $\Delta = \{(f, \vec{a}, b) \mid \overline{f}^Y(\vec{a}) = b \text{ and } \overline{f}^X(\vec{a}) \neq b\}$ of non trivial updates such that $Y = X \oplus \Delta$. This Δ is the **difference** between the two sets and is denoted by $Y \ominus X$.

Let $\Delta(A, X) = \tau_A(X) \ominus X$ be the set of updates done by a sequential algorithm A on the state X.

During an execution, if an increasing number of updates are done (this is the case in example 1 with the parallel λ–calculus) then the algorithm will be said massively parallel, not sequential. The two first postulates cannot ensure that only local and bounded explorations or changes are done at every step. The third postulate states that only a bounded number of terms must be read or updated during one step of the execution:

Postulate 3 (Bounded Exploration). For every algorithm A there exists a finite set T of terms (closed by subterms) such that for every state X and Y, if the elements of T have the same interpretations on X and Y then $\Delta(A, X) = \Delta(A, Y)$.

This T is called the **exploration witness** of A.

Gurevich proved in [12] that if $(f, a_1, \ldots, a_\alpha, b) \in \Delta(A, X)$ then $a_1, \ldots, a_\alpha, b$ are interpretations in X of terms in T. So, since T is finite there exists a bounded number of $a_1, \ldots, a_\alpha, b$ such that the update $(f, a_1, \ldots, a_\alpha, b)$ belongs to $\Delta(A, X)$. Moreover, since $\mathcal{L}(A)$ is finite there exists a bounded number of dynamic symbols f. Therefore, $\Delta(A, X)$ has a bounded number of elements, and for every step of the algorithm only a bounded amount of work is done.

1.2 Fair Simulation

A **model of computation** can be defined as a set of programs given with their operational semantics. In our paper we only study sequential algorithms, which have a step-by-step execution determined by their transition function. So, this operational semantics can be defined by a set of transition rules, as is the case in the following example:

Example 1 (The Lambda-Calculus).
The lambda calculus is defined by of a set of lambda terms (which

are the "programs"), and a set of transformation rules (which are the "operational semantics"):

$$\text{Syntax of Programs:} \quad t =_{def} x \mid \lambda x.t \mid (t_1)t_2$$
$$\beta\text{-reduction:} \quad (\lambda x.t_1)t_2 \rightarrow_\beta t_1[t_2/x]$$

In order to be deterministic the strategy of the transition system must be specified. An example is the call-by-name strategy defined by context:

$$\text{Call-by-Name Context:} \quad C_n\{.\} =_{def} \cdot \mid C_n\{.\}t$$
$$\text{Transition Rule:} \quad C_n\{(\lambda x.t_1)t_2\} \rightarrow_n C_n\{t_1[t_2/x]\}$$

This rule can be implemented in a machine:

$$\text{Operational semantics:} \quad t_1 t_2 \star \quad \pi \succ_0 \quad t_1 \star t_2, \pi$$
$$\lambda x.t_1 \star t_2, \pi \succ_1 t_1[t_2/x] \star \quad \pi$$

These notations and this machine are directly taken from Krivine's [15]. In this machine π is a stack of terms. The symbol $\star$ is a separator between the current program and the current state of the memory. $\succ$ represents one step of computation, where only substitutions have a cost, not explorations inside a term, as is the case in the contextual transition rule. Programs in the machine are closed terms, so final states have the form $\lambda x.t \star \varnothing$.

We will follow the notations $\star$ and $\succ$ in the definition of the operational semantics of imperative programs 7.

Notice that if the substitution is given as an elementary operation this model satisfies the third postulate, because only one term is pushed or popped per step. This is not the case with the lambda-calculus with parallel reductions. For example, with the term $t = \lambda x.(x)x(x)x$ applied to itself:

$$(t)t \rightarrow_p (t)t(t)t \rightarrow_p (t)t(t)t(t)t \rightarrow_p \ldots$$

Indeed, at the step t exactly 2^{t-1} β-reductions are done, which is unbounded.

Sometimes, not only the simulation between two models of computation can be proven, but also their identity. As an example, Serge Grigorieff and Pierre Valarcher proved in [13] that Evolving MultiAlgebras (a variant of the Gurevich's ASMs) can unify common sequential models of computation. For instance, a family of EMAs can not only simulate step-by-step the Turing Machines, it can also be literally identified to them. The same applies for Random Access Machines, or other common models.

But generally it is only possible to prove a simulation between two models of computation. In our framework, a computation model M_1 can simulate another computation model M_2 if for every program P_2

of M_2 there exists a program P_1 of M_1 producing in a "reasonable way" the "same" executions as those produced by P_2. The following two examples will detail what can be used in a "fair" simulation:

Example 2 (Temporary Variables).

In this example a programmer is trying to simulate a `loop` n $\{P\}$ command in an imperative programming language[8] containing `while` commands. The well-known solution is to use a temporary variable i in the new program:

$$i := 0; \texttt{while } i < n \; \{P; i := i + 1; \};$$

This simulation is very natural, but a fresh variable i is necessary.

Another example is to simulate the exchange $x \leftrightarrow y$ between two variables using a temporary variable:

$$v := x; x := y; y := v;$$

In any case, the signature $\mathcal{L}_1$ of the simulating program must be bigger than the signature $\mathcal{L}_2$ of the simulated program.

Notation 3. We follow the notation from [7], where $X|_{\mathcal{L}_2}$ denotes the **restriction** of the $\mathcal{L}_1$-structure X to the signature $\mathcal{L}_2$. The signature of $X|_{\mathcal{L}_2}$ is $\mathcal{L}_2$, its universe is the same than X, and every symbol $s \in \mathcal{L}_2$ has the same interpretation in $X|_{\mathcal{L}_2}$ than in X.

This notation is extended to a set of updates:

$$\Delta|_{\mathcal{L}} =_{def} \{(f, \vec{a}, b) \in \Delta \mid f \in \mathcal{L}\}$$

But fresh function symbols could be "too powerful", for example a dynamical unary symbol env alone would be able to store an unbounded amount of information. In order to obtain a fair simulation, we assume that the difference $\mathcal{L}_1 \setminus \mathcal{L}_2$ between both signatures is a set containing only a bounded number of variables (0-ary dynamical symbols).

The initial values of these **fresh variables** could be a problem if they depend on the inputs. For example, the empty program could compute any $f(\vec{n})$ if we assume that an output variable contains in the initial state the result of the function f on the inputs $\vec{n}$.

So, in this paper we use an initialization which depends[9] only on the constructors[10]. Because this initialization is independent (up to isomorphism) from the initial states, we call it a **uniform initialization**.

---

[8]We will define 6 the precise syntax of imperative programs.

[9]Even the values of fresh variables in the initial states can be irrelevant. See the program P_Π 8 where the variables $\vec{v}$ are explicitly updated with the value of terms $\vec{t}$ before being read.

[10]See the program Π_P 15 where the boolean variable b_P is initialized with $true$ and the others with $false$.

Example 4 (Temporal Dilation).

At every step of a Turing machine, depending on the current state and the symbol in the current cell:

- the state of the machine is updated
- the machine writes a new symbol in the cell
- the head of the machine can move left or right

Usually these actions are considered simultaneous, so only one step of computation is necessary to execute them. This is our classical model M_1 of Turing machines. But if we consider that every action requires one step of computation then we could imagine a model M_3 where three steps are necessary to simulate one step of M_1.

In other words, if we only observe an execution

$$\mathbf{X_0}, X_1, X_2, \mathbf{X_3}, X_4, X_5, \mathbf{X_6}, \dots$$

of M_3 every three steps (the observed states are bolded) then we will obtain an execution defined by $Y_t = X_{3 \times t}$, which is an execution of M_1.

Imagine that M_1 and M_2 are implemented on real machines such that M_3 is three times faster than M_1. In that case if an external observer starts both machines simultaneously and checks their states at every step of M_1 then both machines cannot be distinguished.

In the following a (constant) **temporal dilation** d is allowed. We will say that the simulation is step-by-step, and strictly step-by-step if $d = 1$. Unfortunately, contrary to the previous example this constant may depend on the simulated program.

But this temporal dilation is not sufficient to ensure the termination of the simulation. For example, a simulated execution $Y_0, \dots, Y_t, Y_t, \dots$ could have finished, but the simulating execution

$$X_0, \dots, X_{d \times t}, X_{d \times t+1}, \dots, X_{d \times t+(d-1)}, X_{d \times t}, X_{d \times t+1}, \dots$$

may continue forever. So, an ending condition like $time(A, X) = d \times time(B, X) + e$ is necessary, and corresponds to the usual consideration for asymptotic time complexity.

Definition 2 (Fair Simulation).

Let M_1, M_2 be two models of computation.

M_1 simulates M_2 if for every program P_2 of M_2 there exists a program P_1 of M_1 such that:

1. *$\mathcal{L}(P_1) \supseteq \mathcal{L}(P_2)$, and $\mathcal{L}(P_1) \setminus \mathcal{L}(P_2)$ is a finite set of variables (with a uniform initialization)*

and there exists $d \in \mathbb{N} \setminus \{0\}$ and $e \in \mathbb{N}$ (depending only on P_2) such *that, for every execution $\vec{Y}$ of P_2 there exists an execution $\vec{X}$ of P_1 satisfying:*

2. for every $t \in \mathbb{N}$, $X_{d \times t}|_{\mathcal{L}(P_2)} = Y_t$

3. $time(P_1, X_0) = d \times time(P_2, Y_0) + e$

*If M_1 simulates M_2 and M_2 simulates M_1 then these models of computation are **algorithmically equivalent**, which is denoted by $M_1 \simeq M_2$.*

Remark 3. The second condition $X_{d \times t}|_{\mathcal{L}(P_2)} = Y_t$ implies for $t = 0$ that the initial states are the same, up to temporary variables.

2 Models of Computation

In this section, Gurevich's Abstract State Machines are defined, and we use his theorem Algo = ASM to get a constructive (from an operational point of view) occurrence of sequential algorithms. So, in the rest of the paper, the set of polynomial-time algorithms $\mathtt{Algo}_\mathcal{P}$ will be the set $\mathtt{ASM}_\mathcal{P}$ of ASMs with a polynomial time complexity.

We will define in the second part of the section the language LoopC from [1], and prove that a sublanguage PLoopC has polynomial time. We will prove in the next section that this sublanguage is complete for $\mathcal{P}$-time algorithms. In order to do so, we will prove a bisimulation between $\mathtt{ASM}_\mathcal{P}$ and PLoopC, using the same data structures in these two models of computation.

2.1 Abstract State Machines

Without going into details, the Gurevich's Abstract State Machines (ASM) require only the equality $=$, the constants *true* and *false*, the unary operation $\neg$ and the binary operations $\wedge$.

Definition 3 (ASM programs).

$$\Pi =_{def} f(t_1, \ldots, t_\alpha) := t_0$$
$$| \text{ if } F \text{ then } \Pi_1 \text{ else } \Pi_2 \text{ endif}$$
$$| \text{ par } \Pi_1 \| \ldots \| \Pi_n \text{ endpar}$$

where f is a dynamic α-ary function symbol, $t_0, t_1, \ldots, t_\alpha$ are closed terms, and F is a formula.

Notation 5. For $n = 0$ a par command is an empty program, so let skip be the command par endpar. If the else part of an if is a skip we only write if F then Π endif.

The sets $Read(\Pi)$ of terms read by Π and $Write(\Pi)$ of terms written by Π can be used to define the exploration witness of Π. But we will also use them in the rest of the article, especially to define the μ-formula F_Π p. 124.

$Read(\Pi)$ is defined by induction on Π:

$$Read(f(t_1, \ldots, t_\alpha) := t_0)$$
$$=_{def} \{t_1, \ldots, t_\alpha, t_0\}$$
$$Read(\text{if } F \text{ then } \Pi_1 \text{ else } \Pi_2 \text{ endif})$$
$$=_{def} \{F\} \cup Read(\Pi_1) \cup Read(\Pi_2)$$
$$Read(\text{par } \Pi_1 \| \ldots \| \Pi_n \text{ endpar})$$
$$=_{def} Read(\Pi_1) \cup \cdots \cup Read(\Pi_n)$$

$Write(\Pi)$ is defined by induction on Π:

$$Write(f(t_1, \ldots, t_\alpha) := t_0)$$
$$=_{def} \{f(t_1, \ldots, t_\alpha)\}$$
$$Write(\text{if } F \text{ then } \Pi_1 \text{ else } \Pi_2 \text{ endif})$$
$$=_{def} Write(\Pi_1) \cup Write(\Pi_2)$$
$$Write(\text{par } \Pi_1 \| \ldots \| \Pi_n \text{ endpar})$$
$$=_{def} Write(\Pi_1) \cup \cdots \cup Write(\Pi_n)$$

Remark 4. The exploration witness of Π is the closure by subterms of $Read(\Pi) \cup Write(\Pi)$ and not only $Read(\Pi)$ because the updates of a command could be trivial.

As said p. 97, defining the syntax of programs is not enough to obtain a model of computation, we still have to define their semantics. An ASM program Π determines a transition function $\tau_\Pi(X) =_{def} X \oplus \Delta(\Pi, X)$, where the set of updates $\Delta(\Pi, X)$ done by Π on X is defined by induction:

Definition 4 (Operational Semantics of ASMs).

$$\Delta(f(t_1, \ldots, t_\alpha) := t_0, X) =_{def} \{(f, \overline{t_1}^X, \ldots, \overline{t_\alpha}^X, \overline{t_0}^X)\}$$
$$\Delta(\text{if } F \text{ then } \Pi_1 \text{ else } \Pi_2 \text{ endif}, X) =_{def} \Delta(\Pi_i, X)$$

$$where \ i = \left\{ \begin{array}{l} 1 \ if \ F \ is \ true \ on \ X \\ 2 \ else \end{array} \right.$$

$$\Delta(\text{par } \Pi_1 \| \ldots \| \Pi_n \text{ endpar}, X) =_{def} \Delta(\Pi_1, X) \cup \cdots \cup \Delta(\Pi_n, X)$$

Notice that the semantics of the par is a set of updates done simultaneously, contrary to the imperative language defined in the next subsection, which is strictly sequential.

Remark 5. For every states X and Y, if the terms of $Read(\Pi)$ have the same interpretation on X and Y then $\Delta(\Pi, X) = \Delta(\Pi, Y)$.

We can now define the set ASM of Abstract States Machines:

Definition 5. *An Abstract State Machine M with signature $\mathcal{L}$ is given by:*

- *an* ASM *program* Π *on* $\mathcal{L}$
- *a set* $S(M)$ *of* $\mathcal{L}$-*structures closed by isomorphisms and* τ_{Π}
- *a subset* $I(M) \subseteq S(M)$ *closed by isomorphisms*
- *an application* τ_M, *which is the restriction of* τ_{Π} *to* $S(M)$

For every sequential algorithm A, the finiteness of the exploration witness in the third postulate allows us (see [12]) to write a finite ASM program Π_A, which has the same set of updates than A for every state. Every program Π_A obtained in this way has the same form, which we call the **normal form**:

$$
\begin{array}{ll}
\text{par} & \text{if } F_1 \text{ then } \Pi_1 \\
\| & \text{if } F_2 \text{ then } \Pi_2 \\
& \quad \vdots \\
\| & \text{if } F_c \text{ then } \Pi_c \\
\text{endpar} &
\end{array}
$$

where F_i are "guards", which means that for every state X one and only one F_i is *true*, and the programs Π_i have the form

$$\text{par } u_1 \| \ldots \| u_{m_i} \text{ endpar}$$

where $u_1, \ldots, u_{m_i}$ are update commands.

Remark 6. $\Delta(\Pi_A, X) = \Delta(A, X) = \tau_A(X) \ominus X$, so $\Delta(\Pi_A, X)$ is consistent without trivial updates.

The proof that the set of sequential algorithms is identical to the set of ASMs uses mainly the fact that every ASM has a finite exploration witness. Reciprocally, for every sequential algorithm we can define an ASM with the same transition function:

Theorem 6 (Gurevich, 2000).

$$\text{Algo} = \text{ASM}$$

So, Gurevich proved that his axiomatic presentation for sequential algorithms defines the same objects than his operational presentation of ASMs.

Remark 7. According to this theorem, every ASM is a sequential algorithm and every sequential algorithm can be simulated by an ASM in normal form. So, for every ASM there exists an equivalent ASM in normal form.

2.2 Imperative programming

We know that an imperative language such as Albert Meyer and Dennis Ritchie's Loop defined in [18] can compute any primitive recursive function, but cannot compute some "better" algorithms (see [6] for the *min*

and [19] for the *gcd*). This Loop language has been extended in with
an exit command to obtain every Arithmetical Primitive Recursive
Algorithm[11].

In [16] we generalized this result, by proving that LoopC characterized algorithms with primitive recursive time and data structures. We
use this language because it is minimal. The programs are only sequences of updates, if or loop commands. Notice that the loop commands can be broken if an exception is reached, like in [1]. Moreover,
in the following section we will prove that a sublanguage PLoopC has
polynomial time.

The difference with common models of computation is that the data
structures are not fixed. As is the case for the ASMs, the equality and
the booleans are needed, and the unary integers are necessary for the
loop commands, but the other data structures are seen as oracular.
If they can be implemented in a sequential algorithm then they are
implemented using the same language, universe and interpretation in
this programming language. So, the fair simulation between $\text{ASM}_\mathcal{P}$ and
PLoopC is proven for control structures, up to data structures.

Definition 6 (Syntax of LoopC programs).

$$
\begin{aligned}
c =_{def} \quad & f(t_1, \ldots, t_\alpha) := t_0 \\
& \mid \text{ if } F \ \{P_1\} \text{ else } \{P_2\} \\
& \mid \text{ loop } n \text{ except } F \ \{P\} \\
P =_{def} \quad & \text{end} \\
& \mid c; P
\end{aligned}
$$

*where f is a dynamic α-ary function symbol, $t_0, t_1, \ldots, t_\alpha$ are closed
terms, F is a formula, and n is a variable which is not updated in the
body of the loop.*

Notation 7. As is the case for ASM programs, we write only if F $\{P\}$
for the command if F $\{P\}$ else $\{\text{end}\}$. Following Meyer and Ritchie's
style [18], we write simply loop n $\{P\}$ a command

$$\text{loop } n \text{ except } \mathit{false} \ \{P\}$$

For the sake of clarity, we will omit the end inside curly brackets in the
rest of the paper.

The composition of commands $c; P$ can be generalized by induc-
tion to **composition of programs** $P_1; P_2$ by $\text{end}; P_2 =_{def} P_2$ and
$(c; P_1); P_2 =_{def} c; (P_1; P_2)$. As seen in example 1 p. 97 the operational

---

[11]APRA is defined as the set of the sequential algorithms with a primitive recursive time complexity, using only booleans and unary integers as data structures,
and using only variables as dynamical symbols.

semantics of this LoopC programming language is formalized by a state transition system. A state of the system is a pair $P \star X$ of a LoopC program and a structure. Its transitions are determined only by the head command and the current structure:

Definition 7 (Operational Semantics of LoopC).

$$f(t_1, \ldots, t_\alpha) := t_0; P \star X \succ P \star X \oplus (f, \overline{t_1}^X, \ldots, \overline{t_\alpha}^X, \overline{t_0}^X)$$

$$\texttt{if } F \; \{P_1\} \texttt{ else } \{P_2\}; P_3 \star X \succ P_j; P_3 \star X$$
$$where \; j = \left\{ \begin{array}{l} 1 \; if \; F \; is \; true \; in \; X \\ 2 \; else \end{array} \right.$$
$$\texttt{loop } n \texttt{ except } F \; \{P_1\}; P_2 \star X \succ Q; P_2 \star X \oplus (i, a)$$

$$where \; Q = \left\{ \begin{array}{ll} P_1; \texttt{loop } n \texttt{ except } F \; \{P_1\} & if \; i < n \; and \; \overline{F}^X = false \\ \texttt{end} & else \end{array} \right.$$

$$and \; a = \left\{ \begin{array}{ll} \overline{i}^X + 1 & if \; i < n \; and \; F \; is \; false \; in \; X \\ 0 & else \end{array} \right.$$

i is a dynamical symbol initialized to 0 in the initial states and which does not appear in the program. Each loop has a different counter i.

The successors are unique, so this transition system is deterministic. We denote by $\succ_t$ a succession of t transition steps.

Only the states $\texttt{end} \star X$ have no successor, so they are the terminating states.

Notation 8. P **terminates** on X if there exists t and X' such that:

$$P \star X \succ_t \texttt{end} \star X'$$

Because the transition system is deterministic, t and X' are unique. So X' is denoted $P(X)$ and t is denoted $time(P, X)$. A program is terminal if it terminates for every initial state.

Example 9. This program computes the minimum of two integers m and n in $O(min(m, n))$ steps, and stores the result in the output variable r:

$$P_{min} =_{def} r := 0; \texttt{loop } n \texttt{ except } r = m \; \{r := r + 1; \}; \texttt{end}$$

The execution of this program for $m = 2$ and $n = 3$ on a structure

X is:

$$r := 0; \texttt{loop } 3 \texttt{ except } r = 2 \; \{r := r+1; \}; \star X \oplus (i,0)$$
$$\succ \qquad\qquad \texttt{loop } 3 \texttt{ except } r = 2 \; \{r := r+1; \}; \star X \oplus \{(i,0),(r,0)\}$$
$$\succ r := r+1; \texttt{loop } 3 \texttt{ except } r = 2 \; \{r := r+1; \}; \star X \oplus \{(i,1),(r,0)\}$$
$$\succ \qquad\qquad \texttt{loop } 3 \texttt{ except } r = 2 \; \{r := r+1; \}; \star X \oplus \{(i,1),(r,1)\}$$
$$\succ r := r+1; \texttt{loop } 3 \texttt{ except } r = 2 \; \{r := r+1; \}; \star X \oplus \{(i,2),(r,1)\}$$
$$\succ \qquad\qquad \texttt{loop } 3 \texttt{ except } r = 2 \; \{r := r+1; \}; \star X \oplus \{(i,2),(r,2)\}$$
$$\succ \qquad\qquad\qquad\qquad\qquad\qquad\qquad \textbf{end} \star X \oplus \{(i,0),(r,2)\}$$

So $time(P_{min}, X) = 2 + 2 \times min(\overline{m}^X, \overline{n}^X) = O(min(\overline{m}^X, \overline{n}^X))$

The composition of programs behaves as intended. It is proved in [16] by using only the determinism and the transitivity of the transition system.

Proposition 2 (Composition of Programs).
$P_1; P_2$ terminates on X if and only if P_1 terminates on X and P_2 terminates on $P_1(X)$, such that:

1. $P_1; P_2(X) = P_2(P_1(X))$
2. $time(P_1; P_2, X) = time(P_1, X) + time(P_2, P_1(X))$

As a consequence, we can prove by induction that every LoopC program is terminal.

Because the transition system is deterministic, there exists a unique P' and X' such that $P \star X \succ_t P' \star X'$. Let $\tau_X^t(P)$ be that P' and $\tau_P^t(X)$ be that X':

$$P \star X \succ_t \tau_X^t(P) \star \tau_P^t(X)$$

Remark 8. τ_P^t is not a transition function in the sense of the first postulate 1, because $\tau_P^t(X) \neq \tau_P \circ \cdots \circ \tau_P(X)$.

Definition 8. *The set of updates made by P on X is:*

$$\Delta(P, X) =_{def} \bigcup_{0 \le t < time(P,X)} \tau_P^{t+1}(X) \ominus \tau_P^t(X)$$

Remark 9. In our transition system, one update at most can be done per step, so $\tau_P^{t+1}(X) \ominus \tau_P^t(X)$ is empty or is a singleton. Therefore, the cardinal of $\Delta(P, X)$ is bounded by $time(P, X)$.

In imperative programming languages, an **overwrite** occurs when a variable is updated to a value, then is updated to another value later in the execution. In our framework, this means that there exists in $\Delta(P, X)$ two updates $(f, \vec{a}, b)$ and $(f, \vec{a}, b')$ with $b \neq b'$, which makes $\Delta(P, X)$ inconsistent. So, we say that P is without overwrite on X if $\Delta(P, X)$ is consistent.

Lemma 1 (Updates of a Non-Overwriting Program).

If P is without overwrite on X then $\Delta(P, X) = P(X) \ominus X$.

Proof. The proof is admitted in this paper, but is detailed in [16].　⊣

2.3　Polynomial Time

We proved in [16] that LoopC is algorithmically complete for $\mathcal{PR}$-time algorithms (restricted to $\mathcal{PR}$-space data structures). The purpose of this subsection is to syntactically restrict this language in order to obtain a $\mathcal{P}$-time language PLoopC.

$$
\begin{aligned}
&r := 0 \\
&r := r + 1 \\
&\texttt{loop } n \\
&\quad x := r \\
&\quad \texttt{loop } x \\
&\quad\quad r := r + 1
\end{aligned}
$$

FIGURE 1　A Program for the Exponential Function

The program P_{pow} in figure 1 is a Loop program (see [18]) because it uses only variables, zero, successor, and loops. The command $x := r; \texttt{loop } x\ \{r := r + 1; \}$ computes $r := 2r$, so $\overline{r}^{P_{pow}(X)} = 2^{\overline{n}^X}$. Variables of a Loop program can only be increased by substitution or successor. So, according to the definition 1.1:

$$|P_{pow}(X)| \leq |X| + time(P_{pow}, X)$$

Therefore, P_{pow} is (at least) exponential in time.

In order to obtain a $\mathcal{P}$-time language, Neergaard used in [21] the Bellantoni and Cook's approach (see [2]) separating **safe and normal variables**. Safe variables can be updated, and normal variables are the bounds n in the loop n commands:

Definition 9 (Bounds of Loops).

$$
\begin{aligned}
Bound(\texttt{end}) &=_{def} \{\} \\
Bound(c; P) &=_{def} Bound(c) \cup Bound(P)
\end{aligned}
$$

$$
\begin{aligned}
Bound(f(t_1, \ldots, t_\alpha) := t_0) &=_{def} \{\} \\
Bound(\texttt{if } F\ \{P_1\}\ \texttt{else}\ \{P_2\}) &=_{def} Bound(P_1) \cup Bound(P_2) \\
Bound(\texttt{loop } n\ \texttt{except}\ F\ \{P\}) &=_{def} \{n\} \cup Bound(P)
\end{aligned}
$$

The programming language PLoop of Neergaard contains only programs such that for every loop $n\ \{P\}$ command occurring in the program, $\{n\} \cup Bound(P) \subseteq Stat(P)$. In other words, normal variables and safe variables are distinct sets in loops.

As suggested by the name, Neergaard proved that programs in PLoop are in polynomial time (and space[12]). In order to do so, he used lists as data structures, such that $|(d, e)| = max(|d|, |e|) + 1$. This result can be generalized to "translation functions" (in a geometric sense), those satisfying that:

$$|f(\vec{x})| \leq max|\vec{x}| + c_f$$

Theorem 10 (A $\mathcal{P}$-time Language).

If the operations are translations, then PLoop *is* $\mathcal{P}$-time and $\mathcal{P}$-space.

But the set of translation functions is very restrictive. As an example, the program in figure 2 satisfies Neergaard's condition, but is not $\mathcal{P}$-time because $x \mapsto 2x$ is not a translation.

$$r := 1$$
$$\texttt{loop } n$$
$$r := 2r$$
$$\texttt{loop } r$$

FIGURE 2 A PLoop Program Using more than Translation Functions

As an example, for unary integers the successor is a translation but not the addition, and for binary integers the addition is a translation but not the multiplication. In order to obtain more algorithms, we need to have larger data structures, so we will have to restrict even more our programming language.

The program in figure 2 illustrates that an exponential space in r can be converted into an exponential time using $\texttt{loop } r$ and the composition. This is the reason why the space must be $\mathcal{P}$-time too in Neergaard's theorem 10.

In order to obtain a $\mathcal{P}$-time programming language using every possible data structure, we will simply remove the connection between space and time. So, the bounds will remain static in the whole program, and not only in loops.

Definition 10 ($\mathcal{P}$-time Programming Language).

Let PLoopC *be the set of* LoopC *programs* P *satisfying that:*

$$Bound(P) \subseteq Stat(P)$$

Remark 10. This language is not closed by composition. As an example, the programs $n := pow(n);$ end and $\texttt{loop } n \; \{\};$ end are in PLoopC, not $n := pow(n); \texttt{loop } n \; \{\};$ end.

---

[12]A program P is $\mathcal{C}$-space if there exists $\varphi_P \in \mathcal{C}$ such that, for every initial state X, $|P(X)| \leq \varphi_P(|X|)$.

But this language will be useful anyway. We prove at proposition 3 that programs in PLoopC are $\mathcal{P}$-time, where the degree of the complexity is the depth of the program:

Definition 11 (Depth of a Program).

$$depth(\texttt{end}) =_{def} 0$$
$$depth(c; P) =_{def} max(depth(c), depth(P))$$

$$depth(f(t_1, \ldots, t_\alpha) := t_0) =_{def} 0$$
$$depth(\textit{if } F \ \{P_1\} \ \textit{else} \ \{P_2\}) =_{def} max(depth(P_1), depth(P_2))$$
$$depth(\texttt{loop } n \ \texttt{except } F \ \{P_1\}) =_{def} \delta + depth(P_1)$$

$$where \ \delta = \left\{ \begin{array}{l} 1 \ \textit{if } n \in Dyn(P) \sqcup Init(P) \\ 0 \ \textit{if } n \in Cons(P) \sqcup Oper(P) \end{array} \right.$$

Both cases in the previous definition may be surprising, because the depth is distinguished from the nesting. Indeed, if the bound of a loop is an uniform symbol, we assume that the loop is not a "true" loop, but only a syntactical convention to avoid code duplication, as illustrated at figure 3.

$r := 0$	$r := 0$
$\texttt{loop } 3$	$r := r + 1$
$\quad r := r + 1$	$r := r + 1$
	$r := r + 1$

FIGURE 3 Different nesting, same depth.

Remind that for every program P in PLoopC, $Bound(P) \subseteq Stat(P)$. So, we use only the symbols in $Init(P)$ for the depth. Moreover, if P_1 is a subprogram of P then $Init(P_1) \subseteq Init(P)$. So, in the following proposition and its proof, for the sake of simplicity we use the notation $|X|_{Init}$ for the size in X of the initial symbols of the program and its subprograms:

$$|X|_{Init} =_{def} \max_{f \in Init(P)} \{|f|_X\}$$

Proposition 3 (Polynomial Time).

For every program P in PLoopC there exists $\varphi_P \in \mathcal{P}$ such that for every X:

1. *$time(P, X) \leq \varphi_P(|X|_{Init})$*
2. *$deg(\varphi_P) = depth(P)$*

Remark 11. Initial symbols are static, so their interpretations in $P(X)$ are the same as in X. As a consequence, for every program P,

$|P(X)|_{Init} = |X|_{Init}$.

Proof. The proof is made by induction on P, by using:

- $\varphi_{\text{end}} = 0$
- $\varphi_{P_1;P_2} = \varphi_{P_1} + \varphi_{P_2}$ (according to proposition 2 p. 106)

Using the induction hypothesis, it remains to prove the proposition for commands alone. Both cases for updates and conditionals are straightforward:

- $\varphi_{f(t_1,...,t_\alpha):=t_0} = 1$
- $\varphi_{\text{if } F \text{ then } \{P_1\} \text{ else } \{P_2\}} = 1 + \varphi_{P_1} + \varphi_{P_2}$

So, we focus only on the non-trivial case of loops.

Because a **for** command (where the counter i is read) can be simulated by using a fresh variable, we can assume for the sake of simplicity that counters are not read in the program. So, we can write the execution of the program $P = \text{loop } n \text{ except } F \{P_1\}$ in this way:

$$
\begin{array}{lll}
 & \text{loop } n \text{ except } F \{P_1\}; \text{end} \star & X \oplus (i,0) \\
\succ & P_1; \text{loop } n \text{ except } F \{P_1\}; \text{end} \star & X \oplus (i,1) \\
\succ_{time(P_1,X)} & \text{loop } n \text{ except } F \{P_1\}; \text{end} \star P_1(X) \oplus (i,1) \\
\succ & P_1; \text{loop } n \text{ except } F \{P_1\}; \text{end} \star P_1(X) \oplus (i,2) \\
\vdots & & \\
\succ_{time(P_1,P_1^{a-1}(X))} & \text{loop } n \text{ except } F \{P_1\}; \text{end} \star P_1^a(X) \oplus (i,a) \\
\succ & \text{end} \star P_1^a(X) \oplus (i,0)
\end{array}
$$

where a is the first t such that $0 \le t \le \overline{n}^X$ and F is true in $P_1^t(X)$, or $a = \overline{n}^X$ if F is false in $P_1^t(X)$ for every $0 \le t \le \overline{n}^X$.

1. Time of the execution:
$$
time(P,X) = \sum_{0 \le t \le a-1} (1 + time(P_1, P_1^t(X))) + 1
$$
$$
\le 1 + \overline{n}^X + \sum_{0 \le t \le \overline{n}^X - 1} time(P_1, P_1^t(X))
$$

By induction hypothesis, for every X:
$$
time(P_1, X) \le \varphi_{P_1}(|X|_{Init})
$$

So, $time(P_1, P_1^t(X)) \le \varphi_{P_1}(|P_1^t(X)|_{Init}) = \varphi_{P_1}(|X|_{Init})$. Therefore:

$$
time(P,X) \le 1 + \overline{n}^X + \sum_{0 \le t \le \overline{n}^X - 1} \varphi_{P_1}(|X|_{Init})
$$
$$
= 1 + \overline{n}^X \times (1 + \varphi_{P_1}(|X|_{Init})) \qquad (1)
$$
$$
\le 1 + (\overline{n}^X + 1) \times (1 + \varphi_{P_1}(|X|_{Init})) \qquad (2)
$$

We use both inequalities for the following cases:

- If $n \in Cons(P) \sqcup Oper(P)$, then there exists an integer c_n such that $\overline{n}^X = c_n$.

  So, $\varphi_P = 1 + c_n \times (1 + \varphi_{P_1})$ (1).
- If $n \in Init(P)$, then $\overline{n}^X + 1 = |\overline{n}^X| \leq |X|_{Init}$.

  So, $\varphi_P(x) = 1 + x \times (1 + \varphi_{P_1}(x))$ (2).

2. The degree of the complexity depends on the case:
   - If $n \in Cons(P) \sqcup Oper(P)$, then:

$$
\begin{aligned}
deg(\varphi_P) &= deg(1 + c_n \times (1 + \varphi_{P_1})) \\
&= deg(\varphi_{P_1}) \\
&= depth(P_1) \\
&= depth(P)
\end{aligned}
$$

   - If $n \in Init(P)$, then:

$$
\begin{aligned}
deg(\varphi_P) &= deg(1 + id \times (1 + \varphi_{P_1})) \\
&= 1 + deg(\varphi_{P_1}) \\
&= 1 + depth(P_1) \\
&= depth(P)
\end{aligned}
$$

⊣

The coefficients of φ_P are positive, so if $0 \leq m \leq n$ then $\varphi_P(m) \leq \varphi_P(n)$. More specifically, since $|X|_{Init} \leq |X|$, we have $\varphi_P(|X|_{Init}) \leq \varphi_P(|X|)$. As a consequence, we have $time(P, X) \leq \varphi_P(|X|)$. Therefore, the time complexity of P is at most a polynomial of degree $depth(P)$.

3 ASM Simulates LoopC

3.1 Graphs of Execution

The intuitive idea for translating LoopC programs into ASM programs is to translate separately every command, and to add a variable (for example, the number of the line in the program) to keep track of the current command[13].

Example 11. The imperative program P_{min} of the example 9 p. 105:

$$
\begin{aligned}
&0: \quad r := 0 \\
&1: \quad \textbf{loop } n \textbf{ except } r = m \\
&2: \qquad r := r + 1
\end{aligned}
$$

could be translated into the following ASM program:

Remark 12. The number of a line is between 0 and $length(P)$. So, a finite number of booleans $b_0, b_1, \ldots, b_{length(P)}$ can be used[14] instead of an integer $line$.

---

[13]Programs of this form are called control state ASMs (see [5]).
[14]Remember that booleans must be in the data structure, but integers may not.

```
par   if line = 0 then
           par r := 0 || line := 1 endpar
        endif
    || if line = 1 then
           if (i ≠ n ∧ r ≠ m) then
               par i := i + 1 || line := 2 endpar
           else
               par i := 0 || line := 3 endpar
           endif
        endif
    || if line = 2 then
           par r := r + 1 || line := 1 endpar
        endif
endpar
```

FIGURE 4 Translation of P_{min}

This approach has been suggested in [14], and is fitted for a line-based programming language (for example with goto instructions) but not the structured language LoopC. Indeed, the positions in the program can distinguish two commands even if they are identical for the operational semantics of LoopC:

Example 12. (Labelled LoopC)

To make an easy example, let's compare the two updates $x := x + 1$ in the program of figure 5.

$$\texttt{loop } m \ \{x := x + 1; \ \texttt{loop } n \ \{x := x + 1; \}; \}; \texttt{end}$$

Because their positions are not the same in the program they have different numbers of line. So, we label them with $x := x + 1$ ⓐ and $x := x + 1$ ⓑ to distinguish each one from the other.

We can replace the program P_2 of figure 5 by the program P_4 of figure 6 without changing anything except the labels. The commands $x := x + 1$ ⓐ and $x := x + 1$ ⓑ are the same for the operational semantics, so we should find another way to keep track of the current command.

We will not use booleans $b_0, b_1, \ldots, b_{length(P)}$ indexed by the lines of the program, but booleans indexed by the possible states of the program during the execution. The possible executions of a program will be represented by a graph where the edges are the possible transitions, and the vertices are the possible programs:

Example 13. (Graph of Execution of P_{min})

P_1	$\succ P_2$	$\succ P_3$
loop m	$x := x + 1$ ⓐ	loop n
$x := x + 1$ ⓐ	loop n	$x := x + 1$ ⓑ
loop n	$x := x + 1$ ⓑ	loop m
$x := x + 1$ ⓑ	loop m	$x := x + 1$ ⓐ
	$x := x + 1$ ⓐ	loop n
	loop n	$x := x + 1$ ⓑ
	$x := x + 1$ ⓑ	

FIGURE 5 Loops with Labelling (1)

P_3	$\succ P_4$	$\succ \ldots$
loop n	$x := x + 1$ ⓑ	
$x := x + 1$ ⓑ	loop n	
loop m	$x := x + 1$ ⓑ	
$x := x + 1$ ⓐ	loop m	
loop n	$x := x + 1$ ⓐ	
$x := x + 1$ ⓑ	loop n	
	$x := x + 1$ ⓑ	

FIGURE 6 Loops with Labelling (2)

$$r := 0$$
$$\texttt{loop } n \texttt{ except } r = m$$
$$r := r + 1$$

$$\downarrow$$

$$\texttt{loop } n \texttt{ except } r = m \qquad\qquad r := r + 1$$
$$r := r + 1 \qquad\qquad \texttt{loop } n \texttt{ except } r = m$$
$$r := r + 1$$

$$\downarrow$$

$$\texttt{end}$$

In the following only the vertices of the graph are needed, so the graph of execution of P_{min} will be denoted by the set of possible programs:

$$\mathcal{G}(P_{min}) = \{ \quad r := 0; \texttt{loop } n \texttt{ except } r = m \; \{r := r + 1; \}; \texttt{end},$$
$$\texttt{loop } n \texttt{ except } r = m \; \{r := r + 1; \}; \texttt{end},$$
$$r := r + 1; \texttt{loop } n \texttt{ except } r = m \; \{r := r + 1; \}; \texttt{end},$$
$$\texttt{end}$$
$$\}$$

Notation 14. In order to define graphs of execution we need to introduce the notation:

$$\mathcal{G}; P =_{def} \{P_j; P \mid P_j \in \mathcal{G}\}$$

where $\mathcal{G}$ is a set of imperative programs and P is an imperative program.

Let P be an imperative program. $\mathcal{G}(P)$ is the set of every possible $\tau_X^t(P)$ programs, which does not depend on an initial state X:

Definition 12. *(Graph of Execution)*

$$\mathcal{G}(\texttt{end}) =_{def} \{\texttt{end}\}$$
$$\mathcal{G}(c; P) =_{def} \mathcal{G}(c); P \cup \mathcal{G}(P)$$

$$\mathcal{G}(f(t_1, \ldots, t_\alpha) := t_0) =_{def} \{f(t_1, \ldots, t_\alpha) := t_0; \texttt{end}\}$$
$$\mathcal{G}(\texttt{if } F \; \{P_1\} \texttt{ else } \{P_2\}) =_{def} \{\texttt{if } F \; \{P_1\} \texttt{ else } \{P_2\}; \texttt{end}$$
$$\cup \, \mathcal{G}(P_1) \cup \mathcal{G}(P_2)$$
$$\mathcal{G}(\texttt{loop } n \texttt{ except } F \; \{P\}) =_{def} \mathcal{G}(P); \texttt{loop } n \texttt{ except } F \; \{P\}; \texttt{end}$$

As intended, we can prove (see [16]) that $card(\mathcal{G}(P)) \leq length(P)+1$. So, only a finite number of guards depending only on P are necessary. Notice that for some programs (like P_{min} in example 13 p. 112) which do not follow example 12 p. 112, $card(\mathcal{G}(P)) = length(P) + 1$ can be reached, so the bound is optimal.

Again, to focus on the simulation, we admit in this paper the proof (see [16]) stating that a graph of execution is closed for the operational semantics of the imperative programs:

Proposition 4. *(Operational Closure of Graph of Execution)*

- *If* $f(t_1, \ldots, t_\alpha) := t_0; Q \in \mathcal{G}(P)$
  then $Q \in \mathcal{G}(P)$
- *If* if F $\{P_1\}$ else $\{P_2\}; Q \in \mathcal{G}(P)$
  then $P_1; Q$ *and* $P_2; Q \in \mathcal{G}(P)$
- *If* loop n except F $\{P_1\}; Q \in \mathcal{G}(P)$
  then P_1; loop n except F $\{P_1\}; Q$ *and* $Q \in \mathcal{G}(P)$

3.2 Translation of an Imperative Program

Notation 15. The fresh boolean variables will be denoted b_{P_j}, where $P_j \in \mathcal{G}(P)$. One and only one b_{P_j} will be true for each step of an execution, so in the following we will write $X[b_{P_j}]$ if b_{P_j} is true and the other booleans b_{P_k} are false, where X denotes a $\mathcal{L}(P)$-structure. Notice that $X[b_{P_j}]|_{\mathcal{L}(P)} = X$.

Proposition 4 ensures that the following translation is well-defined:

Definition 13. *(Translation of imperative programs into ASM)*

$$\Pi_P =_{def} \underset{P_j \in \mathcal{G}(P)}{\text{par}} \quad \text{if } b_{P_j} \text{ then } P_j^{tr} \text{ endpar}$$

where P_j^{tr} is defined at the figure 7 p. 116.

Notice that for every $P_j \in \mathcal{G}(P)$, $\Delta(\Pi_P, X[b_{P_j}]) = \Delta(P_j^{tr}, X[b_{P_j}])$. We use this fact in [16] to prove by exhaustion on $\tau_X^t(P)$ that the translation of the imperative program P behaves as intended:

Proposition 5. *(Step-by-Step Simulation)*

For every $t < time(P, X), \tau_{\Pi_P}(\tau_P^t(X)[b_{\tau_X^t(P)}]) = \tau_P^{t+1}(X)[b_{\tau_X^{t+1}(P)}]$

Theorem 16. ASM *fairly simulates* LoopC.

Proof. We prove the three conditions of the fair simulation defined p. 100:

1. $\mathcal{L}(\Pi_P) = \mathcal{L}(P) \cup \{b_{P_j} \mid P_j \in \mathcal{G}(P)\}$
   where $card(\{b_{P_j} \mid P_j \in \mathcal{G}(P)\}) \leq length(P) + 1$.

2. Using proposition 5, we can prove by induction on $t \leq time(P, X)$ that

$$\tau_{\Pi_P}^t(X[b_P]) = \tau_P^t(X)[b_{\tau_X^t(P)}]$$

   Hence $\tau_{\Pi_P}^t(X[b_P])|_{\mathcal{L}(P)} = \tau_P^t(X)$, and the temporal dilation is $\boxed{d = 1}$.

$(\mathbf{end})^{tr} =_{def} \mathbf{par}\ \mathbf{endpar}$

$(f(t_1,\ldots,t_\alpha) := t_0; Q)^{tr}$

$=_{def}$
$\quad\quad \mathbf{par}\ b_{f(t_1,\ldots,t_\alpha):=t_0;Q} := \mathit{false}$
$\quad\quad\quad \|\ f(t_1,\ldots,t_\alpha) := t_0$
$\quad\quad\quad \|\ b_Q := \mathit{true}$
$\quad\quad \mathbf{endpar}$

$(\mathbf{if}\ F\ \mathbf{then}\ \{P_1\}\ \mathbf{else}\ \{P_2\}; Q)^{tr}$

$=_{def}$
$\quad\quad \mathbf{par}\ b_{\mathbf{if}\ F\ \mathbf{then}\ \{P_1\}\ \mathbf{else}\ \{P_2\};Q} := \mathit{false}$
$\quad\quad \|\ \mathbf{if}\ F\ \mathbf{then}$
$\quad\quad\quad\quad b_{P_1;Q} := \mathit{true}$
$\quad\quad \mathbf{else}$
$\quad\quad\quad\quad b_{P_2;Q} := \mathit{true}$
$\quad\quad \mathbf{endif}$
$\quad \mathbf{endpar}$

$(\mathbf{loop}\ n\ \mathbf{except}\ F\ \{\mathbf{end}\}; Q)^{tr}$

$=_{def}$
$\quad\quad \mathbf{if}\ (i \neq n \wedge \neg F)\ \mathbf{then}$
$\quad\quad\quad i := i+1$
$\quad\quad \mathbf{else}$
$\quad\quad\quad\quad \mathbf{par}\ b_{\mathbf{loop}\ n\ \mathbf{except}\ F\ \{\mathbf{end}\};Q} := \mathit{false}$
$\quad\quad\quad\quad \|\ i := 0$
$\quad\quad\quad\quad \|\ b_Q := \mathit{true}$
$\quad\quad\quad \mathbf{endpar}$
$\quad\quad \mathbf{endif}$

$(\mathbf{loop}\ n\ \mathbf{except}\ F\ \{c; P_1\}; Q)^{tr}$

$=_{def}$
$\quad\quad \mathbf{par}\ b_{\mathbf{loop}\ n\ \mathbf{except}\ F\ \{c;P_1\};Q} := \mathit{false}$
$\quad\quad \|\ \mathbf{if}\ (i \neq n \wedge \neg F)\ \mathbf{then}$
$\quad\quad\quad\quad \mathbf{par}\ i := i+1$
$\quad\quad\quad\quad\quad \|\ b_{c;P_1;\mathbf{loop}\ n\ \mathbf{except}\ F\ \{c;P_1\};Q} := \mathit{true}$
$\quad\quad\quad\quad \mathbf{endpar}$
$\quad\quad\quad \mathbf{else}$
$\quad\quad\quad\quad \mathbf{par}\ i := 0$
$\quad\quad\quad\quad\quad \|\ b_Q := \mathit{true}$
$\quad\quad\quad\quad \mathbf{endpar}$
$\quad\quad\quad \mathbf{endif}$
$\quad\quad \mathbf{endpar}$

FIGURE 7 Translation of an Imperative Program

3. If $t = time(P, X)$ then $\tau_X^t(P) = $ **end**.
So, $\Delta(\Pi_P, \tau_P^t(X)[b_{\tau_X^t(P)}]) = \varnothing$, and $\tau_{\Pi_P}^{t+1}(X[b_P]) = \tau_{\Pi_P}^t(X[b_P])$.
Therefore, $time(\Pi_P, X[b_P]) \leq time(P, X)$. (1)
Let $t < time(P, X)$.
According to the operational semantics p. 105:
If $\tau_X^t(P) \star \tau_P^t(X) \succ \tau_X^{t+1}(P) \star \tau_P^{t+1}(X)$ then

$$\tau_X^t(P) \neq \tau_X^{t+1}(P) \text{ or } \tau_X^t(P) = \texttt{loop } n \texttt{ except } F \text{ \{\};} Q$$

In the first case, we have $b_{\tau_X^t(P)} \neq b_{\tau_X^{t+1}(P)}$, and in the second
case $\tau_P^t(X) \neq \tau_P^{t+1}(X)$, since $\overline{i}^{\tau_P^{t+1}(X)} = \overline{i}^{\tau_P^t(X)} + 1$.
In any case, $\tau_P^{t+1}(X)[b_{\tau_X^{t+1}(P)}] \neq \tau_P^t(X)[b_{\tau_X^t(P)}]$.
So, $\tau_{\Pi_P}^{t+1}(X[b_P]) \neq \tau_{\Pi_P}^t(X[b_P])$.
Therefore, $time(\Pi_P, X[b_P]) \geq time(P, X)$. (2)
According to (1) and (2), we have $time(\Pi_P, X[b_P]) = time(P, X)$,
so $\boxed{e = 0}$.

$\dashv$

Therefore, as stated in the conclusion, ASMs in polynomial time can
fairly simulate programs of PLoopC, since (according to proposition 3
p. 109) they are in polynomial time, and PLoopC is a sublanguage of
LoopC.

4 PLoopC Simulates Polynomial-Time ASM

Let Π be an ASM program with a polynomial-time complexity $\varphi_\Pi \in \mathcal{P}$.
The purpose of this section is to find a PLoopC program P_{step} simulating
the same executions as Π. We construct this program P_{step} in three
steps:

1. Translate Π into an imperative program P_{step} simulating one step
   of the ASM.

2. Repeat P_{step} a sufficient number of times, depending on φ_Π, the
   complexity of Π.

3. Ensure that the final program stops at the same time as the ASM,
   up to temporal dilation.

4.1 Translation of one Step

Remember that Π contains only updates, **if** and **par** commands. The
intuitive solution is to translate the commands directly, without paying
attention to the parallelism:

Definition 14 (Syntactical Translation of the ASM programs).

$(f(t_1, \ldots, t_\alpha) := t_0)^{tr}$
$$=_{def} f(t_1, \ldots, t_\alpha) := t_0; \mathtt{end}$$
$(\mathtt{if}\ F\ \mathtt{then}\ \Pi_1\ \mathtt{else}\ \Pi_2\ \mathtt{endif})^{tr}$
$$=_{def} \mathtt{if}\ F\ \mathtt{then}\ \{\Pi_1^{tr}\}\ \mathtt{else}\ \{\Pi_2^{tr}\}; \mathtt{end}$$
$(\mathtt{par}\ \Pi_1 \| \ldots \| \Pi_n\ \mathtt{endpar})^{tr}$
$$=_{def} \Pi_1^{tr}; \ldots; \Pi_n^{tr}$$

Updates and $\mathtt{if}$ commands are the same in these two models of computation, but the simultaneous commands of ASM must be sequentialized in LoopC, so this translation does not respect the semantics of the ASM programs:

Example 17. Let X be a structure such that $\bar{x}^X = 0$ and $\bar{y}^X = 1$, and Π be the program:

$$\Pi = \mathtt{par}\ x := y \| y := x\ \mathtt{endpar}$$

Since both updates are done simultaneously, the semantics of Π is to exchange the value of x and y. In that case $\Delta(\Pi, X) = \{(x, 1), (y, 0)\}$, so $\tau_\Pi(X) = X \oplus \{(x, 1), (y, 0)\}$.

$$\Pi^{tr} = x := y; y := x; \mathtt{end}$$

But the semantics of Π^{tr} is to replace the value of x by the value of y and leave y unchanged. In that case, we have the following execution:

$$x := y; y := x; \mathtt{end} \star X$$
$$\succ \qquad y := x; \mathtt{end} \star X \oplus \{(x, 1)\}$$
$$\succ \qquad \mathtt{end} \star X \oplus \{(x, 1), (y, 1)\}$$

So $\tau_\Pi(X) = X \oplus \{(x, 1), (y, 0)\} \neq X \oplus \{(x, 1), (y, 1)\} = \Pi^{tr}(X)$.

In order to capture the simultaneous behavior of the ASM program, we need to store the values of the variables read in the imperative program. As an example, if $v_x = x$ and $v_y = y$ in X then:

$$x := v_y; y := v_x; \mathtt{end} \star X$$
$$\succ \qquad y := v_x; \mathtt{end} \star X \oplus \{(x, 1)\}$$
$$\succ \qquad \mathtt{end} \star X \oplus \{(x, 1), (y, 0)\}$$

Indeed, even if x has been updated, its old value is still in v_x.

Definition 15 (Substitution of a Term by a Variable).

$(f(t_1, \ldots, t_\alpha) := t_0)[v/t]$
$$=_{def} f(t_1[v/t], \ldots, t_\alpha[v/t]) := t_0[v/t]$$
$(\mathtt{if}\ F\ \mathtt{then}\ \Pi_1\ \mathtt{else}\ \Pi_2\ \mathtt{endif})[v/t]$
$$=_{def} \mathtt{if}\ F[v/t]\ \mathtt{then}\ \Pi_1[v/t]\ \mathtt{else}\ \Pi_2[v/t]\ \mathtt{endif}$$
$(\mathtt{par}\ \Pi_1 \| \ldots \| \Pi_n\ \mathtt{endpar})[v/t]$
$$=_{def} \mathtt{par}\ \Pi_1[v/t] \| \ldots \| \Pi_n[v/t]\ \mathtt{endpar}$$

$$where\ t_1[v/t_2] =_{def} \begin{cases} v & if\ t_1 = t_2 \\ t_1 & else \end{cases}$$

Remark 13. Since the temporary variables are fresh, if t_1 and t_2 are distinct terms then $\Pi[v_{t_1}/t_1][v_{t_2}/t_2] = \Pi[v_{t_2}/t_2][v_{t_1}/t_1]$ As a consequence, since the substitutions can be made in any order, for the terms $t_1, \ldots, t_r$ read by Π (see the definition p. 102), the notation $\Pi[\vec{v_t}/\vec{t}]$ is not ambiguous.

But using $\Pi[\vec{v_t}/\vec{t}]^{tr}$ for P_{step} is not sufficient, because two issues remain:

1. The variables $\vec{v_t}$ must be initialized with the value of the terms $\vec{t}$. Because the fresh variables must have a uniform initialization (see p. 99), we have to update the variables $\vec{v_t}$ explicitly at the beginning of the program by using a sequence of updates:

$$v_{t_1} := t_1; \ldots; v_{t_r} := t_r;$$

2. The execution time depends on the current initial state.
   This is an issue because, according to our definition of the fair simulation p. 100, every step of the ASM Π must be simulated by d steps, where d depends only on Π. In order to obtain a uniform temporal dilation, we will add skip commands[15] to the program:

$$\begin{aligned} \text{skip } 0 \quad &=_{def} \text{ end} \\ \text{skip } n+1 &=_{def} \text{ if } true\ \{\}; \text{skip } n \end{aligned}$$

According to the Gurevich's Theorem, every ASM is equivalent to an ASM in normal form, so we can assume that Π is in normal form (see p. 103). Therefore, its translation has the form:

$$\begin{aligned} &\text{if } F_1 \text{ then } \{\Pi_1^{tr}\}; \\ &\text{if } F_2 \text{ then } \{\Pi_2^{tr}\}; \\ &\quad\vdots \\ &\text{if } F_c \text{ then } \{\Pi_c^{tr}\}; \\ &\text{end} \end{aligned}$$

Remind that every F is a guard, which means that one and only one F_i is true for the current state X. The block of updates $\Pi_i^{tr} = u_1; \ldots; u_{m_i}; \text{end}$ requires m_i steps to be computed by the imperative program, so we add skip $m - m_i$ at the end of the block, where m is

---

[15]It may seem strange in an algorithmic purpose to lose time, but these skip commands do not change the asymptotic behavior and are necessary for our strict definition of the fair simulation. It is possible to weaken the definition of the simulation to simulate one step with $\leq d$ steps and not $= d$ steps, but we wanted to prove the result for the strongest definition possible.

defined by:

$$m =_{def} max\{m_i \mid 1 \le i \le c\}$$

$P_{step} =_{def}$
  $v_{t_1} := t_1;$
  $v_{t_2} := t_2;$
  $\vdots$
  $v_{t_r} := t_r;$
  if v_{F_1} **then** {
    $f_1^1(\vec{v}_{t_1^1}) := v_{t_1^1};$
    $f_2^1(\vec{v}_{t_2^1}) := v_{t_2^1};$
    $\vdots$
    $f_{m_1}^1(\vec{v}_{t_{m_1}^1}) := v_{t_{m_1}^1};$
    skip $m - m_1;$
  };
  if v_{F_2} **then** {
    $f_1^2(\vec{v}_{t_1^2}) := v_{t_1^2};$
    $f_2^2(\vec{v}_{t_2^2}) := v_{t_2^2};$
    $\vdots$
    $f_{m_2}^2(\vec{v}_{t_{m_2}^2}) := v_{t_{m_2}^2};$
    skip $m - m_2;$
  };
  $\vdots$
  if v_{F_c} **then** {
    $f_1^c(\vec{v}_{t_1^c}) := v_{t_1^c};$
    $f_2^c(\vec{v}_{t_2^c}) := v_{t_2^c};$
    $\vdots$
    $f_{m_c}^c(\vec{v}_{t_{m_c}^c}) := v_{t_{m_c}^c};$
    skip $m - m_c;$
  };
end

FIGURE 8 Translation P_{step} of one Step of Π

We obtain at figure 8 p. 120 the translation P_{step} of one step of the ASM program Π. Let X be a state of the ASM with program Π, extended with the variables $\vec{v}_t$. As intended, we prove that P_{step} simulates one step of Π in a constant time t_Π:

Proposition 6 (Semantical Translation of the ASM programs).

There exists t_Π, depending only on Π, such that for every state X of P_{step}:

- $(P_{step}(X) \ominus X)|_{\mathcal{L}(\Pi)} = \Delta(\Pi, X|_{\mathcal{L}(\Pi)})$
- $time(P_{step}, X) = t_\Pi$

Proof. The sequence of updates $v_{t_1} := t_1; \ldots; v_{t_r} := t_r$; requires r steps. Because the variables $\vec{v_t}$ are fresh they don't appear in the terms $\vec{t}$. So, in the state Y after these updates, $\overline{v_{t_k}}^Y = \overline{t_k}^X$. Moreover, in the rest of the program the variables $\vec{v_t}$ are not updated, so for every following state Y, $\overline{v_{t_k}}^Y = \overline{t_k}^X$.

In particular, for every $1 \le j \le c$, $\overline{v_{F_j}}^Y = \overline{F_j}^X$. Since these conditionals are guards, one and only one is *true* in X. Let F_i be this formula. Therefore, in every following state Y, $\overline{v_{F_i}}^Y = true$, and for every $j \ne i$, $\overline{v_{F_j}}^Y = false$.

$i - 1$ steps are required to erase the conditionals before F_i, one step is required to enter the block of F_i, and after the commands in that block $c - i$ steps are required to erase the conditionals after F_i. So, $(i - 1) + 1 + (c - i) = c$ steps are required for the conditionals.

Since for every following state Y, $\overline{v_{t_k}}^Y = \overline{t_k}^X$, the set of updates done in the block of F_i is $\Delta(\Pi, X|_{\mathcal{L}(\Pi)})$. These updates require m_i steps, then the `skip` command requires $m - m_i$ steps. So the commands in the block require $m_i + (m - m_i) = m$ steps, and the execution time depends only on Π:

$$time(P_{step}, X) = r + c + m = t_\Pi$$

The updates done by P_{step} are the initial updates and the updates done in the block of F_i:

$$\Delta(P_{step}, X) = \{(v_{t_1}, \overline{t_1}^X), \ldots, (v_{t_r}, \overline{t_r}^X)\} \cup \Delta(\Pi, X|_{\mathcal{L}(\Pi)})$$

The fresh variables are updated only once, and since Π is in normal form, $\Delta(\Pi, X|_{\mathcal{L}(\Pi)})$ is consistent. So, P_{step} is without overwrite on X, and according to proposition 1 p. 106:

$$\Delta(P_{step}, X) = P_{step}(X) \ominus X$$

So $(P_{step}(X) \ominus X)|_{\mathcal{L}(\Pi)} = \Delta(\Pi, X|_{\mathcal{L}(\Pi)})$. ⊣

More generally, we can use this result to prove by induction on t that:

Corollary 1. $P^t_{step}(X)|_{\mathcal{L}(\Pi)} = \tau^t_\Pi(X|_{\mathcal{L}(\Pi)})$

4.2 Translation of the Complexity

P_{step} simulates in constant time one step of the ASM program Π, so we want to repeat it a sufficient number of times in order to simulate every execution of the ASM. In this paper we focus on the polynomial time algorithms, so we assume that there exists a polynomial function φ_Π such that for every initial state X:

$$time(\Pi, X) \leq \varphi_\Pi(|X|)$$

Since φ_Π is a polynomial function, there exists $a_0, \dots, a_{deg(\varphi_\Pi)} \in \mathbb{Z}$ such that:

$$\varphi_\Pi(|X|) = \sum_{0 \leq n \leq deg(\varphi_\Pi)} a_n |X|^n \leq \left(\sum_{0 \leq n \leq deg(\varphi_\Pi)} max(0, a_n) \right) |X|^{deg(\varphi_\Pi)}$$

Therefore, there exists $c \in \mathbb{N}$ depending only of φ_Π, such that:

$$time(\Pi, X) \leq c \times |X|^{deg(\varphi_\Pi)}$$

We assume that the program has access to the size of its inputs, so it has access to $|X|$, which is the maximum (or the sum) of these values. Therefore, the following program has an execution time greater than Π on X, where c and $size$ are fresh variables initialized respectively with $\sum_{0 \leq n \leq deg(\varphi_\Pi)} max(0, a_n)$ and $|X|$:

$$
\begin{array}{l}
\texttt{loop } c \\
\quad \texttt{loop } size \\
\qquad \ddots \quad deg(\varphi_\Pi) \text{ times} \\
\qquad \texttt{loop } size
\end{array}
$$

Notice that according to the definition 11 p. 109, the depth of this program is $deg(\varphi_\Pi)$. The intuitive program repeating P_{step} is:

$$
\begin{array}{l}
\texttt{loop } c \\
\quad \texttt{loop } size \\
\qquad \ddots \\
\qquad \texttt{loop } size \\
\qquad \quad P_{step}
\end{array}
$$

In that case, between two executions of P_{step} the number of steps depends on the actual depth in the program, so the simulation will not have a constant temporal dilation. We want the program to execute one step of a loop then to execute P_{step} , then to execute another step of a loop and so on... So, we need to duplicate[16] P_{step} before each body of a

---

[16]Like in [1], with the difference that we choose to have every execution of P_{step}

loop (when the execution enters a loop) and after each `loop` command (when the execution erases a loop):

$$
\begin{aligned}
&\texttt{loop } c \\
&\quad P_{step} \\
&\quad \texttt{loop } size \\
&\qquad P_{step} \\
&\qquad\qquad \ddots \\
&\qquad\qquad\quad \texttt{loop } size \\
&\qquad\qquad\qquad P_{step} \\
&\qquad\qquad\quad P_{step} \\
&\qquad\qquad \ddots \\
&\qquad P_{step} \\
&\quad P_{step}
\end{aligned}
$$

As a consequence, our candidate is

$$\texttt{loop } c \ \{P_{step}; \texttt{loop}^{deg(\varphi_\Pi)} \ size \ \{P_{step}\}\}; P_{step}$$

where $\texttt{loop}^i \ n \ \{P\}$ is defined by induction:

$$
\begin{aligned}
\texttt{loop}^0 \ n \ \{P\} &=_{def} \texttt{end} \\
\texttt{loop}^{i+1} \ n \ \{P\} &=_{def} \texttt{loop } n \ \{P; \texttt{loop}^i \ n \ \{P\}\}; P
\end{aligned}
$$

The temporal dilation is $d = t_\Pi + 1$, since the program alternates between `loop` commands and executions of P_{step}. But we can't ensure that the program stops at the same time as Π, so we need to detect the end of the execution.

For any initial state of the program P_{step}, the fresh variables $\vec{v_t}$ store the value of the interpretation of the terms $\vec{t}$, then the terms $\vec{t}$ are updated. This means that at the end of P_{step} the variables $\vec{v_t}$ have the old values of the terms. In particular, if the initial state is $P_{step}^t(X)$, after one execution of P_{step} we have:

$$\overline{v_{t_k}}^{P_{step}^{t+1}(X)} = \overline{t_k}^{P_{step}^t(X)}$$

Π terminates when no more updates are done. In that case, the old values of the terms read by Π are the same as the new values. Therefore, since the old values are stored in the variables $\vec{v_t}$, every v_{t_k} is equal to t_k in the terminating state:

$$F_\Pi =_{def} \bigwedge_{t \in Read(\Pi)} v_t = t$$

_____

after every command of the program, not before. This will make sense when we will add one occurrence of P_{step} before the program in order to initialize the μ-formula F_Π.

We call it the "μ-formula" because it is similar to the minimization operator μ from recursive functions (see [7]):

Lemma 2 (The μ-formula).

$$time(\Pi, X|_{\mathcal{L}(\Pi)}) = min\{t \in \mathbb{N} \mid \overline{F_\Pi}^{P_{step}^{t+1}(X)} = true\}$$

Proof. $time(\Pi, X|_{\mathcal{L}(\Pi)}) = min\{t \in \mathbb{N} \mid \tau_\Pi^t(X|_{\mathcal{L}(\Pi)}) = \tau_\Pi^{t+1}(X|_{\mathcal{L}(\Pi)})\}$, so all that is left to prove (see [16]) is that $\tau_\Pi^t(X|_{\mathcal{L}(\Pi)}) = \tau_\Pi^{t+1}(X|_{\mathcal{L}(\Pi)})$ if and only if $\overline{F_\Pi}^{P_{step}^{t+1}(X)}$ is true, by using the remark p. 102 on $Read(\Pi)$. $\dashv$

So, the current candidate to simulate the ASM program Π is the program:

> loop c except F_Π
>   if $\neg F_\Pi$ $\{P_{step}\}$
>   loop $size$ except F_Π
>   if $\neg F_\Pi$ $\{P_{step}\}$
>
>   ⋱
>
>   loop $size$ except F_Π
>   if $\neg F_\Pi$ $\{P_{step}\}$
>   if $\neg F_\Pi$ $\{P_{step}\}$
>
>   ⋱
>
>   if $\neg F_\Pi$ $\{P_{step}\}$
>   if $\neg F_\Pi$ $\{P_{step}\}$

The temporal dilation becomes $d = t_\Pi + 2$, since entering the conditionals costs one more step. But two issues remain:

1. The variables $\vec{v}_t$ must be properly initialized to obtain a correct value for the μ-formula F_Π. We do that simply by adding an occurrence of P_{step} at the beginning of the program. F_Π becomes true after $time(\Pi, X|_{\mathcal{L}(\Pi)}) + 1$ steps, so we execute the program P_{step} one more time after the end of Π. This is not an issue because the execution time of P_{step} is t_Π, as required by the third condition of the fair simulation.

2. The simulation is correct until F_Π becomes true, and after that the remaining steps consist to erase the last loop commands. But their number depends on the current depth, determined by the initial state. This number is bounded by $deg(\varphi_\Pi) + 1$, so the current ending time can be bounded too. In fact, for every remaining loop commands two steps are done: erase the loop then

erase the following if $\neg F_\Pi \{P_{step}\}$. Therefore, the ending time is bounded[17] by $max_{end} = 2 \times (deg(\varphi_\Pi) + 1)$.

By using a fresh variable i_{end} which counts the number of steps done after F_Π became true, we can add at the end of the program the program $\text{skip } i_{end} \rightarrow max_{end}$ defined by:

$$\text{skip } i \rightarrow 0 \qquad =_{def} \text{ end}$$
$$\text{skip } i \rightarrow m + 1 =_{def} \text{ if } i = m + 1 \text{ \{end\} else \{skip } i \rightarrow m\}; \text{end}$$

For every state X, we can prove by induction on $0 \leq \overline{i_{end}}^X \leq \overline{max_{end}}^X$ that:

$$timc(\text{skip } i_{end} \rightarrow max_{end}, X) = \overline{max_{end}}^X - \overline{i_{end}}^X + 1$$

It remains to set the correct value for i_{end}. This variable is initialized to 0 and for each remaining loop commands three steps are done: erase the loop, enter the if and update i_{end}. So, we replace in our candidate program each if $\neg F_\Pi \{P_{step}\}$ by if $\neg F_\Pi \{P_{step}\}$ else $\{i_{end} := i_{end} + 3; \text{end}\}$, and max_{end} becomes $3 \times (deg(\varphi_\Pi) + 1)$.

4.3 The Simulation

For every ASM program Π we obtain at figure 9 its translation P_Π simulating the execution of Π:

Theorem 18. PLoopC *fairly simulates* ASM$_P$.

Proof. We prove the three conditions of the fair simulation defined p. 100:

1. $\mathcal{L}(P_\Pi) = \mathcal{L}(\Pi) \sqcup \{v_t \mid t \in Read(\Pi)\} \sqcup \{c, size, i_{end}\}$
   So, there is a finite number of fresh variables, depending only on Π.

2. Until F_Π becomes true the execution alternates between:
   (a) t_Π steps of P_{step}, which simulates one step of Π, according to proposition 6 p. 120.
   (b) Then one step to enter the body of a loop, or erase a loop command.
   (c) Then one step to enter the conditional if $\neg F_\Pi \{P_{step}\}$, then repeat from the beginning.

   So, each step of Π is simulated by exactly $\boxed{d = t_\Pi + 2}$ steps of its translation P_Π.

---

[17] max_{end} does not depend of the initial state so we can use constructors instead of a variable, and define $\text{skip } i \rightarrow m$ by induction on m. We cannot do that for c because contrary to conditionals, loop commands can only be bounded by a variable, not a term.

P_{step}
loop c except F_Π
    if $\neg F_\Pi$ $\{P_{step}\}$ else $\{i_{end} := i_{end} + 3; \text{end}\}$
    loop $size$ except F_Π
       if $\neg F_\Pi$ $\{P_{step}\}$ else $\{i_{end} := i_{end} + 3; \text{end}\}$

$\therefore$ $deg(\varphi_\Pi)$ times
       loop $size$ except F_Π
          if $\neg F_\Pi$ $\{P_{step}\}$ else $\{i_{end} := i_{end} + 3; \text{end}\}$
       if $\neg F_\Pi$ $\{P_{step}\}$ else $\{i_{end} := i_{end} + 3; \text{end}\}$

$\therefore$ $deg(\varphi_\Pi)$ times
    if $\neg F_\Pi$ $\{P_{step}\}$ else $\{i_{end} := i_{end} + 3; \text{end}\}$
if $\neg F_\Pi$ $\{P_{step}\}$ else $\{i_{end} := i_{end} + 3; \text{end}\}$
skip $i_{end} \to max_{end}$

FIGURE 9 Translation P_Π of the ASM program Π

Moreover, the execution is sufficiently long. Indeed, if c and $size$ are initialized respectively with $\sum_{0 \leq n \leq deg(\varphi_\Pi)} max(0, a_n)$ and $|X_0|$ in an initial state X_0 then:

$$time(\Pi, X) \leq \bar{c}^{X_0} \times (\overline{size}^{X_0})^{deg(\varphi_\Pi)}$$

Notice that c and $size$ are never updated, so this inequality holds for every state of the execution. Moreover, since c does not depend on the chosen initial state, according to definition 11 p. 109:

$$depth(P_\Pi) = deg(\varphi_\Pi)$$

3. Therefore, $time(\Pi, X|_{\mathcal{L}(\Pi)})$ repetitions of these steps simulate the ASM program. Then, according to lemma 2 p. 124, t_Π more steps for the last iteration of P_{step} make F_Π true[18].

Then, until skip $i_{end} \to max_{end}$ is reached, the execution alternates between:

(a) One step to erase a loop command.

(b) Then one step to enter the else part of the conditional

$$\text{if} \neg F_\Pi.$$

(c) Then one step $i_{end} := i_{end} + 3$, then repeat from the beginning.

---

[18]Even if $time(\Pi, X|_{\mathcal{L}(\Pi)}) = 0$, in which case the initial P_{step} executes all these steps.

Because the variable i_{end} is initialized to 0, when $\texttt{skip } i_{end} \rightarrow max_{end}$ is reached at the state X_{final} the value $\overline{i_{end}}^{X_{final}}$ is the number of steps done since F_Π is true. Then $\overline{max_{end}}^{X_{final}} - \overline{i_{end}}^{X_{final}} + 1$ steps are done by $\texttt{skip } i_{end} \rightarrow max_{end}$. Therefore, the ending time is:

$$e = t_\Pi + \overline{i_{end}}^{X_{final}} + \overline{max_{end}}^{X_{final}} - \overline{i_{end}}^{X_{final}} + 1 = t_\Pi + \overline{max_{end}}^{X_{final}} + 1$$

$$\text{So } \boxed{e = t_\Pi + 3 \times (deg(\varphi_\Pi) + 1) + 1}$$

$\dashv$

5   Conclusion and Discussion

We proved p. 125 that ASM fairly simulates LoopC. So, because PLoopC is a sublanguage of LoopC with polynomial time (see p. 109), ASM$_\mathcal{P}$ fairly simulates PLoopC. Reciprocally, we proved p. 115 that PLoopC fairly simulates ASM$_\mathcal{P}$. Therefore, according to the definition p. 100 of the algorithmic equivalence, PLoopC characterizes polynomial time algorithms:

Theorem 19. PLoopC $\simeq$ Algo$_\mathcal{P}$.

This result can be seen as an end of the quest for an *algorithmically complete* language for the set of PTIME algorithms.

Moreover, this language does not require constraints on data structures to be PTIME, unlike PLoop (see [21]).

Nevertheless, PLoopC is not very practicable:

- It is not fully compositional. Indeed, in $P_1; P_2$, we must ensure that the inputs of P_2 are not outputs of P_1.
- And moreover, it is difficult to program in this language because the complexity must be anticipated before writing the program.

In order to obtain the better of both worlds, it would be pleasant to construct an intermediary language, between our PLoopC and Neergaard's PLoop.

We do not need every possible first order structures, only common data structures, which are stronger than Neergaard's translations. So, we are looking for a compromise, by using a restriction on data structures, in order to gain more flexibility from a programmer's perspective.

But, the fact remains that other PTIME languages can be compared to PLoopC for the algorithmic completeness.

References

[1] Andary P., Patrou B. and Valarcher P., A theorem of representation for primitive recursive algorithms, *Fundamenta Informaticae*, XX (2010), pp. 118.

[2] Bellantoni S., and Cook S., A new recursion-theoretic characterisation of the polytime functions, *Computational complexity*, vol. 2 (1992), pp. 97110.

[3] Blass A., Dershowitz N., and Gurevich Y., When are two algorithms the same?, *Bull. Symbolic Logic* Volume 15, Issue 2 (2009), 145-168.

[4] Bonfante G., Some programming languages for LOGSPACE and PTIME, *11th International Conference*, AMAST 2006, Kuressaare, Estonia, July 5-8, 2006.

[5] Borger E., Abstract State Machines: A Unifying View of Models of Computation and of System Design Frameworks, *Annals of Pure and Applied Logic* (2005).

[6] Colson L., About primitive recursive algorithms, *Theoretical Computer Science*, 83 (1991) 5769.

[7] Cori R., Lascar D., and Pelletier D., Mathematical Logic: A Course With Exercises: Part I and II, Paris, *Oxford University Press* (2000, 2001).

[8] Dershowitz N. and Yuri Gurevich, A natural axiomatization of Church's thesis. *Bulletin of symbolic logic*, 2008.

[9] Doyle P., Dexter S., and Gurevich Y., Gurevich abstract state machine and schonhage storage modification machines. *J. Universal Computer Science*, 3(4):279–303, 1997.

[10] Gurevich Y., Sequential abstract state machines capture sequential algorithms. *ACM Transactions on Computational Logic*, 1:77–111, 2000.

[11] Gurevich Y., Evolving Algebras 1993: Lipari Guide. In *Specification and Validation Methods*, pages 9–36. Oxford University Press, 1993.

[12] Gurevich, Yuri, Sequential Abstract State Machines Capture Sequential Algorithms, *ACM Transactions on Computational Logic* (2000).

[13] Grigorieff S and Valarcher P., Evolving Multialgebras unify all usual models for computation in sequential time, *27th International Symposium on Theoretical Aspects of Computer Science*, STACS (2010).

[14] Grigorieff S and Valarcher P., Classes of Algorithms: Formalization and Comparison, *Bulletin of the EATCS* 107 (2012).

[15] Krivine J.-L., A call-by-name lambda-calculus machine, *Higher Order and Symbolic Computation* 20 (2007) 199-207.

[16] Marquer Y., Caractérisation impérative des algorithmes séquentiels en temps quelconque, primitif récursif ou polynomial, *dr-apeiron.net/doku.php/en:research:thesis-defense*

(thesis defended in 2015).

[17] Marquer Y., Algorithmic Completeness of Imperative Programming Languages, *dr-apeiron.net/doku.php/en:research:fi-while* (submitted to Fundamenta Informaticae).

[18] Meyer A. R. and Ritchie D. M., The complexity of loop programs. In *Proc. ACM Nat. Meeting*, 1976.

[19] Moschovakis Y. N., On primitive recursive algorithms and the greatest common divisor function. *Theor. Comput. Sci.*, 301(1-3):1–30, 2003.

[20] Moschovakis Y. N., What is an algorithm? In Springer, editor, *Mathematics unlimited – 2001 and beyond*, pages 919–936. B. Enqquist and W. Schmid, 2001.

[21] Neergaard P. M., Ploop: A Language For Polynomial Time, Manuscript note, *www.cs.brandeis.edu/ cs117a/ploop.pdf*, 2003.

[22] Niggl K.-H., Control structures in programs and computational complexity, *Annals of Pure and Applied Logic*, Volume 133, Issues 13, May 2005, Pages 247–273.

[23] Michel D., and Valarcher P., A total functional programming language that computes APRA, *New Studies in Weak Arithmetic*, Stanford, CSLI Lecture Notes, No. 196, Sep. 2009.

[24] Vinar T., Biedl T., Buss J., Demaine E. D., Demaine M. L., and Hajiaghayi M., Palindrome recognition using a multidimensional tape, *Theoretical Computer Science* 302 (2003).

6

Near-Linearity and Open Induction

Mojtaba Moniri

Abstract:

The union of the set of all spectra $(\lfloor n\alpha \rfloor)_{n \in \mathbb{N}^{\geq 1}}$, for $\alpha \in \mathbb{R}^{\geq 0}$, and that of rational upper spectra minus one $(\lceil n\alpha \rceil - 1)_{n \in \mathbb{N}^{\geq 1}}$, for $\alpha \in \mathbb{Q}^{>0}$, is characterized in $\mathbb{N}^{(\mathbb{N}\setminus\{0\})}$ by additive, equivalently multiplicative near-linearity. There are counterparts for finite initial segments of a sequence to be extendible to some inhomogeneous spectrum $(\lfloor n\alpha + \gamma \rfloor)_{n \in \mathbb{N}^{\geq 1}}$. The equivalences of both pairs of criteria are limit-based in one direction, and rely on induction in the other. We turn to weak arithmetic and compare these properties for functions over models M of Open Induction in the language $\mathcal{L} = \{+, \cdot, <, 0, 1\}$. We establish that homogeneous multiplicative near-linearity continues to characterize the union of the two types of functions in these models, but the multiplicative formulation is now stronger than the additive in both homogeneous and inhomogeneous cases. Next we prove that for any $M \models$ IOpen, the spectrum of $\varphi = \frac{1+\sqrt{5}}{2} \in \mathrm{RC}(M)$ has jumps 2 precisely on the range of the upper spectrum minus one of φ, regardless of its (ir)rationality. We obtain some independence results using rationality of all real algebraic numbers in Shepherdson's model M_0. Here the range of the golden Beatty spectrum is open-definable in $\mathcal{L}_{\lfloor \frac{x}{y} \rfloor}$. The same holds for $\lfloor \frac{a}{b} \lfloor \frac{c}{d} n \rfloor \rfloor$ in any Euclidean division ring. In the standard setting, if $\alpha > 0$, $0 < \beta < 1$ and $\alpha\beta$ is irrational, then $\lfloor \lfloor n\alpha \rfloor \beta \rfloor = \lfloor n\alpha\beta \rfloor$ on a set of positive density. If in addition the standard α and β are algebraic, this also holds for any $\mathbb{Z}$-chain of M_0. Examples of further variants in M and M_0 are brought up.

2010 *Mathematics Subject Classification*: Primary 03F30; Secondary 11U10, 03C62.

Studies in Weak Arithmetics, Volume 3.
Patrick Cégielski, Ali Enayat,
Roman Kossak.
Copyright © 2016, CSLI Publications

Keywords: Beatty Sequence; Shepherdson's Model; Golden Spectrum; Iterated Floor Function.

1 Spectra in the Standard Model of Arithmetic

There has been a good deal of work in number theory in characterizing the class of spectra $(\lfloor r\alpha \rfloor)_{r \in \mathbb{N}^{\geq 1}}$, for $\alpha \in \mathbb{R}^{\geq 0}$, by near-linearity properties. The same is true for the inhomogeneous case $(\lfloor n\alpha + \gamma \rfloor)_{n \in \mathbb{N}^{\geq 1}}$ as well. This section includes several notions of near-linearity and elaborates on the extent to which they characterize various spectra. The paper grew out of an analysis of the implications between some near-linearity properties of spectra but in weak systems of arithmetic, specifically Open Induction, IOpen. In section 2 we will be concerned with provability (or non-provability) in IOpen of certain such standard implications.

1.1 Irrational and Rational-Slope Spectra, and Notions of Near-Linearity

For $\alpha \in \mathbb{R}^{\geq 0}$, the (lower) spectrum of α is the sequence $(\lfloor r\alpha \rfloor)_{r \in \mathbb{N}^{\geq 1}}$. When the slope $\alpha > 0$, we also consider its upper spectrum minus one $(\lceil r\alpha \rceil - 1)_{r \in \mathbb{N}^{\geq 1}}$. If $\alpha \in \mathbb{Q}^{>0}$, these sequences disagree. The sequences in our discussions would have nonnegative integer values, so we exclude the upper spectrum minus one with slope 0, i.e. the constant sequence -1 (although it would satisfy the main properties we will deal with). Beatty sequences are spectra of irrationals $\alpha > 0$.

For $(a_r)_{r \in \mathbb{N}^{\geq 1}}$ in $\mathbb{N}^{(\mathbb{N} \setminus \{0\})}$, let us call the property $(\forall r, s)(a_{r+s} = a_r + a_s \vee a_{r+s} = a_r + a_s + 1)$ additive near-linearity, abbreviated ANL. It was shown in [8] that for a sequence with ANL, every finite initial segment is extendible to various spectra with intervals for slopes indicated. The review of that paper in Mathematical Reviews published by AMS included the misleading sentence *"It is shown that this [ANL] holds if and only if the sequence is the spectrum of a real number"*. Also, in [15] it was mistakenly claimed that the spectra are characterized among sequences $(a_r) \in \mathbb{N}^{(\mathbb{N} \setminus \{0\})}$ via $(\forall r, s)(\frac{a_r}{r} < \frac{a_s+1}{s})$. [But the last page therein missed the inclusiveness of the second inequality in the argument for our next Fact 1, see (*) below.] We call this condition multiplicative near-linearity MNL, a consequence of the conjunction of MNL_{left}: $(\forall s, r)(ra_s \leq a_{rs})$ and $\text{MNL}_{\text{right}}$: $(\forall s, r)(a_{rs} \leq ra_s + r - 1)$. But the upper spectra minus one satisfy ANL and MNL too, see Facts 2 and 4 (the correct formulation was already given in [5]). The rational slope ones constitute all non-spectra whose finite initial segments are extendible to various Beatty sequences (of irrational, hence non-original, slope).

Fact 1. *In* $\mathbb{N}^{(\mathbb{N}\setminus\{0\})}$, *any MNL sequence is either a spectrum or an upper spectrum minus one.*

Proof. For all $n \in \mathbb{N}^{\geq 1}$, $c_n := \max_{r \leq n} \frac{a_r}{r} < d_n := \min_{r \leq n} \frac{a_r+1}{r}$. So $(c_n)_{n \in \mathbb{N}^{\geq 1}}$ and $(d_n)_{n \in \mathbb{N}^{\geq 1}}$ converge (resp. increasingly and decreasingly) to the same limit $\alpha \in \mathbb{R}^{\geq 0}$. Therefore $(\forall r)(\frac{a_r}{r} \leq \alpha \leq \frac{a_r+1}{r})$ (∗, this is where the paragraph above referred to). Hence $(\forall r)(a_r = \lfloor r\alpha \rfloor \vee a_r = r\alpha - 1)$. If $\alpha \notin \mathbb{Q}$, the second disjunct never occurs. For $\alpha \in \mathbb{Q}$ and r a multiple of the denominator of α, it would either never or always hold, that is $(\forall r)(a_r = \lfloor r\alpha \rfloor) \vee (\forall r)(a_r = \lceil r\alpha \rceil - 1)$ (for the latter $\alpha > 0$). The reason is that if $a_r = r\alpha$ but $a_s = s\alpha - 1$, then $\frac{a_r}{r} < \frac{a_s+1}{s}$ would imply $\alpha < \alpha$. ⊣

Fact 2. *For any nonnegative slope, its spectrum and upper spectrum minus one satisfy ANL.*

Proof. For the former, consider whether $\{r\alpha\} + \{s\alpha\} < 1$, or $\{r\alpha\} + \{s\alpha\} \geq 1$. Here $\{\cdot\}$ denotes the fractional part. For the latter, if $\{r\alpha\} + \{s\alpha\} \leq 1$ while $\{r\alpha\}, \{s\alpha\} > 0$, then $\lceil (r+s)\alpha \rceil = \lceil r\alpha \rceil + \lceil s\alpha \rceil - 1$. Otherwise, $\lceil (r+s)\alpha \rceil = \lceil r\alpha \rceil + \lceil s\alpha \rceil$. (The constant -1 is also ANL.) ⊣

Fact 3. *For sequences in $\mathbb{N}$, ANL implies both MNL_{left} and MNL_{right} (hence it implies MNL).*

Proof. Assuming ANL for $(a_r)_{r \in \mathbb{N}^{\geq 1}}$, one shows $(\forall s, r)(ra_s \leq a_{rs} \leq ra_s + r - 1)$ by induction on r. ⊣

We will be considering a weak arithmetic context with limited induction where the corresponding Fact 1 holds, but both components of Fact 3 fail. However, the combination of the not-carrying-over Fact 3 and the other prevailing Fact 2 (induction-free) would still hold via:

Fact 4. *Spectra and upper spectra minus one (slopes in $\mathbb{R}^{\geq 0}$, resp. $\mathbb{R}^{>0}$) satisfy $MNL_{left \& right}$.*

Proof. This argument is also induction-free, just use $\lfloor u \rfloor \leq u < \lfloor u \rfloor + 1$ and $\lceil u \rceil - 1 < u \leq \lceil u \rceil$. ⊣

Corollary 1. (i) In $\mathbb{N}^{(\mathbb{N}\setminus\{0\})} \setminus \{0\}$, (a_r) is a spectrum (resp. upper spectrum minus one) with slope in $\mathbb{Q}^{>0}$ iff the same sequence and, for some $m \in \mathbb{N}^{\geq 1}$, the sequence obtained by replacing all a_{mr} by $a_{mr} - 1$ (resp. $a_{mr} + 1$) are nearly-linear (any of the equivalent ANL or MNL).

(ii) Beatty sequences are those which are nearly-linear unlike any of their such variants.

1.2 Corrigendum to Our "Beatty Sequences and the Arithmetical Hierarchy"

The rational upper spectra minus one were overlooked also in a joint publication [7] by the present author, the second half of page 132 there had some inaccuracies. In particular, $\lim_{k\to\infty}\lfloor\frac{xf(k)}{k}\rfloor$ need not exist, and when it does, pushing the limit inside brackets *could* add one:

Remark 1. One can see that an $f \in \mathbb{N}^{(\mathbb{N}\setminus\{0\})}$ satisfies $(\forall x)(f(x) = \lim_y\lfloor x\frac{f(y)}{y}\rfloor)$ iff it is either the zero function or an upper spectrum minus one of positive slope. The mentioned limit does not exist for the lower spectra of nonzero rational slopes. E.g., with the slope $\frac{1}{2}$, for fixed even x, every other term of $\lfloor x\frac{\lfloor\frac{y}{2}\rfloor}{y}\rfloor$ will be $\frac{x}{2}$, resp. $\frac{x}{2} - 1$ (for large y). Also note that if $f(x) = \lceil x\alpha\rceil - 1$ and $\alpha \in \mathbb{Q}^{>0}$, then for x a multiple of the denominator of α, $\lim_y\lfloor x\frac{f(y)}{y}\rfloor = \alpha x - 1 = \lfloor x\lim_y\frac{f(y)}{y}\rfloor - 1$.

1.3 The Two Formulations of Inhomogeneous Near-Linearity

Recall the homogeneous near-linearity properties already mentioned:

$$(\forall x, y)[f(x + y) - f(x) - f(y) \in \{0, 1\}] \quad \text{(ANL)},$$
$$(\forall x, y)[yf(x) - xf(y) < x] \quad \text{(MNL)}.$$

In this paper we also consider two notions of inhomogeneous (meaning not necessarily homogeneous throughout this paper) near-linearity abbreviated as indicated:

$$(\forall x, y, z)[(f(x + y) - f(x)) - (f(z + y) - f(z)) \in \{-1, 0, 1\}] \quad \text{(IANL)},$$
$$(\forall x, y, u, v)[v(f(x + y) - f(x) - 1) < y(f(u + v) - f(u) + 1)] \quad \text{(IMNL)}.$$

Sequences satisfying additive near-linearity are sometimes called balanced sequences. Multiplicative near-linearity could also be called slope near-linearity. The IMNL property can be remembered as $\frac{\Delta_y f - 1}{\Delta x} < \frac{\Delta' f + 1}{\Delta' x}$. It was proved in [2] that IANL and IMNL are equivalent over $\mathbb{N}$, both characterizing sequences all whose finite initial segments are extendible to an inhomogeneous spectrum $(\lfloor n\alpha + \gamma\rfloor)_{n\in\mathbb{N}^{\geq 1}}$.

Fact 5. *We have the following diagram concerning the near-linearity properties for a function $f : \mathbb{N}^{\geq 1} \to \mathbb{N}$, with the arrow strictly one-way (the abbreviation LSORUSM1 stands for lower spectra or rational upper spectra minus one).*

$$MNL \equiv LSORUSM1 \equiv ANL \equiv MNL_{left \ \& \ right}$$

$$\downarrow$$

$$IMNL \equiv IANL$$

Proof. ANL implies IANL: The values $f(x + y)$ and $f(z + y)$ are expanded either with or without the 1 in the statement of ANL. If both or neither come with 1, then the IANL instance is verified with 0. Otherwise, it is verified with ± 1.

IMNL does not imply ANL (so IMNL does not imply MNL and IANL does not imply ANL either): Consider, e.g., $(n + 2)_{n \in \mathbb{N}^{\geq 1}}$. ⊣

2 Comparing Near-Linearity Notions in IOpen

We will be dealing with the weak arithmetic system of Open Induction, IOpen. Here are some of its basics and Shepherdson's model M_0 for this theory. Recall that the instance of induction corresponding to a formula $\varphi(x, \overline{y})$ with respect to the distinguished free variable x is $I_x(\varphi) : \forall \overline{y}(\varphi(0, \overline{y}) \wedge \forall x(\varphi(x, \overline{y}) \rightarrow \varphi(x + 1, \overline{y})) \rightarrow \forall x \varphi(x, \overline{y}))$. Open Induction domains are those discretely ordered commutative rings with 1 whose nonnegative part satisfy the scheme of induction for open (i.e. quantifier-free) formulas in the language $\{+, \cdot, <, 0, 1\}$. These nonnegative parts are called models of IOpen. A closely related notion is that of an integer part (IP) for an ordered field F, a discrete subring which approximates every element of F within 1 (equivalently within a finite distance). Not all ordered fields have IP's (see [4]), but every real closed field has such (see [13]).

Fact 6. *[Shepherdson's characterization theorem, [18]] Open Induction domains are precisely the integer parts of real closed fields (equivalently, IP's of their real closure).*

This implies every open formula with one (free) variable in a model of IOpen defines a finite union of intervals (which may be unbounded above or reduced to singletons) and so the scheme of Least Number Principle for open formulas would also hold in a model of IOpen. The latter properties crucially depend on the language of arithmetic as above and the proof of the equivalence would not work for extended languages.

Shepherdson also introduced a computable nonstandard model for IOpen, he used non-infinitesimal parts of those Puiseux series over the real algebraic numbers whose constant term is in $\mathbb{N}$, and the dominating term is positive for nonzero Puiseux series. Throughout the paper, let M_0 denote Shepherdson's model of IOpen in the usual language of ordered semi-rings (also M would denote an arbitrary model of IOpen). To be specific:

Fact 7. *[Shepherdson's end-extension theorem applied to the standard model $\mathbb{N}$] Let M_0 be the set of finite sums $\sum_{i=1}^{n} a_i t^{-\frac{n+1-i}{q}} + a_0$, where*

$n \in \mathbb{N}$, $q \in \mathbb{N}^{>0}$, $a_i \in \tilde{\mathbb{Q}}$ *(real algebraic numbers) with* $a_n > 0$ *if* $n > 0$, $a_0 \in \mathbb{Z}$, *and* $a_0 \in \mathbb{N}$ *if* $n = 0$. *Equip* M_0 *with the obvious addition, multiplication, 0, and 1 and the anti-lexicographic order determined by making* t *a positive infinitesimal. Then* $M_0 \models IOpen$.

Puiseux series over any real closed field, and in particular over $\tilde{\mathbb{Q}}$, form a real closed field and the value of the corresponding Shepherdson integer part *function* applied to such a series is obtained by cutting off the positive exponent terms and replacing the constant term by its integer part or the latter minus one if the constant term of the series is an integer and the cut off infinitesimal part is negative. The range of this function is the model M_0.

For a function $f : M^{\geq 1} \to M$, where $M \models$ IOpen, we continue to abbreviate the properties ANL, MNL, MNL_{left}, $\text{MNL}_{\text{right}}$, IANL, and IMNL above in the same way. In section 1 we dealt with them over the standard model $\mathbb{N}$, we now deal with how these are related in the fragment IOpen. In one direction we use the Scott completion of ordered fields, see [16]. A cut C in an ordered field is called regular if it is of zero distance to its complement: $(\forall \epsilon \in F^{>0})(C + \epsilon \not\subset C)$.

Fact 8. *[Scott's theorem] Any ordered field can be densely embedded in another where in the latter all regular cuts have a least upper bound. Furthermore if the original field is dense in its real closure, then its (Scott) completion as above is real closed.*

2.1 IOpen Proves $\text{MNL}_{\text{left \& right}} \to$ ANL, but Not Either Half of the Converse

We will now see that the argument for MNL implying ANL can be generalized to the system IOpen, but both halves of the converse, ANL $\to$ MNL_{left} and ANL $\to$ $\text{MNL}_{\text{right}}$, can be violated for $f : M_0^{\geq 1} \to M_0$.

Proposition 1. Let $M \models IOpen$, F be the fraction field of the ring generated by M, and K the Scott completion of F. Then any MNL function $f : M^{\geq 1} \to M$ is either the spectrum or upper spectrum minus one, for some slope in K. (Of course the two possibilities agree if $\alpha \notin F$.)

Proof. By Fact 6, F is dense in its real closure R. So by Fact 8, K is real closed. Therefore K contains R as a dense subfield. Note also that the integer part M of R would remain an integer part for K. We now argue in the latter field K. Let us use the notation a_r for $f(r)$. The downward closure C of the set of all $\frac{a_r}{r}$'s when $r \in M$ is a cut in K. By MNL, for all $r, s \in M$, $\frac{a_r}{r} < \frac{a_s+1}{s}$. Therefore the cut C is regular since for every $r \in M$, the elements $\frac{a_r}{r}$ and $\frac{a_r+1}{r}$ are respectively in C and $K \setminus C$

and their difference $\frac{1}{r}$ can become as small in K as we like (since M is cofinal in K). So the cut would have a (nonnegative) least upper bound α in K and the rest of the proof is similar to the standard case. Namely, $(\forall r \in M)(\frac{a_r}{r} \leq \alpha \leq \frac{a_r+1}{r})$. Hence $(\forall r \in M)(a_r = \lfloor r\alpha \rfloor \vee a_r = r\alpha - 1)$ and by the strict inequalities $\frac{a_r}{r} < \frac{a_s+1}{s}$, we have $(\forall r \in M)(a_r = \lfloor r\alpha \rfloor) \vee (\forall r \in M)(a_r = \lceil r\alpha \rceil - 1)$. (The argument would go through even when, like in Shepherdson's model M_0, GCD's do not necessarily exist and there is no reduced representation of rationals using relatively prime numerator and denominator: if $\alpha = \frac{p}{q} = \frac{k}{\ell}$ is in the fraction field F of M_0, and we assume $a_q = q\alpha - 1$, but $a_\ell = \ell\alpha$, then $\frac{\ell\alpha}{\ell} < \frac{q\alpha-1+1}{q}$, which is impossible.) ⊣

Now the fact that over M, all lower spectra and all upper spectra minus one of positive slope satisfy MNL$_{\text{left \& right}}$ is seen as in the standard case, see Fact 4. Also, MNL$_{\text{left \& right}}$ trivially implies MNL in this general context too. The spectra and upper spectra minus one in the present context also satisfy ANL, the argument for Fact 2 goes through.

However, the situation is different for the converse of the last statement as we shall now see. Indeed, ANL would not necessarily (i.e. not for all models of IOpen) imply MNL for functions $M^{\geq 1} \to M$ (note that the proof given for Fact 3 would not work in IOpen as it involved an induction not available there). The fact that there is an additive, indeed ring, homomorphism from the ring generated by Shepherdson's model M_0 into $\mathbb{Z}$ plays an important role in the next two proofs.

Proposition 2. For a function $f : M_0^{\geq 1} \to M_0$, the property ANL does not imply MNL$_{\text{left}}$.

Proof. Define $f : M_0^{\geq 1} \to M_0$ by mapping $\sum_{i=1}^{n} a_i t^{-r_i} + a_0$ to the integer part of half of its Archimedean (i.e., constant) term, that is to $\lfloor \frac{a_0}{2} \rfloor$ (the $\frac{1}{2}$ is nothing special here). This function satisfies ANL since the Archimedean term of the sum of two elements in M_0 is the sum of their Archimedean terms and standard spectra satisfy ANL, even if one or both of the constant terms is/are negative. But it fails MNL$_{\text{left}}$, $(\forall s, r)(rf(s) \leq f(rs))$, e.g. when $r = t^{-1}$, and $s = 2$. ⊣

Proposition 3. For a function $f : M_0^{\geq 1} \to M_0$, the property ANL does not imply MNL$_{\text{right}}$.

Proof. Define $f : M_0^{\geq 1} \to M_0$ by $f(a_0) = 0$ and $f(\sum_{i=1}^{n} a_i t^{-r_i} + a_0) = \sum_{i=1}^{n} a_i t^{-r_i} - 1$ if $n > 0$. Note that if at most one of $r, s \in M_0$ is infinite, then $f(r + s) = f(r) + f(s)$, while if they are both infinite, then $f(r+s) = f(r) + f(s) + 1$. Therefore f satisfies ANL. But it fails MNL_{right}, $(\forall s, r)(f(rs) \leq rf(s) + r - 1)$, as, e.g., $r = s = t^{-1} + 1$ shows. ⊣

2.2 IOpen ⊢ (IMNL → IANL), Converse is Independent There

Proposition 4. Let $M \models IOpen$. Then any IMNL function $f : M^{\geq 1} \to M$ satisfies IANL.

Proof. As in the proof in the homogeneous case, we can assume without loss of generality that the real closure K of the fraction field of the ring generated by M is Scott complete. Consider the cut $C \subset K$ defined as the downward closure of the set of all elements of the form $\frac{f(x+y)-f(x)-1}{y}$, for $x, y \in M^{\geq 1}$ (IMNL implies these quotients are bounded above). Note that C is regular since for every $x, y \in M^{\geq 1}$, $\frac{f(x+y)-f(x)+1}{y} \in K \setminus C$ and so there are elements in the complement and the cut itself which are $\frac{2}{y}$ apart and the latter can take as small values in K as we like. So C would have a (nonnegative) least upper bound α in K. For any $x, y \in M^{\geq 1}$, we will have $\frac{f(x+y)-f(x)-1}{y} \leq \alpha \leq \frac{f(x+y)-f(x)+1}{y}$ and so $f(x+y) - f(x) \in [y\alpha - 1, y\alpha + 1]$. Of course with the same α, for any $z, y \in M$, we have $f(z+y) - f(z) \in [y\alpha - 1, y\alpha + 1]$. It is impossible for one of $f(x + y) - f(x)$ and $f(z + y) - f(z)$ to take the maximum possible value and the other one the minimum such. To see that, assume $f(x + y) - f(x)$ takes the highest possible value and $f(z + y) - f(z)$ takes its lowest possible value. We would then have $\frac{f(x+y)-f(x)-1}{y} = \frac{y\alpha+1-1}{y} = \alpha \not< \alpha = \frac{y\alpha-1+1}{y} = \frac{f(z+y)-f(z)+1}{y}$, contradicting IMNL for f. Now since $f(x + y) - f(x), f(z + y) - f(z) \in [y\alpha - 1, y\alpha + 1] \cap M$ cannot be 2 units apart, they are at most 1 unit apart and this shows IANL holds for f. ⊣

Similar to the homogeneous case, the standard proof for the converse involved an induction not available in IOpen. We now show there could be no proof of that converse valid in IOpen. Once again, the Archimedean part homomorphism from the ring generated by M_0 into $\mathbb{Z}$ is going to play a role.

Proposition 5. For a function $f : M_0^{\geq 1} \to M_0$, the property IANL is strictly weaker than IMNL.

Proof. By the last proposition, if IMNL holds for f, then so does IANL. To show the failure of the converse consider, e.g., $f : M_0^{\geq 1} \to M_0$ by $f(\sum_{i=1}^{n} a_i t^{-r_i} + a_0) = \lfloor \frac{a_0}{2} + \frac{2}{3} \rfloor$. Again the Archimedean part of $x + y$

plus that of z equals the Archimedean part of $(x + y) + z$, and the same holds for the Archimedean part of x plus that of $y + z$. So the mentioned inhomogeneous spectrum of the sum of their Archimedean terms satisfy IANL, even if any of the three constant terms are negative. On the other hand f fails IMNL: $(\forall x, y, u, v)[v(f(x + y) - f(x) - 1) < y(f(u + v) - f(u) + 1)]$ e.g. when $x = u = 1$, $y = 4$, and $v = t^{-1}$. ⊣

2.3 A Diamond of the Near-Linearity Properties in IOpen

We have already done most of the work for the following result.

Theorem 1. The following provability diagram holds among the near-linearity properties we considered for a function f over a model of IOpen, with all four arrows strictly one-way and the sides incomparable:

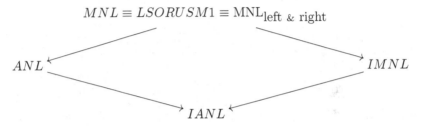

$$MNL \equiv LSOR USM1 \equiv MNL_{\text{left \& right}}$$

$$ANL \qquad\qquad\qquad\qquad IMNL$$

$$IANL$$

Proof. Observe that:

- MNL implies IMNL. This is seen as follows: $\frac{f(x+y)-f(x)-1}{y} \leq \frac{f(y)}{y} < \frac{f(v)+1}{v} \leq \frac{f(u+v)-f(u)+1}{v}$. The middle inequality is directly due to MNL and the other two inequalities due to MNL implying ANL for functions over models of IOpen.

- ANL does not imply IMNL. Earlier we considered $f : M_0^{\geq 1} \to M_0$ defined by $f(\sum_{i=1}^{n} a_i t^{-r_i} + a_0) = \lfloor \frac{a_0}{2} \rfloor$ and said it satisfies ANL but not MNL. Now we note that it even fails IMNL: $(\forall x, y, u, v)[v(f(x + y) - f(x) - 1) < y(f(u + v) - f(u) + 1)]$, since v can be greater than any finite multiple of y. E.g. take $x = u = 1$, $y = 3$, and $v = t^{-1}$.

Now using the results of the preceding subsections, the proof is complete. ⊣

3 The Golden Slope and Models of IOpen

Consider the golden ratio $\varphi = \frac{1+\sqrt{5}}{2}$, and the function $M^{\geq 1} \to M$ defined by $x \mapsto \lfloor x\varphi \rfloor$ (it satisfies MNL). If $M = M_0$, then φ is rational.

3.1 A Look-Back Formula for $\lfloor x\varphi \rfloor$ Insensitive to Rationality/Irrationality

In the standard case over $\mathbb{N}$, several interesting algorithms are known which generate the terms of the golden Beatty sequence $(\lfloor n\varphi \rfloor)_{n \in \mathbb{N}^{\geq 1}}$, e.g. see [20] and [21]. Below we prove a look-back formula for $\lfloor x\varphi \rfloor$ in the general setting of a model M of IOpen (its standard version is apparently not directly mentioned in the references). Note that φ is in the real closure of the fraction field of any M. Let us begin by something general, not just for φ.

Lemma 1. Fix a positive element σ in an ordered field extension of the fraction field of a model M of IOpen (e.g. σ could be in the real closure of the fraction field). Assume $x \in M$ is such that for some $u \in M$, we have $x = \lceil u\sigma \rceil - 1$. Then $\lfloor \frac{x+1}{\sigma} \rfloor > \frac{x}{\sigma}$.

Proof. We will have $\lfloor \frac{x+1}{\sigma} \rfloor = \lfloor \frac{\lceil u\sigma \rceil}{\sigma} \rfloor \geq \lfloor \frac{u\sigma}{\sigma} \rfloor = \frac{u\sigma}{\sigma} > \frac{\lceil u\sigma \rceil - 1}{\sigma} = \frac{x}{\sigma}$. ⊣

Proposition 6. For any $x \in M \models IOpen$, for the jumps of the golden spectrum we have:

$$(\exists u)(x = \lceil u\varphi \rceil - 1) \longleftrightarrow \lfloor (x+1)\varphi \rfloor - \lfloor x\varphi \rfloor = 2.$$

Proof. If $x = \lceil u\varphi \rceil - 1$ for some $u \in M$, then by the lemma above $\frac{x}{\varphi} < \lfloor \frac{x+1}{\varphi} \rfloor \leq \frac{x+1}{\varphi}$. Now using $\frac{1}{\varphi} = \varphi - 1$, we get $x(\varphi - 1) < \lfloor \frac{x+1}{\varphi} \rfloor \leq (x+1)(\varphi - 1)$. Therefore $x\varphi < \lfloor \frac{x+1}{\varphi} \rfloor + x \leq (x+1)\varphi - 1 < (x+1)\varphi$. Notice that $\lfloor \frac{x+1}{\varphi} \rfloor + x \in M$, and furthermore the right-closed interval $((x+1)\varphi - 1, (x+1)\varphi]$ includes an element in M. This shows that $\lfloor (x+1)\varphi \rfloor - \lfloor x\varphi \rfloor > 1$. Now since $\lfloor (x+1)\varphi \rfloor - \lfloor x\varphi \rfloor$ is either 1 or 2, it must be 2.

Conversely, assume x is such that $\lfloor (x+1)\varphi \rfloor - \lfloor x\varphi \rfloor = 2$. Let $m \in M$ be such that $m - 1 = \lfloor x\varphi \rfloor$ and $m + 1 = \lfloor (x+1)\varphi \rfloor$, so $x\varphi < m < m + 1 \leq (x+1)\varphi$. Now using $\varphi = \frac{1}{\varphi} + 1$, we get $x(\frac{1}{\varphi} + 1) < m < m + 1 \leq (x+1)(\frac{1}{\varphi} + 1)$. This shows $\frac{x}{\varphi} < m - x < m + 1 - x \leq \frac{x+1}{\varphi} + 1$. So $\frac{x}{\varphi} < m - x \leq \frac{x+1}{\varphi}$. Therefore $x < (m - x)\varphi \leq x + 1$, and hence $x + 1 = \lceil (m-x)\varphi \rceil$, i.e. $x = \lceil (m-x)\varphi \rceil - 1$. This shows that u can be taken $m - x$. ⊣

3.2 Some Open-Definable Subsets of M_0 in $\mathcal{L}_{\lfloor \frac{x}{y} \rfloor}$

The language extending $\mathcal{L}$ by $\lfloor \frac{x}{y} \rfloor$ (to be total, let it be 0 when $y = 0$) is denoted $\mathcal{L}_{\lfloor \frac{x}{y} \rfloor}$. As Kaye observed in [12], M_0 does not satisfy open induction in this extended language as the standard cut $\mathbb{N}$ is open-definable. E.g., as in his second proof, $y \in \mathbb{N}$ if and only if y and $y + 1$ both divide t^{-1}, meaning $y \lfloor \frac{t^{-1}}{y} \rfloor = t^{-1}$ and similarly for $y + 1$.

A note might be in order as to why that argument coexists with M_0 being a model of IOpen in the original language $\mathcal{L}$. Over any M, the *graph* of the function $\lfloor \frac{x}{y} \rfloor$ is open-definable in $\mathcal{L}$, since $z = \lfloor \frac{x}{y} \rfloor$ is equivalent to $y = z = 0 \vee zy \leq x < (z+1)y$. It is also true that $PA^- + I_z(zy \leq x)$ proves the Euclidean Division Ring axiom EDR. Namely, for all x, y, with $y \neq 0$, there exists (a unique) z such that $zy \leq x < (z+1)y$ and so the function $\lfloor \frac{x}{y} \rfloor$ (with value 0 when $y = 0$) is provably total in IOpen in $\mathcal{L}$. Therefore for any x and nonzero y in any model M of IOpen, $\lfloor \frac{x}{y} \rfloor$ exists in M, in particular for M_0. (Using all of IOpen, we can also infer this from Shepherdson criterion since $\frac{x}{y}$ for nonzero y is in the fraction field and so in the real closure of the fraction field of the model.) We cannot refer to it in a quantifier-free way but we can in a bounded-existential way: $u = \lfloor \frac{x}{y} \rfloor \equiv (\exists z \leq x)[(y = z = 0 \vee zy \leq x < (z+1)y) \wedge z = u]$. If we were working in a class Σ_n of formulas, then as usual we could refer to the complexity of a function by the complexity of its graph. But that is not the case for more restricted classes of formulas like open formulas.

Over two decades after Kaye's remarks and his interesting results, apparently almost nothing more is known about IOpen($\lfloor \frac{x}{y} \rfloor$), as pointed out in [11]. That perhaps includes whether M_0 can be extended to a model of IOpen($\lfloor \frac{x}{y} \rfloor$).

Not to meet any of that challenge, here are a few small observations. The formula for the golden spectrum implies other consequences in this regard (not in the context of IOpen($\lfloor \frac{x}{y} \rfloor$), rather M_0 in the signature $\mathcal{L}_{\lfloor \frac{x}{y} \rfloor}$ and still with open induction in the original language $\mathcal{L}$).

Lemma 2. For $u \in M_0$, the constant term of u is nonzero if and only if $\lfloor (\lfloor u\varphi \rfloor + 1)\varphi \rfloor - \lfloor \lfloor u\varphi \rfloor \varphi \rfloor = 2$ (as in the standard case).

Proof. Only if: Here $u\varphi \notin M_0$, so $\lceil u\varphi \rceil - 1 = \lfloor u\varphi \rfloor$ and Proposition 6 tells us $\lfloor (\lfloor u\varphi \rfloor + 1)\varphi \rfloor - \lfloor \lfloor u\varphi \rfloor \varphi \rfloor = 2$. *If:* We show the contrapositive.

If the constant term of $u \in M_0$ is 0, then $u\varphi, u\varphi^2 \in M_0$ equal their integer parts and so $\lfloor (\lfloor u\varphi \rfloor + 1)\varphi \rfloor - \lfloor \lfloor u\varphi \rfloor \varphi \rfloor = 1$. $\dashv$

Corollary 2. In M_0, the range of the golden Beatty spectrum is open-definable in $\mathcal{L}_{\lfloor \frac{x}{y} \rfloor}$.

Proof. An element $z \in M_0$ is in the mentioned range if and only if it satisfies $\lfloor \frac{z}{\varphi} \rfloor \varphi = z \vee (\lfloor \frac{z}{\varphi} \rfloor \varphi \neq z \wedge \lfloor (z+1)\varphi \rfloor - \lfloor z\varphi \rfloor = 2)$. Note that φ would actually be viewed as $\frac{\varphi t^{-1}}{t^{-1}}$. $\dashv$

Corollary 3. There is an $\mathcal{L}_{\lfloor \frac{x}{y} \rfloor}$-open definable subset of M_0 consisting of exactly one element from each $\mathbb{Z}$-chain.

Proof. The $\mathcal{L}_{\lfloor \frac{x}{y} \rfloor}$-formula $\lfloor (\lfloor u\varphi \rfloor + 1)\varphi \rfloor - \lfloor \lfloor u\varphi \rfloor \varphi \rfloor = 1$ defines the set of elements in M_0 whose constant term is 0. [This happens exactly when $(\exists y \leq u)(u^2 = y^2 + uy)$, a bounded existential definition in $\mathcal{L}$.] One could take the conjunction with $u \neq 0$, and then the standard cut would not be represented as in the statement. ⊣

3.3 Further Independence Results Implied by Rationality of φ in M_0

In the next result we observe that replacing the ceiling minus one by the integer part as in the standard case when φ is irrational, neither direction in our look-back formula is saved.

Proposition 7. In M_0, none of the properties $(\exists u)(x = \lfloor u\varphi \rfloor)$ or $\lfloor (x+1)\varphi \rfloor - \lfloor x\varphi \rfloor = 2$ in x implies the other.

Proof. By Proposition 6, the second property is equivalent to $(\exists u)(x = \lceil u\varphi \rceil - 1)$ and this is not implied by, and does not imply, the first property: Take $v \in M_0$ with constant term 0. Then $v\varphi \in M_0$ also. Now $v\varphi$ is of the form $\lfloor u\varphi \rfloor$ (with $u = v$), but not $\lceil u\varphi \rceil - 1$ for any u. On the other hand, $v\varphi - 1$ is of the form $\lceil u\varphi \rceil - 1$ (with $u = v$), but not $\lfloor u\varphi \rfloor$ for any u. ⊣

It is easy to see that the famous Beatty theorem (stating that the Beatty sequences of two irrationals whose reciprocals add up to 1 make a partition of $\mathbb{N}$) fails for rationals and that failure is also the case for M_0. Here we see that a couple of standard consequences also fail, and in contrast to the way we saved a fact above for the golden ratio, this time the failure is persistent even with ceiling minus one instead of the integer part.

Proposition 8. Neither of the standard facts $(\forall x, y)(\lfloor x\varphi \rfloor \neq y + \lfloor y\varphi \rfloor)$ or $(\forall y)(\exists x)(y = \lfloor x\varphi \rfloor \vee y = x + \lfloor x\varphi \rfloor)$ hold in M_0.

Proof. We know these indeed hold in the standard model $\mathbb{N}$ since φ is irrational and $\frac{1}{\varphi} + \frac{1}{\varphi+1} = 1$.

Let us observe that the two ranges intersect. E.g., t^{-1} is divisible by both φ and $\varphi + 1$. So $\lfloor \frac{t^{-1}}{\varphi}\varphi \rfloor = \lfloor \frac{t^{-1}}{\varphi+1}(\varphi + 1) \rfloor = t^{-1}$.

Next, let $x = \varphi t^{-1} - 1$. Clearly $x \in M_0$ and $x \neq \lfloor y\varphi \rfloor$, for any $y \in M_0$, since $\lfloor (t^{-1} - 1)\varphi \rfloor = \varphi t^{-1} - 2$. Let us show that it does not belong to the spectrum of $\varphi + 1$ in M_0 either. Assume by way of contradiction that $\varphi t^{-1} - 1 \leq u(\varphi + 1) < \varphi t^{-1}$. Dividing by $\varphi^2 = \varphi + 1$, we would have $\frac{t^{-1}}{\varphi} - \frac{1}{\varphi^2} \leq u < \frac{t^{-1}}{\varphi}$. But this is impossible since $\frac{t^{-1}}{\varphi}, u \in M$ and $\frac{1}{\varphi^2} < 1$. ⊣

Proposition 9. Replacing the integer part by the ceiling minus one, neither are saved: the ranges of the functions $\lceil x\varphi \rceil - 1$ and $\lceil x(1+\varphi) \rceil - 1$ intersect and their union does not cover M_0.

Proof. For the intersection being nonempty, same example as above works: $\lceil \frac{t^{-1}}{\varphi}\varphi \rceil - 1 = \lceil \frac{t^{-1}}{\varphi+1}(\varphi + 1) \rceil - 1 = t^{-1} - 1$. For the union being proper in M_0, we may consider the example φt^{-1}. First, we cannot have $\lceil x\varphi \rceil - 1 = \varphi t^{-1}$, since otherwise $\varphi t^{-1} < \varphi x \leq \varphi t^{-1} + 1$ and then $t^{-1} < x \leq t^{-1} + \frac{1}{\varphi}$. But that is a contradiction since $x, t^{-1} \in M_0$ and $\frac{1}{\varphi} < 1$. To show that it is not in the range of $\lceil x(\varphi + 1) \rceil - 1$ either, assume by way of contradiction that $\varphi t^{-1} < x(\varphi+1) \leq \varphi t^{-1} + 1$. Then $\frac{1}{\varphi}t^{-1} < x \leq \frac{1}{\varphi}t^{-1} + \frac{1}{\varphi^2}$. But $x, \frac{1}{\varphi}t^{-1} \in M_0$ and $\frac{1}{\varphi^2} < 1$, so we cannot have $x \in M_0$. $\dashv$

4 Iterated Rational Spectra

Studies on iterated spectra $(\lfloor \lfloor n\alpha \rfloor \beta \rfloor)_{n \in \mathbb{N}^{\geq 1}}$ have included issues on "hearing the shape." E.g., [9] dealt with recovering the parameters uniquely. A theorem on recovering the real parameters, in the irrational case, for $\lfloor \lfloor n\alpha \rfloor \beta \rfloor$ was proved and an example was given, in the rational parameters case, where that uniqueness of parameters $\frac{a}{b}$ and $\frac{c}{d}$ in $\lfloor \frac{a}{b} \lfloor \frac{c}{d} n \rfloor \rfloor$ fail, namely $\lfloor \frac{2}{9} \lfloor \frac{3}{7} n \rfloor \rfloor = \lfloor \frac{2}{7} \lfloor \frac{1}{3} n \rfloor \rfloor$ for all $n \in \mathbb{N}^{\geq 1}$ (we verify this below). Although if a quadruple (a, b, c, d) fails this type of identity, it may still fail to be recoverable from the range of an iterated rational spectrum, nevertheless it motivates the considerations in this section which includes some related issues as well.

4.1 Iterated Rational Spectra and EDR

In this subsection we present some facts on iterated rational spectra which hold in the induction free arithmetic PA^- (i.e., axioms for nonnegative parts of discretely ordered commutative rings with 1), augmented by the Euclidean division axiom. Let EDR denote this extension of PA^-. [As mentioned in subsection 3.2, over PA^- the Euclidean division axiom is a consequence of a single instance of induction on an open formula $I_z(zy \leq x)$, so IOpen proves EDR.] Let E denote an EDR.

Lemma 3. For positive (here and in the sequel) a, b, c, d, n in E, we have:

(i) $\lfloor \frac{a}{b} \lfloor \frac{c}{d} n \rfloor \rfloor = ac\lfloor \frac{n}{bd} \rfloor + \lfloor \frac{a}{b} \lfloor \frac{c}{d} \text{Rem}(n, bd) \rfloor \rfloor$.

(ii) $\lfloor \frac{a}{b} \lfloor \frac{c}{d} \text{Rem}(n, bd) \rfloor \rfloor < ac$.

Proof. Let $n = kbd + r$, with $0 \leq r < bd$, and $cr = qd + s$, with $0 \leq s < d$.

(i) We have $k = \lfloor \frac{n}{bd} \rfloor$ and $q = \lfloor \frac{cr}{d} \rfloor$. Note that $\lfloor \frac{c}{d} n \rfloor = \lfloor kbc + q + \frac{s}{d} \rfloor = kbc + q$. Therefore $\lfloor \frac{a}{b} \lfloor \frac{c}{d} n \rfloor \rfloor = kac + \lfloor \frac{aq}{b} \rfloor = \lfloor \frac{n}{bd} \rfloor ac + \lfloor \frac{a}{b} \lfloor \frac{c}{d} r \rfloor \rfloor = \lfloor \frac{n}{bd} \rfloor ac + \lfloor \frac{a}{b} \lfloor \frac{c}{d} \mathrm{Rem}(n, bd) \rfloor \rfloor$.

(ii) We have $0 \leq \lfloor \frac{cr}{d} \rfloor < bc$, and so $0 \leq \frac{a}{b} \lfloor \frac{cr}{d} \rfloor < ac$. Therefore $0 \leq \lfloor \frac{a}{b} \lfloor \frac{cr}{d} \rfloor \rfloor < ac$. $\dashv$

So the quotients of $\lfloor \frac{a}{b} \lfloor \frac{c}{d} n \rfloor \rfloor$ and $\lfloor \frac{a}{d} \lfloor \frac{c}{b} n \rfloor \rfloor$ when divided by ac are the same, they both equal the quotient of n divided by bd. The remainder of $\lfloor \frac{a}{b} \lfloor \frac{c}{d} n \rfloor \rfloor$ when divided by ac is the same as the similar expression when n is replaced by its remainder by bd. Of course a similar statement also holds for $\lfloor \frac{a}{d} \lfloor \frac{c}{b} n \rfloor \rfloor$. When the said remainder is 0, say $n = kbd$, then obviously $\lfloor \frac{a}{b} \lfloor \frac{c}{d} n \rfloor \rfloor = \lfloor \frac{a}{d} \lfloor \frac{c}{b} n \rfloor \rfloor = ack$.

Corollary 4. If $\lfloor \frac{a}{b} \lfloor \frac{c}{d} n \rfloor \rfloor = \lfloor \frac{a}{d} \lfloor \frac{c}{b} n \rfloor \rfloor$ holds for $1 \leq n \leq bd - 1$ in E, then it holds for all $n \in E$.

We continue to be in the context of EDR. Part of the standard version of the next lemma is a consequence of the right-continuity of the integer part function on $\mathbb{R}$. Recall that the generalized inverse of a non-decreasing function f from $\mathbb{N} \to \mathbb{N}$ or $\mathbb{R} \to \mathbb{R}$ is the function g defined as the minimum, respectively infimum, of the set of values $\{z | f(z) \geq m\}$. Then $f(n) = m$ iff $g(m) \leq n < g(m + 1)$. For $g(m)$ to be equal to n, this latter element would need to be the least such that $f(n) = m$. These special instances of the least number principle over E that we consider in the lemma are indeed satisfied in E. We obtain a formula for the generalized inverse of $\lfloor \frac{a}{b} \lfloor \frac{c}{d} n \rfloor \rfloor$ on E.

Lemma 4. With all parameters in E, the least $n \in E$ at which $\lfloor \frac{a}{b} \lfloor \frac{c}{d} n \rfloor \rfloor$ reaches values not less than m is $\lceil \frac{d}{c} \lceil \frac{mb}{a} \rceil \rceil$.

Proof. We will argue in the fraction field of E with the obvious order extended from E, and the floor and ceiling functions from such fractions to E. We also use the equivalence of $\lfloor u \rfloor < v$ with $u < \lceil v \rceil$, and $v \leq \lfloor u \rfloor$ with $\lceil v \rceil \leq u$. The inequality $m \leq \lfloor \frac{a}{b} \lfloor \frac{c}{d} n \rfloor \rfloor$ is successively equivalent to $m \leq \frac{a}{b} \lfloor \frac{c}{d} n \rfloor$, $\frac{mb}{a} \leq \lfloor \frac{c}{d} n \rfloor$, $\lceil \frac{mb}{a} \rceil \leq \frac{c}{d} n$, $\frac{d}{c} \lceil \frac{mb}{a} \rceil \leq n$, and to $\lceil \frac{d}{c} \lceil \frac{mb}{a} \rceil \rceil \leq n$. So the least n in the proposition is $\lceil \frac{d}{c} \lceil \frac{mb}{a} \rceil \rceil$. (Note that if $\lceil \frac{d}{c} \lceil \frac{mb}{a} \rceil \rceil = \lceil \frac{d}{c} \lceil \frac{(m+1)b}{a} \rceil \rceil$, then m is not in the range of $\lfloor \frac{a}{b} \lfloor \frac{c}{d} n \rfloor \rfloor$.) $\dashv$

Corollary 5. Suppose $ac < bd$ in E. Then the identity $\lfloor \frac{a}{b} \lfloor \frac{c}{d} n \rfloor \rfloor \equiv \lfloor \frac{a}{d} \lfloor \frac{c}{b} n \rfloor \rfloor$ holds if and only if for all $m \in E$ with $1 \leq m \leq ac - 1$, we have $\lceil \frac{d}{c} \lceil \frac{mb}{a} \rceil \rceil = \lceil \frac{b}{c} \lceil \frac{md}{a} \rceil \rceil$.

Corollary 6. The range of $\lfloor \frac{a}{b} \lfloor \frac{c}{d} n \rfloor \rfloor$, where $n \in E$ (and so are the parameters), is quantifier-free $\mathcal{L}_{\lfloor \frac{x}{y} \rfloor}$-definable and bounded-existential $\mathcal{L}$-definable.

Proof. The mentioned range equals $\{m \mid \lceil \frac{d}{c} \lceil \frac{mb}{a} \rceil \rceil < \lceil \frac{d}{c} \lceil \frac{(m+1)b}{a} \rceil \rceil\}$, or $\{m \mid \lfloor \frac{a}{b} \lfloor \frac{c}{d} \lceil \frac{d}{c} \lceil \frac{mb}{a} \rceil \rceil \rfloor \rfloor = m\}$. Note that any property $P(\cdots, \lceil \frac{u}{v} \rceil, \cdots)$ in terms of expressions like $\lceil \frac{u}{v} \rceil$ (so four, respectively two here) can equivalently be replaced by $(v \lfloor \frac{u}{v} \rfloor = u \wedge P(\cdots, \lfloor \frac{u}{v} \rfloor, \cdots)) \vee (v \lfloor \frac{u}{v} \rfloor < u \wedge P(\cdots, \lfloor \frac{u}{v} \rfloor + 1, \cdots))$. So we get a disjunction of 16, respectively 4, disjuncts of subformulas as mentioned. The disjunction will be equivalent to the original formula, still quantifier-free, and now in the language $\mathcal{L}_{\lfloor \frac{x}{y} \rfloor}$. The assertion in $\mathcal{L}$ is obtained as in Subsection 3.2. ⊣

Corollary 7. The predicate $(\mu n)[\lfloor \frac{a}{b} \lfloor \frac{c}{d} n \rfloor \rfloor \geq m]$ of variables a, b, c, d, and m ranging in E is likewise quantifier-free $\mathcal{L}_{\lfloor \frac{x}{y} \rfloor}$-definable and bounded-existential $\mathcal{L}$-definable.

4.2 Some Positive and Negative Examples of a Certain "Not Hearing"

When the first parameter above $a = 1$, we always have an identity:

Example 1. We have $\lfloor \frac{1}{b} \lfloor \frac{c}{d} n \rfloor \rfloor = \lfloor \frac{1}{d} \lfloor \frac{c}{b} n \rfloor \rfloor$, for any $b, c, d, n \geq 1$ (it holds in any EDR).

To verify this, note that $\lceil \frac{d}{c} \lceil bm \rceil \rceil = \lceil \frac{b}{c} \lceil dm \rceil \rceil$ for any m (in particular for $1 \leq m \leq c - 1$).

Example 2. We have $\lfloor \frac{2}{9} \lfloor \frac{3}{7} n \rfloor \rfloor = \lfloor \frac{2}{7} \lfloor \frac{1}{3} n \rfloor \rfloor$ $(= \lfloor \frac{2}{7} \lfloor \frac{3}{9} n \rfloor \rfloor)$ for all n (this also holds in EDR).

This is the example that Graham-O'Bryant gave as we mentioned at the beginning of section 4. We can see it holds by observing that $\lceil \frac{7}{3} \lceil \frac{9m}{2} \rceil \rceil = \lceil \frac{9}{3} \lceil \frac{7m}{2} \rceil \rceil$, which is obvious for even m and immediate for odd m.

Now we mention some negative examples.

Example 3. $\lfloor \frac{10}{11} \lfloor \frac{7}{15} n \rfloor \rfloor \neq \lfloor \frac{2}{3} \lfloor \frac{7}{11} n \rfloor \rfloor$, for $n \in \mathbb{N}$, e.g. at $n = 4$.

Example 4. A negative example could be obtained if we reduce one of the inner fractions in the Graham-O'Bryant example by $\frac{3}{3}$, namely we have $\lfloor \frac{2}{7} \lfloor \frac{1}{3} n \rfloor \rfloor \neq \lfloor \frac{2}{3} \lfloor \frac{1}{7} n \rfloor \rfloor$, e.g. at $n = 12$.

Observe that $\lceil \frac{3}{1} \lceil \frac{7m}{2} \rceil \rceil \neq \lceil \frac{7}{1} \lceil \frac{3m}{2} \rceil \rceil$, indeed at $m = 1$, the values are 12 and 14. The latter tells us the values are different at $n = 12, 13$. Direct inspection shows that at all numbers different than 12 and 13 mod 21, the equation holds.

Example 5. Here are some further examples of the sensitivity of the issue with respect to the presentation of a rational number, compare with the above.

(i) $\lfloor \frac{2}{8} \lfloor \frac{3}{5} n \rfloor \rfloor \neq \lfloor \frac{2}{5} \lfloor \frac{3}{8} n \rfloor \rfloor$, e.g. at $n = 7$.

(ii) $\lfloor \frac{2}{9} \lfloor \frac{6}{14} n \rfloor \rfloor \neq \lfloor \frac{2}{14} \lfloor \frac{6}{9} n \rfloor \rfloor$, e.g. at $n = 11$.

4.3 Iterated Rational Spectra and M_0

In Shepherdson's model M_0, the GCD property fails: the greatest common divisors need not exist (e.g. it does not exist for $\sqrt{2}t^{-1}$ and t^{-1}), and so elements in the fraction field do not necessarily have a reduced form (e.g. $\frac{\varphi t^{-1}}{t^{-1}} = \frac{\varphi t^{-\frac{1}{2}}}{t^{-\frac{1}{2}}} = \cdots$). So the above Remark makes more sense here when we consider iterated rational spectra over M_0.

The next statement belongs to this paper and this section since in the special case when the reals considered are algebraic, they would be rational in M_0. In this subsection we will be writing the inner integer part before the other factor.

Lemma 5. (i) Let $\alpha, \beta \in \mathbb{R}^{>0}$, with $\beta < 1$. Suppose $n \in \mathbb{N}$ is such that $\{n\alpha\beta\} \geq \beta$. Then $\lfloor \lfloor n\alpha \rfloor \beta \rfloor = \lfloor n\alpha\beta \rfloor$.

(ii) Assume furthermore that $\alpha\beta$ is irrational. Then the density of the set of n's where $\lfloor \lfloor n\alpha \rfloor \beta \rfloor = \lfloor n\alpha\beta \rfloor$ is at least $1 - \beta$ (so density is positive).

Proof. (i) Suppose not. Then, together with the always true $\lfloor \lfloor n\alpha \rfloor \beta \rfloor \leq \lfloor n\alpha\beta \rfloor$, we would have $\lfloor n\alpha \rfloor \beta < \lfloor n\alpha\beta \rfloor \leq n\alpha\beta$. Hence

$$\{n\alpha\beta\} < \{n\alpha\}\beta < \beta,$$

a contradiction.

(ii) The sufficient condition above is $\{n\alpha\beta\} \geq \beta$, i.e. $\{n\alpha\beta\} \in [\beta, 1)$. Now use the uniform distribution of the fractional parts of multiples of irrationals, see, e.g., [14]. ⊣

Example 6. Suppose α, β are algebraic reals with $\alpha > 0$, $0 < \beta < 1$, and $\alpha\beta$ is irrational. Then the statement in part (ii) above when considered for $n \in M_0$ comes down to the constant term of n since the non-Archimedean part times any algebraic would be a Shepherdson integer. So the conclusion would hold in any $\mathbb{Z}$-chain of M_0 as well (when the density would be relative to the $\mathbb{Z}$-chain).

The rationality in M_0 of all real algebraic numbers produces independence statements related to variants of iterated spectra, here is an example.

Example 7. For $n \in \mathbb{N}^{\geq 1}$, we have $\lfloor \lfloor n\sqrt{2} \rfloor \sqrt{2} + n\sqrt{2} \rfloor \equiv 2n + \lfloor n\sqrt{2} \rfloor - 1$.

To see this, let $m = \lfloor n\sqrt{2} \rfloor$, and $f = \{n\sqrt{2}\}$. From $f \times (\sqrt{2} - 1) < 1$, we get $m\sqrt{2} + f\sqrt{2} + m - 1 < m\sqrt{2} + m + f$, that is $2n + \lfloor n\sqrt{2} \rfloor - 1 < \lfloor n\sqrt{2} \rfloor \sqrt{2} + n\sqrt{2}$. So far this was valid in both $\mathbb{N}$ and M_0. Next, in $\mathbb{N}$ but not in M_0, for all n we also have $f > 0$ and so $f < f\sqrt{2}$. Therefore $m\sqrt{2} + m + f < m\sqrt{2} + f\sqrt{2} + m$, i.e. $\lfloor n\sqrt{2} \rfloor \sqrt{2} + n\sqrt{2} < 2n + \lfloor n\sqrt{2} \rfloor$.

The latter fails in M_0, since n can be taken so that $n\sqrt{2} \in M_0$, e.g., $n = \frac{\sqrt{2}}{2}t^{-1}$. However, we have the following.

Example 8. In any model of IOpen, for any element n we have

$$\lfloor \lfloor n\sqrt{2} \rfloor \sqrt{2} + n\sqrt{2} \rfloor = 2n + \lfloor n\sqrt{2} \rfloor + D,$$

where $D = -1$ if $n\sqrt{2}$ does not belong to the model and $D = 0$ if it does. In Shepherdson's model the conditions are respectively equivalent to $a \neq 0$ and $a = 0$, where a is the constant term of the element in M_0.

5 Some Related Comments and Questions

The above sections motivate the following areas of pursuit:

5.1 Characterization of Infinite IMNL Sequences in $\mathbb{N}$?

To possibly get more than what the proof of IOpen $\vdash$ (IMNL $\to$ IANL) showed, it is worth trying to use a stronger notion of completeness for ordered fields, namely symmetric completeness as in [17]. Picking up from $|f(x+y) - f(x) - y\alpha| \leq 1$, relabel to get $f(x) - f(y) - (x-y)\alpha \leq 1$ and therefore $f(x) - x\alpha \leq f(y) - y\alpha + 1$. This shows that any pair of intervals of the form $[f(x) - x\alpha, f(x) - x\alpha + 1]$ intersect. We can assume K is already symmetrically complete (this means that any decreasing sequence, of any ordinal length, of closed bounded intervals in K has a nonempty intersection), otherwise we would embed it in such a real closed field. We leave it to the reader to see if the setting would then be enough to assure the existence of a β in the intersection of all such intervals. If so, then for every $x \in M^{\geq 1}$, we would have $f(x) \leq x\alpha + \beta \leq f(x) + 1$ (*) and so $(\forall x \in M^{\geq 1})(f(x) = \lfloor x\alpha + \beta \rfloor \vee f(x) = x\alpha + \beta - 1)$. Here one could not use the uniformity that was established at the similar step in the homogeneous case (in the next paragraph we see why), rather could copy (*) for a few other elements and get $f(x+y) + f(z) \leq (x+y+z)\alpha + 2\beta \leq f(x+y) + f(z) + 2$ and the same for $f(x) + f(y+z)$. Similar to what we did earlier in the proof of Proposition 2.4, it would then be impossible that $f(x+y) + f(z) = (x+y+z)\alpha + 2\beta$ and $f(x) + f(y+z) = (x+y+z)\alpha + 2\beta - 2$, and IANL for f follows.

In the argument in the paragraph above, for suitable α and potential β we reached $(\forall x \in M^{\geq 1})(f(x) = \lfloor x\alpha + \beta \rfloor \vee f(x) = x\alpha + \beta - 1)$. We indicated not to always expect the comprehensiveness, like in the homogeneous case, that f would have to be either a spectrum (inhomogeneous $\lambda n.\lfloor \alpha n + \beta \rfloor$ here), or inhomogeneous upper spectrum minus one ($\lambda n.\lceil \alpha n + \beta \rceil - 1$). In fact, such an f could experience the minus 1 on some but not all occasions when $\alpha n + \beta$ belongs to M. Indeed, a sequence $f : \mathbb{N}^{\geq 1} \to \mathbb{N}$ can satisfy IMNL and hence have a slope α and inhomogeneity β for the whole infinite sequence f as mentioned above (and the algorithm in [3] for finite subsequences will succeed in finding non-empty domains for possible α and β), yet for some $n, m \in \mathbb{N}$, with $n\alpha + \beta, m\alpha + \beta - 1 \in \mathbb{N}$, we have $f(n) = n\alpha + \beta$ but $f(m) = m\alpha + \beta - 1$. Such an α would have to be rational (since with irrational slope, the inside of the brackets would be an integer at most once). In particular, consider the sequence $r_n = n + 2$. If for $n = 1$, we reduce the value by 1, the resulting sequence would still satisfy IMNL. This can be checked via the standard-equivalent IANL. This raises the following.

Question 1. What is the inhomogeneous counterpart of the standard LSORUSM1?

It would also be nice to formulate inhomogeneous versions of properties MNL_{left} and $\text{MNL}_{\text{right}}$.

5.2 Near-Linearity Notions and Reverse Mathematics

The following observations are based on communications with Denis Hirschfeldt [10]. We leave the details and any extensions to the reader.

The proof of Fact 1 was based on "the intersection of a nested sequence of closed bounded intervals where the diameters tend to 0 is nonempty." This is provable in RCA_0, see [19], therefore Fact 1 is also provable in RCA_0. The proof of Fact 3 was just an application of $I\Delta_0^0$, and hence provable in RCA_0 (or even RCA_0^*). Hence "ANL, MNL, and LSORUSM1 are equivalent" is provable in RCA_0.

For the inhomogeneous case, assuming IMNL, existence of α would be similar, and our first treatment did not rely on any β. Alternatively, "the intersection of an infinite sequence of closed bounded intervals any two of which intersect is nonempty" was used in showing the existence of β such that $f(x) \leq x\alpha + \beta \leq f(x) + 1$. The quoted fact can be seen to be provable in RCA_0. Hence "IANL and IMNL are equivalent" is provable in RCA_0.

One can see that "every finite sequence of nonnegative integers satisfying ANL is extendible to an infinite Beatty sequence", and also the inhomogeneous counterpart "any finite sequence of nonnegative integers satisfying IANL is extendible to some infinite inhomogeneous Beatty sequence" are also provable in RCA_0 (or even RCA_0^*).

5.3 Hearing the Shape of (ir)rational Beatty in M_0?

To show some independence results in this paper, we used M_0. Here we formulate something whose status in M_0 is left undetermined. We dealt with the spectrum of φ in M and in M_0, where this slope was rational in the latter. Consider $\varphi\sqrt{1-t}$ in the real closure of fraction field of M_0 (one can see that it is not in the fraction field itself). This element is infinitely close and to the left of φ. It is easy to find $x \in M_0$ such that $\lfloor x\varphi\sqrt{1-t}\rfloor \neq \lfloor x\varphi\rfloor$. E.g., we could take $x = \lfloor \frac{1}{\varphi(1-\sqrt{1-t})}\rfloor + 1$. Along the results and questions in [1], it would be nice to know things like:

Question 2. How are $\{ \lfloor x\varphi\rfloor \mid x \in M_0 \}$ and $\{ \lfloor x\varphi\sqrt{1-t}\rfloor \mid x \in M_0 \}$ compared?

Also of interest would be whether in M_0, the latter set is dense mod 1.

5.4 Both φ and $\sqrt{2}$ Irrational, or Rational with Same Possible Denominators?

What we saw in sections 3 and 4 imply that both $\mathbb{N}$ and M_0 model

$$(\forall x)(\lfloor(\lfloor x\varphi\rfloor + 1)\varphi\rfloor - \lfloor\lfloor x\varphi\rfloor\varphi\rfloor + \lfloor\lfloor x\sqrt{2}\rfloor\sqrt{2} + x\sqrt{2}\rfloor - 2x - \lfloor x\sqrt{2}\rfloor = 1)$$

(for $x \in \mathbb{N}$ and most elements $x \in M_0$, the expression on the left of the equality simplifies to $2 + (-1)$, while for the exceptional case when the constant term of $x \in M_0$ is 0, it equals $1 + 0$).

Question 3. IOpen $\vdash^?$ $(\forall x)(\lfloor(\lfloor x\varphi\rfloor + 1)\varphi\rfloor - \lfloor\lfloor x\varphi\rfloor\varphi\rfloor + \lfloor\lfloor x\sqrt{2}\rfloor\sqrt{2} + x\sqrt{2}\rfloor - 2x - \lfloor x\sqrt{2}\rfloor = 1)$?

Also, recalling in M_0 all real algebraic numbers are rational, is there a dichotomy among models $M \models$ IOpen of being either normal (integrally closed in its fraction field like in the standard case $\mathbb{N}$, see [6]), or this extremely non-normal like M_0? (I.e., couldn't there be a model M where some but not all irrational real algebraic numbers remain irrational in M?)

Acknowledgement. The author thanks the anonymous referee for very helpful comments, the editors for their handling of the volume, and the JAF34/MAMLS organizers for travel support. Feedback received from Mehdi Ghasemi, Denis Hirschfeldt, and Jeffrey Lagarias improved the paper.

References

[1] S.M. Ayat, The Skolem-Bang Theorems in Ordered Fields with an IP, *J. Algebra* 319 (2008), 4869-4890.

[2] M. Boshernitzan and A.S. Fraenkel, Nonhomogeneous Spectra of Numbers, *Discrete Math.* 34 (1981), 325-327.

[3] M. Boshernitzan and A.S. Fraenkel, A Linear Algorithm for Nonhomogeneous Spectra of Numbers, *J. Algorithms* 5 (1984), 187-198.

[4] S. Boughattas, Résultats Optimaux sur l'existence d'une Partie Entière dans les Corps Ordonnés, *J. Symbolic Logic* 58 (1993), 326-333.

[5] J.L. Davison, Some Remarks on Beatty-type Sequences, *Congr. Numer.* 30 (1981), 257-263.

[6] L. van den Dries, Some Model Theory and Number Theory for Models of Weak Systems of Arithmetic, in: L. Pacholski, J. Wierzejewski, and A.J. Wilkie (Eds.), *Model Theory of Algebra and Arithmetic*, Lect. Notes Math., 834, pp. 346-362, Springer, Berlin, 1980.

[7] M. Ghasemi and M. Moniri, Beatty Sequences and the Arithmetical Hierarchy, in: A. Enayat, I. Kalantari, and M. Moniri (Eds.), *Logic in Tehran*, Lect. Notes Log. 26, pp. 126-133, Assoc. Symbol. Logic, La Jolla, CA (and A.K. Peters, Ltd., Wellesley, MA), 2006.

[8] R.L. Graham, S. Lin, and C.-S. Lin, Spectra of Numbers, *Math. Mag.*, 51 (1978) 174-176.

[9] R. Graham and K. O'Bryant, Can You Hear the Shape of a Beatty Sequence?, in: *Additive Number Theory*, Festschrift in Honor of the Sixtieth Birthday of Melvyn B. Nathanson, D. Chudnovsky and G. Chudnovsky (Eds.), pp. 39-52, Springer, New York, 2010.

[10] D. Hirschfeldt, Personal Communication.

[11] E. Jeřábek and L.A. Kołodziejczyk, Real Closures of Models of Weak Arithmetic, *Arch. Math. Logic* 52 (2013), 143-157.

[12] R. Kaye, Open Induction, Tennenbaum Phenomena, and Complexity Theory, in: P. Clote and J. Krajíček (Eds.), *Arithmetic, Proof Theory, and Computational Complexity*, Oxford Logic Guides 23, pp. 222-237, Oxford Univ. Press, New York, 1993.

[13] M. H. Mourgues and J. P. Ressayre, Every Real Closed Field Has an Integer Part, *J. Symbolic Logic* 58 (1993), 641-647.

[14] I. Niven, *Diophantine Approximations*, Reprint of the 1963 Original, Dover Publications, Inc., Mineola, NY, 2008.

[15] M.A. Nyblom, On the Spectrum of Real Numbers Revisited, *Fibonacci Quart.*, 43 (2005) 299-301.

[16] D. Scott, On Completing Ordered Fields, in: W.A.J. Luxemburg (Ed.), *Applications of Model Theory to Algebra, Analysis, and Probability*, pp. 274-278, Holt, Rinehart and Winston, New York, 1969.

[17] S. Shelah, Quite Complete Real Closed Fields, *Israel J. Math.*, 142 (2004), 261-272.

[18] J.C. Shepherdson, A Non-standard Model for a Free Variable Fragment of Number Theory, *Bull. Acad. Polon. Sci. Sér. Sci. Math. Astronom. Phys.*, 12 (1964) 79-86.

[19] S.G. Simpson, *Subsystems of Second Order Arithmetic*, Springer-Verlag, Berlin, 1999.

[20] Th. Skolem, On Certain Distributions of Integers in Pairs with Given Differences, *Math. Scand.* 5 (1957), 57-68.

[21] K.B. Stolarsky, Beatty Sequences, Continued Fractions, and Certain Shift Operators, *Canad. Math. Bull.*, 19 (1976) 473-482.

7

Submodel Lattices of Existentially Closed Models of Arithmetic

James H. Schmerl

Abstract: Only models $\mathcal{N}$ of TA_2 are considered here. If $\mathcal{N}$ is such a model, then its submodel lattice $\mathcal{L}(\mathcal{N})$ consists of those of its submodels that are also models of TA_2. The main result of this paper is that if L is a finite distributive lattice, then there is an e.c. $\mathcal{N}$ such that $\mathcal{L}(\mathcal{N}) \cong \mathbf{2} \oplus L$.

True Arithmetic (TA) is the theory of the standard model $\mathbb{N} = (\omega, +, \times, 0, 1, \leq)$ of arithmetic. For $n < \omega$, let $\mathsf{TA}_n = \mathsf{TA} \cap \Pi_n$. The focus of this paper will be on TA_2, which, by the MRDP Theorem, is logically equivalent to the set of $\forall\exists$ sentences in TA. The study of models of TA_2 was initiated by Hirschfeld [2],[3],[4].

Suppose that $\mathcal{N} = (N, +, \times, 0, 1, \leq)$ is a model of TA_2. A submodel $\mathcal{M} = (M, +, \times, 0, 1, \leq) = \mathcal{N}|M$ is a model of TA_2 iff M is closed under all functions $f : N \longrightarrow N$ that are parameter-free Δ_1-definable in $\mathcal{N}$. Consequently, those submodels of $\mathcal{N}$ that are models of TA_2 form a lattice, which we denote by $\mathcal{L}(\mathcal{N})$ and refer to as the **submodel lattice** of $\mathcal{N}$. This lattice is algebraic so, in particular, it is complete.

From now on (unless indicated otherwise) $\mathcal{M}, \mathcal{N}, \ldots$ will denote models of TA_2 having universes $M, N, \ldots$.

A model $\mathcal{M}$ is **existentially closed** (or, briefly, **e.c.**) if whenever

I wish to thank Volodya Shavrukov on two counts. First, the results herein came about following some stimulating email correspondence with him. Second, his comments on some earlier versions of this manuscript contributed beneficially to the present one.

Studies in Weak Arithmetics, Volume 3.
Patrick Cégielski, Ali Enayat,
Roman Kossak.

$\mathcal{M} \in \mathcal{L}(\mathcal{N})$, then $\mathcal{M} \prec_1 \mathcal{N}$. (Just to be sure, $\mathcal{M} \prec_1 \mathcal{N}$ iff whenever $\varnothing \neq X \subseteq N$ and X is defined in $\mathcal{N}$ by an $\exists_1$-formula with only parameters from M allowed, then $X \cap M \neq \varnothing$.) The standard model is e.c. It was proved in [p.151, 4] that there is a single $\exists\forall$ formula that defines the set of standard elements in every e.c. model.

The question that is investigated here is:

Which finite lattices can appear as $\mathcal{L}(\mathcal{N})$, where $\mathcal{N}$ is an e.c. model?

This question arose out of some work of Thomas McLaughlin. Notice that we can refer to a subset $X \subseteq N$ as a *set of generators* of $\mathcal{N}$ if $\mathcal{N} = \bigcap \{\mathcal{M} \in \mathcal{L}(\mathcal{N}) : X \subseteq M\}$. A model $\mathcal{N}$ is a *Nerode semiring* if it is nonstandard and finitely generated. First of all, McLaughlin showed [Theorem 2.6, 6] that not every e.c. Nerode semiring is minimal or, in other words, there is an e.c. Nerode semiring $\mathcal{N}$ such that $\mathcal{L}(\mathcal{N})$ is not a 2-element lattice. Much later, McLaughlin [7] initiated the study of possible $\mathcal{L}(\mathcal{N})$ for $\mathcal{N}$ a not necessarily e.c. model, the most significant result for us being that all finite chains can appear as some $\mathcal{L}(\mathcal{N})$. To be more specific, for $1 \leq n < \omega$, let **n** be a chain of length n. Then McLaughlin proved [Theorem 3.1, 7] proved that if $1 \leq n < \omega$, then there is a model $\mathcal{N}$ such that $\mathcal{L}(\mathcal{N}) \cong \mathbf{n}$. These results lead naturally to the question posed above.

If K, L are finite lattices (so that both 1_K and 0_L exist), then their **linear sum** $K \oplus L$ is the lattice that is the disjoint union of K and L (except that 1_K and 0_L are identified) such that $r \leq s$ whenever $r \in K$ and $s \in L$. For example, $\mathbf{2} \oplus \mathbf{3} \cong \mathbf{4}$. It is proved in [8] that for any (not necessarily e.c.) nonstandard model $\mathcal{N}$, if $\mathcal{L}(\mathcal{N})$ is finite, then there is a lattice L such that $\mathcal{L}(\mathcal{N}) \cong \mathbf{2} \oplus L$.

The following theorem, referred to hereinafter as *the Theorem*, answers the question for distributive lattices.

Theorem. *Let D be a finite distributive lattice. There is an e.c. model $\mathcal{N}$ such that $\mathcal{L}(\mathcal{N}) \cong \mathbf{2} \oplus D$.*

The rest of this paper comprises three sections devoted to proving the Theorem. We begin in §1 with a discussion of the method to be used for proving theorems of this type. In §2 we prove a weakening of the Theorem with D being restricted to Boolean lattices. The proof of the Theorem is completed in §3.

A comment about notation: Both ω and $\mathbb{N}$ will be used for the set of all natural numbers. But $\mathbb{N}$ will also be used for the standard model.

1 The Method

How does one go about proving that for a given finite lattice L there is an e.c. model $\mathcal{N}$ such that $\mathcal{L}(\mathcal{N}) \cong \mathbf{2} \oplus L$? In this section, we take a look at one approach, which is the one we will use to prove the Theorem.

We begin with a review of some definitions used in studying the lattices of elementary substructures of models of Peano Arithmetic as presented, for example, in [Chapter 4, 5]. Not all of this will be used here, but it should help with motivating this section.

For any set A, we let $\mathrm{Eq}(A)$ be the set of equivalence relations on A, considered as a lattice. The two extreme equivalence relations are $\mathbf{0}_A = \bigwedge \mathrm{Eq}(A)$, which is the equality relation, and $\mathbf{1}_A = \bigvee \mathrm{Eq}(A) = A^2$, which is the trivial equivalence relation having exactly one equivalence class. If L is a finite lattice, then a function $\alpha : L \longrightarrow \mathrm{Eq}(A)$ is a **representation** of L provided that α is one-to-one and

$$\alpha(0_L) = \mathbf{1}_A, \ \alpha(1_L) = \mathbf{0}_A, \ \alpha(r \vee s) = \alpha(r) \wedge \alpha(s)$$

for each $r, s \in L$. (We do not require that $\alpha(r \wedge s) = \alpha(r) \vee \alpha(s)$.) The representation α is **finite** if A is finite. If $\alpha : L \longrightarrow \mathrm{Eq}(A)$ is a representation and $B \subseteq A$, then $\alpha|B$ is the function $\beta : L \longrightarrow \mathrm{Eq}(B)$ such that if $r \in L$, then $\beta(r) = \alpha(r) \cap B^2$. It may happen that $\alpha|B$ is not a representation. We define when a representation $\alpha : L \longrightarrow \mathrm{Eq}(A)$ is n-CPP by recursion on $n < \omega$:

- α is 0-CPP if for no $r \in L$ does $\alpha(r)$ have exactly 2 equivalence classes;
- α is $(n + 1)$-CPP if whenever $\Theta \in \mathrm{Eq}(A)$, then there are $B \subseteq A$ and $r \in L$ such that $\alpha|B$ is an n-CPP representation and $\Theta \cap B^2 = \alpha(r) \cap B^2$.

Observe that if α is $(n + 1)$-CPP, then it is n-CPP. The definition of 0-CPP guarantees that whenever $\alpha : L \longrightarrow \mathrm{Eq}(A)$ is $(n + 1)$-CPP and $A = B \cup C$, then $\alpha|B$ or $\alpha|C$ is n-CPP.

These notions are useful when studying the lattice $\mathrm{Lt}(\mathcal{N})$ of elementary substructures of a model $\mathcal{N}$ of PA. They are also useful when studying the possible $\mathcal{L}(\mathcal{N})$, where $\mathcal{N} \models \mathsf{TA}_2$, as presented in [8]. For example, the following are equivalent for any finite lattice L:

- For each $n < \omega$, L has an n-CPP representation.
- For each $n < \omega$, L has a finite n-CPP representation.
- There is $\mathcal{N} \models \mathsf{TA}$ such that $\mathrm{Lt}(\mathcal{N}) \cong \mathbf{2} \oplus L$.
- For every completion T of PA, if $T \neq \mathsf{TA}$, then there is $\mathcal{N} \models T$ such that $\mathrm{Lt}(\mathcal{N}) \cong L$.
- There is $\mathcal{N} \models \mathsf{TA}_2$ such that $\mathcal{L}(\mathcal{N}) \cong \mathbf{2} \oplus L$.

Suppose that $\alpha : L \longrightarrow \mathrm{Eq}(A)$ and $\beta : L \longrightarrow \mathrm{Eq}(B)$ are representations of the finite lattice L. Then α and β are **isomorphic** (in symbols: $\alpha \cong \beta$) if there is a bijection $f : A \longrightarrow B$ such that whenever $r \in L$

and $x, y \in A$, then $\langle x, y \rangle \in \alpha(r)$ iff $\langle f(x), f(y) \rangle \in \beta(r)$. We say that α **arrows** β (in symbols: $\alpha \longrightarrow \beta$) if whenever $\Theta \in \mathrm{Eq}(A)$, then there are $C \subseteq A$ and $r \in L$ such that $\alpha | C \cong \beta$ and $\Theta \cap C^2 = (\alpha | C)(r)$. Notice that if, for each $n < \omega$, there are representations $\alpha_0, \alpha_1, \ldots, \alpha_n$ of L such that α_0 is 0-CPP and $\alpha_n \longrightarrow \alpha_{n-1} \longrightarrow \cdots \longrightarrow \alpha_0$, then L has n-CPP representations for every $n < \omega$. A seemingly stronger situation would be that there are representations $\alpha_0, \alpha_1, \alpha_2, \ldots$ such that α_0 is 0-CPP and $\alpha_{n+1} \longrightarrow \alpha_n$ for all $n < \omega$.

Returning to our current topic, we make some new definitions.

Definition 1.1. (*Part 1*) Fix a finite lattice L, and let $\langle \alpha_i : i < \omega \rangle$ be a sequence of representations of L, where $\alpha_i : L \longrightarrow \mathrm{Eq}(A_i)$ for each $i < \omega$ and $A = \bigcup_{i < \omega} A_i$. We will say that this sequence is **precanonical** if the following hold:

(C1) Each α_i is a finite 0-CPP representation of L.

(C2) Whenever $i < j < \omega$, then $A_i \cap A_j = \varnothing$.

(C3) The sequence $\langle \alpha_i : i < \omega \rangle$ is computable.

(C4) For each $i < \omega$ there is $j < \omega$ such that $\alpha_j \longrightarrow \alpha_i$.

By (C3), we mean: the sequence $\langle c_i : i < \omega \rangle$ is computable, where c_i is the canonical code of α_i, and also that A is computable.

If $\langle \alpha_i : i < \omega \rangle$ is a precanonical sequence of representations of L, then for each $n < \omega$, there is i_n such that α_{i_n} is an n-CPP representation of L. In fact, there is a subsequence $\langle \alpha_{i_n} : n < \omega \rangle$ such that $\alpha_{i_{n+1}} \longrightarrow \alpha_{i_n}$ for all $n < \omega$. Moreover, there is such a subsequence which itself is precanonical.

If $\langle \alpha_i : i < \omega \rangle$ is a precanonical sequence of representations of L (as in Definition 1.1 (Part 1)), then the set B is **large** if $B \subseteq A$, B and there is an increasing sequence $i_0 < i_1 < i_2 < \cdots < \omega$ such that $\langle \alpha_{i_j} | (B \cap A_{i_j}) : j < \omega \rangle$ is a precanonical sequence of representations of L. Usually, the large sets that we consider will be computable.

We note the following easy proposition that helps put into perspective the condition (C5) in the upcoming Part 2 of Definition 1.1.

Proposition 1.2. *Suppose that $\langle \alpha_i : i < \omega \rangle$ is a precanonical sequence of representations of the finite lattice L. Whenever X is computable, then there is a large computable B such that either $B \subseteq X$ or $B \cap X = \varnothing$.* $\dashv$

Definition 1.1. (*Part 2*) We say that the sequence $\langle \alpha_i : i < \omega \rangle$ is **canonical** if it is precanonical (as in Definition 1.1(Part 1)) and the following holds:

(C5) Whenever $X \subseteq A$ is computably enumerable (or, briefly, c.e.), then there is a large computable $B \subseteq A$ such that either $B \subseteq X$ or $B \cap X = \varnothing$.

Theorem 1.3. *If L is a finite lattice and there is a canonical sequence of representations of L, then there is an e.c. model $\mathcal{N}$ such that $\mathcal{L}(\mathcal{N}) \cong \mathbf{2} \oplus L$.*

Before proving this theorem, we review the method for constructing finitely generated models as presented in [4] and [7]. We will call an ultrafilter $\mathcal{F}$ of the Boolean algebra of all computable subsets of $\mathbb{N}$ a Δ_1-ultrafilter. Now let $\mathcal{F}$ be a nonprincipal Δ_1-ultrafilter. Let $\mathcal{U} \supseteq \mathcal{F}$ be an ultrafilter (of $\mathcal{P}(\mathbb{N})$). Let $\mathcal{M}$ be the ultrapower of the standard model relative to $\mathcal{U}$. Of course, $\mathcal{M}$ is a model of TA, so it is a model of TA_2. Consider, the element $[id]_{\mathcal{U}} \in M$ that is the equivalence class to which the identity function belongs. Then the submodel $\mathcal{N}$ of $\mathcal{M}$ that $[id]_{\mathcal{U}}$ generates is a Nerode semiring. The model $\mathcal{N}$ depends on $\mathcal{F}$, but it is independent of which ultrafilter $\mathcal{U}$ was chosen. In fact, we can avoid the ultrafilter $\mathcal{U}$ and simply let $\mathcal{N}$ consist all $[f]_{\mathcal{F}}$, where f is a computable function from $\mathbb{N}$ to $\mathbb{N}$. Here, $[f]_{\mathcal{F}}$ is the set of all computable g that agree with f modulo $\mathcal{F}$. We will write that $\mathcal{N} = \mathbb{N}/\mathcal{F}$.

This type of construction, a Δ_1-ultrapower, yields all Nerode semirings. Let $\mathcal{N}$ be a Nerode semiring and let $a \in N$ be a generator. Let $\mathcal{F}$ consist of those $X \subseteq \mathbb{N}$ such that $a \in X^{\mathcal{N}}$, where $X^{\mathcal{N}}$ is the unique subset of $\mathbb{N}$ that is defined by any Σ_1 (or Π_1) formula that defines X in the standard model. Then $\mathcal{F}$ is a Δ_1-ultrafilter, and $\mathcal{N} \cong \mathbb{N}/\mathcal{F}$.

Lemma 1.4. *If L is a finite lattice and there is a precanonical sequence of representations of L, then there is $\mathcal{N}$ such that $\mathcal{L}(\mathcal{N}) \cong \mathbf{2} \oplus L$.*

Proof. Let L be a finite lattice and let $\langle \alpha_i : i < \omega \rangle$ be a precanonical sequence of representations of L. Suppose that $\alpha_i : L \longrightarrow \mathrm{Eq}(A_i)$ and $A = \bigcup_{i < \omega} A_i \subseteq \mathbb{N}$.

Let $\Theta_0, \Theta_1, \Theta_2, \ldots$ be an enumeration of all computable equivalence relations on $\mathbb{N}$. By recursion, we will obtain a decreasing sequence $X_0 \supseteq X_1 \supseteq X_2 \supseteq \cdots$ of large computable subsets of $\mathbb{N}$.

We start by letting $X_0 = A$.

Proceeding inductively, suppose that we have X_n. Consider the equivalence relation Θ_n. We can effectively get three sequences: an increasing sequence $j_0 < j_1 < j_2 < \cdots$ of elements of ω, a sequence $B_0, B_1, B_2, \ldots$ of subsets of $\mathbb{N}$ and a sequence $r_0, r_1, r_2, \ldots$ of elements of L such that for each $i < \omega$, $B_i \subseteq X_n \cap A_{j_i}$, $\alpha_{j_i}|B_i \cong \alpha_i$ and $\alpha_{j_i}(r_i) \cap B_i^2 = \Theta_n \cap B_i^2$. Since L is finite, we get $r \in L$ and an infinite $I \subseteq \omega$ such that $r_i = r$ for every $i \in I$. Now let $X_{n+1} = \bigcup_{i \in I} B_i$.

We claim that the sets $X_0, X_1, X_2, \ldots$ generate a Δ_1-ultrafilter $\mathcal{F}$.

To see this, consider, a computable $Y \subseteq \mathbb{N}$ and let $n < \omega$ be such that Θ_n has no more than 2 equivalence classes and one of them is Y. Then, either $X_{n+1} \subseteq Y$ or $X_{n+1} \cap Y = \varnothing$.

Let $\mathcal{N} = \mathbb{N}/\mathcal{F}$.

We next show that $\mathcal{L}(\mathcal{N}) \cong \mathbf{2} \oplus L$. For each $r \in L$, let $g_r : \mathbb{N} \longrightarrow \mathbb{N}$ be a computable function such that whenever $i < \omega$ and $x \in A_i$, then $g_r(x)$ is the least $y \in A_i$ such that $\langle x, y \rangle \in \alpha_i(r)$. For $r \in \mathbf{2} \oplus L$, let $\mathcal{N}_r$ be the submodel of $\mathcal{N}$ generated by $[g_r]_{\mathcal{F}}$ if $r \in L$ and $\mathcal{N}_r$ is the standard submodel of $\mathcal{N}$ if $r = 0_{\mathbf{2} \oplus L}$. We claim that the function $r \mapsto \mathcal{N}_r$ is an isomorphism from L onto $\mathcal{L}(\mathcal{N})$. The proof of this is fairly straightforward, generally following the proof of [Theorem 4.5.8, 5]. We leave the details for the reader to work out. $\quad\dashv$

To get an e.c. Nerode semiring, it is both necessary and sufficient to use a Σ_1-ultrafilter[1], where a Δ_1-ultrafilter is a Σ_1-ultrafilter if for every c.e. $X \subseteq \mathbb{N}$ there is $Y \in \mathcal{F}$ such that either $Y \subseteq X$ or $Y \cap X = \varnothing$. To complete the proof of Theorem 1.3, we show how to modify the proof of Lemma 1.4. We now are assuming that $\langle \alpha_i : i < \omega \rangle$ is a canonical sequence of representations of L. We interleave into the construction of the sequence $X_0 \supseteq X_1 \supseteq X_2 \supseteq \cdots$ some additional steps.

Let $Z_0, Z_1, Z_2, \ldots$ be an enumeration of all the c.e. subsets of $\mathbb{N}$. Suppose that we have X_n. Obtain X_{n+1} as in the proof of Lemma 1.4, but now let's call that set Y_n. By (C5), we now let $X_{n+1} \subseteq Y_n$ be large and computable such that either $X_{n+1} \subseteq Z_n$ or $X_{n+1} \cap Z_n = \varnothing$. With this new construction, we get a Σ_1-ultrafilter $\mathcal{F}$, thereby proving Theorem 1.3. $\quad\dashv$

The lattice $\mathbf{1}$ has, up to isomorphism, exactly one representation. It is easy to see that there is a canonical sequence of representations of $\mathbf{1}$, so Theorem 1.3 has the following immediate corollary, which was proved in [7].

Corollary 1.5. *There is an e.c. model $\mathcal{N}$ such that $\mathcal{L}(\mathcal{N}) \cong \mathbf{2}$.*

2 Boolean Lattices

Fix $n < \omega$, and let $\mathbf{B}_n$ be a Boolean lattice with exactly n atoms. To be definitive, let $\mathbf{B}_n = \mathcal{P}(n) = \mathcal{P}(\{0, 1, \ldots, n-1\})$, where $0_{\mathbf{B}_n} = \varnothing$, $1_{\mathbf{B}_n} = n$ and $\wedge$ and $\vee$ are $\cap$ and $\cup$, respectively. A finite lattice is a Boolean lattice iff it is isomorphic to some $\mathbf{B}_n$. Every Boolean lattice is distributive; moreover, a finite lattice is distributive iff it is a sublattice of a Boolean lattice. In this section, we prove the following weakened

---

[1]Warning: This terminology is used here just for convenience, and is not at all standard, conflicting with both [4] and [7].

version of the Theorem.

Theorem 2.1. *For each $n < \omega$, there is an e.c. model $\mathcal{N}$ such that $\mathcal{L}(\mathcal{N}) \cong \mathbf{2} \oplus \mathbf{B}_n$.*

Because of Theorem 1.3, it suffices to prove the following more combinatorial lemma.

Lemma 2.2. *For each $n < \omega$, there is a canonical sequence of representations of $\mathbf{B}_n$.*

As was commented on at the end of §1, the case when $n = 0$ is immediate since $\mathbf{B}_0 \cong \mathbf{1}$. Thus, for the remainder of this section, we assume that $1 \le n < \omega$.

Before starting the proof, let's settle on some more terminology and notation. As already mentioned, both ω and $\mathbb{N}$ will be used for the set of all natural numbers. When referring to the first infinite ordinal, we always use ω. If $n < \omega$, then ω^n is an ordinal whereas $\mathbb{N}^n$ is a set of n-tuples. If $x \in \mathbb{N}^n$, then $x = \langle x_0, x_1, \dots, x_{n-1} \rangle$. If $A \subseteq \mathbb{N}^m$ and $B \subseteq \mathbb{N}^n$, then $A \times B \subseteq \mathbb{N}^{m+n}$. If $a \le b < \omega$, then $[a, b) = \{m \in \mathbb{N} : a \le m < b\}$.

Let $\{I_0, I_1, I_2, \dots\}$ be a computable partition of $\mathbb{N}$ such that $|I_i| = i+3$ for $i < \omega$. We could, for example, let $I_i = [i(i+5)/2, (i+1)(i+6)/2)$ for $i < \omega$. We define $\beta_{n,i} : \mathbf{B}_n \longrightarrow \mathrm{Eq}(I_i^n)$ so that if $r \in \mathbf{B}_n$ and $x, y \in I_i^n$, then

$$\langle x, y \rangle \in \beta_{n,i}(r) \iff r \subseteq \{j < n : x_j = y_j\}.$$

Let $A_n = \bigcup_{i < \omega} I_i^n$.

We will show that $\langle \beta_{n,i} : i < \omega \rangle$ is a canonical sequence of representations.

It is easy to see that each $\beta_{n,i}$ is a finite representation of $\mathbf{B}_n$. Although it's not relevant here, notice that

$$\beta_{n,i}(r \wedge s) = \beta_{n,i}(r) \vee \beta_{n,i}(s)$$

for every $r, s \in \mathbf{B}_n$. Each $\beta_{n,i}$ is 0-CPP since $|I_i| \ge 3$. Thus, (C1) of Definition 1.1 is satisfied. Clearly, $\langle \beta_{n,i} : i < \omega \rangle$ is computable so (C3) is satisfied. (C2) is obvious. Observe that (C4) is essentially just the finite version of the Canonical Ramsey Theorem of Erdős & Rado [1]. Thus, $\langle \beta_{n,i} : i < \omega \rangle$ is precanonical. To complete the proof that $\langle \beta_{n,i} : i < \omega \rangle$ is a canonical, we need to verify (C5) of Definition 1.1.

Some more definitions will be useful. Suppose that $1 \le n < \omega$ and that $A, B \subseteq \mathbb{N}^n$. A function $f : A \longrightarrow B$ is an **embedding** of A into B if whenever $x, y \in A$ and $i < n$, then $x_i = y_i$ iff $f(x)_i = f(y)_i$. If there is an embedding $f : A \longrightarrow B$, then A is **embeddable** into B. We define an n-**pattern** to be a finite subset of $\mathbb{N}^n$. We let $\mathcal{P}_n$ be the set of n-patterns. Clearly, a subset $B \subseteq A_n$ is large if every n-pattern is embeddable into B. We extend the notion of a large set to all subsets of $\mathbb{N}^n$: a subset $B \subseteq \mathbb{N}^n$ is **large** every n-pattern is embeddable into B.

It is an immediate consequence of the next lemma that $\langle \beta_{n,i} : i < \omega \rangle$ satisfies (C5). This lemma, which is of independent interest, is the principal new fact of this section.

Lemma 2.3. *Suppose that $X \subseteq A \subseteq \mathbb{N}^n$ are such that X is c.e. and A is large and computable. Then there is a large computable $B \subseteq A$ such that either $B \subseteq X$ or $B \cap X = \varnothing$.*

The remainder of this section is devoted to proving Lemma 2.3.

Suppose that $1 \leq n < \omega$. We let $\pi : \mathbb{N}^n \longrightarrow \mathbb{N}^{n-1}$ be the projection such that if $x \in \mathbb{N}^n$, then $\pi(x) = y \in \mathbb{N}^{n-1}$, where $y_i = x_i$ for every $i < n - 1$. If $A, B \subseteq \mathbb{N}^n$ and $f : A \longrightarrow B$ is a bijective embedding, then f is an **isomorphism** from A to B and we say that A is **isomorphic** to B and write $A \cong B$. If $m < \omega$, then an n-pattern P is an m^n-**cube** if there are subsets $J_0, J_1, \ldots, J_{n-1} \subseteq \mathbb{N}$ such that $|J_0| = |J_1| = \cdots = |J_{n-1}| = m$ and $P = J_0 \times J_1 \times \cdots \times J_{n-1}$. If P, Q are n-patterns and P is an m^n-cube, then Q is an m^n-cube iff $P \cong Q$. Obviously, A_n is a large subset of $\mathbb{N}^n$. Observe that if $A \subseteq \mathbb{N}^n$ is large and c.e., then there is a computable $B \subseteq A$ and a computable isomorphism $f : A_n \longrightarrow B$.

We mention three simple facts about embeddings. The first is that the composition of two embeddings is an embedding. The second is that if $P, Q \in \mathcal{P}_n$, then $f : P \longrightarrow Q$ is an embedding iff there are injections $f_0, f_1, \ldots, f_{n-1} : \mathbb{N} \longrightarrow \mathbb{N}$ such that whenever $x = \langle x_0, x_1, \ldots, x_{n-1} \rangle \in P$, then $f(x) = \langle f_0(x_0), f_1(x_1), \ldots, f_{n-1}(x_{n-1}) \rangle$. The last is that if $P, Q \in \mathcal{P}_n$, Q is an m^n-cube and $f, g : P \longrightarrow Q$ are embeddings, then there is an automorphism h of Q (that is, $h : Q \longrightarrow Q$ is an isomorphism) such that $f = hg$.

The following definition introduces a key concept used in the proof of Lemma 2.3.

Definition 2.4. An n-pattern P is **prevalent** if whenever $X \subseteq A_n$, X is c.e., and there is no large computable $B \subseteq A_n \backslash X$, then there is $m < \omega$ such that P is embeddable into $X \cap I_m^n$.

It is immediate from the definition that every singleton pattern is prevalent. Also, if $P, Q \in \mathcal{P}_n$ are such that P is prevalent and Q is embeddable into P, then Q is prevalent. The next proposition shows that prevalent n-patterns have a property that is stronger than what is given in the definition.

Proposition 2.5. *Suppose that $P \in \mathcal{P}_n$ is prevalent, $X \subseteq A_n$ is c.e., and there is no large computable $B \subseteq A_n \backslash X$. Then, for every infinite c.e. $I \subseteq \omega$, there is $m \in I$ such that P is embeddable into $X \cap I_m^n$.*

Proof. Let $I \subseteq \omega$ be infinite and c.e. Then there are a large computable $A \subseteq A_n$, a computable isomorphism $g : A_n \longrightarrow A$ and a computable injection $h : \omega \longrightarrow I$ such that $g(x) \in I_{h(k)}^n$ whenever $k < \omega$ and $x \in I_k^n$. Clearly, there is no large computable subset of $A \backslash X$, so there is no large computable subset of $A_n \backslash g^{-1}(X)$. Since $g^{-1}(X)$ is c.e., there are $k < \omega$ and an embedding f of P into $g^{-1}(X) \cap I_k^n$. Then, gf embeds P into $X \cap I_{h(k)}^n$. ⊣

The significance of the notion of prevalence is that if every n-pattern is prevalent, then the Theorem is true when restricted to that n. This is made precise in Lemma 2.6.

Lemma 2.6. *Suppose that $1 \leq n < \omega$. If every n-pattern is prevalent, then whenever $X \subseteq A \subseteq \mathbb{N}^n$ are such that X is c.e. and A is large and computable, then there is a large computable $B \subseteq A$ such that either $B \subseteq X$ or $B \cap X = \varnothing$.*

Proof. Let $X \subseteq A \subseteq \mathbb{N}^n$ be such that X is c.e. and A is large and computable. Without loss of generality, let $A = A_n$. Suppose that there is no large computable $B \subseteq A_n \backslash X$. We will show that there is a large computable $B \subseteq X$. Since X is c.e., it suffices to show that X is large. Thus, it suffices to show that for every n-pattern P there is $m < \omega$ such that P is embeddable into $X \cap I_m^n$. Indeed, this is so since P is prevalent. ⊣

The next definition introduces another key concept used in the proof.

Definition 2.7. The relation $\mathsf{rank}(P) \leq \alpha$, where $P \in \mathcal{P}_n$ and α is an ordinal, is defined by ordinal-recursion as follows:

- if $P = \varnothing$, then $\mathsf{rank}(P) \leq \alpha$ for all α;
- if $|P| = 1$, then $\mathsf{rank}(P) \leq \alpha$ iff $\alpha \geq 1$;
- if $|P| \geq 2$, then $\mathsf{rank}(P) \leq \alpha$ iff there are an n-pattern Q, a computable sequence $\langle Q_m : m < \omega \rangle$ of n-patterns and a sequence $\langle \beta_m : m < \omega \rangle$ such that $\mathsf{rank}(Q) \leq \beta_0$, $\mathsf{rank}(Q_m) \leq \beta_m < \alpha$ for each $m < \omega$, and whenever $m < \omega$ and $Q \cong Q' \subseteq I_m^n$, then P is embeddable in $Q' \cup Q_m$.

Notice that in Definition 2.7, the n-pattern Q can be replaced by any $Q' \cong Q$ (that is, Q can be thought of as an isomorphism type of an n-pattern), whereas each Q_m is a specific n-pattern. Also in this same case, there is a smallest $k < \omega$ such that Q is embeddable into I_k^n. Thus, Q_m is irrelevant for $m < k$. On the other hand, if $m > k$, then there are many Q' such that $Q \cong Q' \subseteq I_m^n$.

The next proposition is easily proved by induction on ordinals. It will be used without specific mention.

Proposition 2.8. *Suppose that $P, R \in \mathcal{P}_n$, $\alpha \leq \beta$, R is embeddable into P, and $\mathsf{rank}(P) \leq \alpha$. Then $\mathsf{rank}(R) \leq \beta$.* ⊣

We say that the n-pattern P is **ranked** if there is an ordinal α such that $\mathsf{rank}(P) \leq \alpha$. For a ranked n-pattern P, we let $\mathsf{rank}(P) = \alpha$ if α is the least ordinal for which $\mathsf{rank}(P) \leq \alpha$. The next lemma relates ranked and prevalent n-patterns.

Lemma 2.9. *If $1 \leq n < \omega$, then every ranked n-pattern is prevalent.*

Proof. Fix n such that $1 \leq n < \omega$. We will prove, by ordinal-induction on α, that if $P \in \mathcal{P}_n$ and $\mathsf{rank}(P) \leq \alpha$, then P is prevalent.

As an inductive hypothesis, assume that α is an ordinal such that whenever $Q \in \mathcal{P}_n$ and $\mathsf{rank}(Q) < \alpha$, then Q is prevalent. Now consider an n-pattern P such that $\mathsf{rank}(P) = \alpha$. We will prove that P is prevalent.

If $\alpha \leq 1$, then $|P| \leq 1$ and, as was already noted, P is prevalent.

Now, suppose that $\alpha \geq 2$. Consider some c.e. $X \subseteq A_n$ such that $A_n \backslash X$ does not have a large computable subset. We will show that P is embeddable into some $I_k^n \cap X$. Let Q and the sequence $\langle Q_m : m < \omega \rangle$ be as in Definition 2.7. In particular, $\mathsf{rank}(Q) < \alpha$ and $\mathsf{rank}(Q_m) < \alpha$ for each $m < \omega$, so by the inductive hypothesis, Q and each Q_m are prevalent. Thus, for each $m < \omega$, there are $k_m < \omega$ and Q'_m such that $Q_m \cong Q'_m \subseteq I_{k_m}^n \cap X$. Moreover, by Proposition 2.5, we can require that the following also hold:

- Both sequences $\langle k_m : m < \omega \rangle$ and $\langle Q'_m : m < \omega \rangle$ are computable.
- $k_0 < k_1 < k_2 < \cdots$.
- For each $m < \omega$, $k_m > |Q_m| + m$.

The first of the above requirements is essential. The second is not, but it is included since it seems natural. The third requirement allows that any embedding $f : Q_m \longrightarrow I_{k_m}^n$ can be extended to an embedding $f_m : Q_m \cup I_m^n \longrightarrow I_{k_m}^n$.

We further assume that $\langle f_m : m < \omega \rangle$ is a computable sequence such that for each $m < \omega$, f_m is an embedding of $Q_m \cup I_m^n$ into $I_{k_m}^n$ and $f_m[Q_m] = Q'_m$.

Since Q is prevalent, by the inductive hypothesis we can let $m < \omega$ and Q' be such that $Q \cong Q' \subseteq f_m[I_m^n] \cap X$. Thus, P is embeddable into $Q' \cup Q'_m \subseteq I_{m_k}^n \cap X$. ⊣

Thus, to prove Lemma 2.3, it suffices to prove the following lemma.

Lemma 2.10. *If $1 \leq n < \omega$, then every n-pattern is ranked.*

This lemma will be proved by induction on n. To maintain this induction, we will have to prove something stronger for which another definition is needed.

Definition 2.11. Suppose that $1 \leq n < \omega$ and that $\mathcal{P}$ is a computable set of n-patterns. We will say that the quadruple $\langle \alpha, \Phi, \Psi, \Theta \rangle$ is an **effective ranking** of $\mathcal{P}$ if α is a computable ordinal and $\Phi : \mathcal{P} \longrightarrow \alpha$, $\Psi : \mathcal{P} \longrightarrow \mathcal{P}$ and $\Theta : \mathcal{P} \times \omega \longrightarrow \mathcal{P}$ are computable functions such that for each $P \in \mathcal{P}$ the following hold:

(1) If $P = \varnothing$, then $\Phi(P) = 0$, $\Psi(P) = \varnothing$ and $\Theta(P, m) = \varnothing$ for each $m < \omega$.

(2) If $P \neq \varnothing$, then $\Phi(\Psi(P)) < \Phi(P)$ and $\Phi(\Theta(P, m)) < \Phi(P)$ for all $m < \omega$.

(3) If $|P| \geq 2$, then whenever $m < \omega$ and $\Psi(P) \cong Q \subseteq I_m^n$, then P is embeddable into $Q \cup \Theta(P, m)$.

(4) For each $m < \omega$, $\Theta(P, m) \cap I_m^n = \varnothing$.

If there is an effective ranking of $\mathcal{P}$, then $\mathcal{P}$ is **effectively rankable**. Moreover, if there are Φ, Ψ, Θ such that $\langle \alpha, \Phi, \Psi, \Theta \rangle$ is an effective ranking of $\mathcal{P}$, then we say that $\mathcal{P}$ is **effectively α-rankable**.

Obviously, if $\mathcal{P}$ is an effectively rankable set of n-patterns, then every $P \in \mathcal{P}$ is ranked. (Clearly, if $\mathcal{P}$ is effectively α-rankable, then $\mathsf{rank}(P) < \alpha$ for all $P \in \mathcal{P}$. Moreover, if $\langle \alpha, \Phi, \Psi, \Theta \rangle$ is an effective ranking of $\mathcal{P}$ and $P \in \mathcal{P}$, then $\mathsf{rank}(P) \leq \Phi(P)$.) Thus, it will suffice to prove the following Lemma 2.12 instead of Lemma 2.9.

Lemma 2.12. $\mathcal{P}_n$ *is effectively rankable.*

Our goal for the rest of this section is to prove Lemma 2.12. In the hope of making the proof more transparent, we reduce it to a series of sublemmas. We start with the simplest instance of Lemma 2.12.

Sublemma 2.12.1. $\mathcal{P}_1$ *is effectively rankable.*

Proof. We will show that $\mathcal{P}_1$ is effectively ω-rankable. Let $\Phi : \mathcal{P}_1 \longrightarrow \omega$, $\Psi : \mathcal{P}_1 \longrightarrow \mathcal{P}_1$ and $\Theta : \mathcal{P}_1 \times \omega \longrightarrow \mathcal{P}_1$ be such that whenever $P \in \mathcal{P}_1$, then:

- $\Phi(P) = |P|$,
- if $|P| \leq 1$, then $\Psi(P) = \Theta(P, m) = \varnothing$ for all $m < \omega$,
- if $|P| \geq 2$, then $|\Psi(P)| = 1$ and, for all $m < \omega$, $|\Theta(P, m)| = |P| - 1$ and $\Theta(P, m) \cap I_m = \varnothing$.

Clearly, there are such Φ, Ψ and Θ. Moreover, there are computable such functions, so we choose them to be computable.

One easily verifies that $\langle \omega, \Phi, \Psi, \Theta \rangle$ is an effective ranking of $\mathcal{P}_1$. ⊣

Definition 2.12.2. If $P \in \mathcal{P}_n$, then R is a **section** of P if:

- $R \in \mathcal{P}_{n+1}$,
- $\pi[R] = P$,
- whenever $x, y \in R$ are distinct, then $\pi(x) \neq \pi(y)$ and $x_n \neq y_n$.

Obviously, every n-pattern has a section. If R is a section of some P, then R is a section of $\pi[R]$. Sections are unique up to isomorphism; that is, if $P \cong P' \in \mathcal{P}_n$ and R, R' are sections of P, P' respectively, then $R \cong R'$. We let $\mathcal{R}_{n+1,1}$ be the set of those $R \in \mathcal{P}_{n+1}$ that are sections of $\pi[R]$. (For $2 \leq k < \omega$, we will define $\mathcal{R}_{n+1,k}$ later.) It is obvious that $\mathcal{R}_{n+1,1}$ is a computable set of $(n+1)$-patterns.

Sublemma 2.12.3. *If $\mathcal{P}_n$ is effectively rankable, then $\mathcal{R}_{n+1,1}$ is effectively rankable.*

Proof. Let $\langle \alpha_n, \Phi_n, \Psi_n, \Theta_n \rangle$ be an effective ranking of $\mathcal{P}_n$. Thus, $\mathcal{P}_n$ is effectively α_n-rankable. We will prove that $\mathcal{R}_{n+1,1}$ is also effectively α_n-rankable.

Let $\Phi : \mathcal{R}_{n+1,1} \longrightarrow \alpha_n$, $\Psi : \mathcal{R}_{n+1,1} \longrightarrow \mathcal{R}_{n+1,1}$ and $\Theta : \mathcal{R}_{n+1,1} \times \omega \longrightarrow \mathcal{R}_{n+1,1}$ be such that whenever $R \in \mathcal{R}_{n+1,1}$, then:

- $\Phi(R) = \Phi_n(\pi[R]) < \alpha_n$;
- $\pi[\Psi(R)] = \Psi_n(\pi[R])$;
- $\pi[\Theta(R,m)] = \Theta_n(\pi[R], m)$ and $\Theta(R,m) \cap (\mathbb{N}^n \times I_m) = \varnothing$
  for all $m < \omega$.

Clearly, there are such Φ, Ψ and Θ. Moreover, there are computable such functions, so we choose them to be computable.

We prove that $\langle \alpha_n, \Phi, \Psi, \Theta \rangle$ is an effective ranking of $\mathcal{R}_{n+1,1}$. Consider $R \in \mathcal{R}_{n+1,1}$. Conditions (1), (2) and (4) of Definition 2.11 are immediate. To prove (3), suppose that $|R| \geq 2$. Let $P = \pi[R]$, so that $P \in \mathcal{P}_n$ and $|P| \geq 2$. Suppose that $\Psi(R) \cong Q \subseteq I_m^{n+1}$. Then, P is embeddable into $\Psi_n(\pi[Q]) \cup \Theta_n(P, m)$. Let $P' \cong P$ be such that $P' \subseteq \Psi_n(\pi[Q]) \cup \Theta_n(P, m)$. Then there is $R' \subseteq Q \cup \Theta(R, m)$ such that $R' \cong R$. ⊣

Incidentally, the same proof can be used to prove: *If $R \in \mathcal{P}_{n+1}$, R is a section of $P \in \mathcal{P}_n$ and P is ranked, then R is ranked and* $\mathrm{rank}(R) \leq \mathrm{rank}(P)$.

Definition 2.12.4. If $P \in \mathcal{P}_n$ and $1 \leq k < \omega$, then we say that R is a k-**fold section** of P if:

- $R \in \mathcal{P}_{n+1}$,
- $\pi[R] = P$,
- $|\pi^{-1}(x) \cap R| = k$ for every $x \in P$,
- whenever $y, z \in R$ are distinct, then $y_n \neq z_n$.

Sections and 1-fold sections are exactly the same. Obviously, every n-pattern has a k-fold section for every $k \geq 1$. For $1 \leq k < \omega$, we note that that k-fold sections are unique up to isomorphism; that is, if $P \cong P' \in \mathcal{P}_n$ and R, R' are k-fold sections of P, P' respectively, then $R \cong R'$. We let $\mathcal{R}_{n+1,k}$ be the set of those $R \in \mathcal{P}_{n+1}$ for which there is ℓ such that $1 \leq \ell \leq k$ and R is an ℓ-fold section of $\pi[R]$. We let $\mathcal{R}_{n+1} = \bigcup_{1 \leq k < \omega} \mathcal{R}_{n+1,k}$. Obviously, $\mathcal{R}_{n+1}$ and every $\mathcal{R}_{n+1,k}$, are computable.

Sublemma 2.12.5. *If $\mathcal{P}_n$ is effectively rankable and $1 \leq k < \omega$, then $\mathcal{R}_{n+1,k}$ is effectively rankable.*

Proof. Let $\langle \alpha, \Phi, \Psi, \Theta \rangle$ be an effective ranking of $\mathcal{P}_n$. We will show by induction on k that $\mathcal{R}_{n+1,k}$ is effectively $(\alpha + k - 1)$-rankable. The basis step, when $k = 1$, is just Lemma 2.12.3.

For the inductive step, suppose that $1 \leq k < \omega$ and that $\mathcal{R}_{n+1,k}$ is effectively $(\alpha + k - 1)$-rankable. Let $\langle \alpha + k - 1, \Phi_k, \Psi_k, \Theta_k \rangle$ be an effective ranking of $\mathcal{R}_{n+1,k}$. We will get $\langle \alpha + k, \Phi_{k+1}, \Psi_{k+1}, \Theta_{k+1} \rangle$ that is an effective ranking of $\mathcal{R}_{n+1,k+1}$.

Let $\Phi_{k+1} : \mathcal{R}_{n+1,k+1} \longrightarrow \alpha + k$, $\Psi_{k+1} : \mathcal{R}_{n+1,k+1} \longrightarrow \mathcal{R}_{n+1,k+1}$ and $\Theta_{k+1} : \mathcal{R}_{n+1,k+1} \times \omega \longrightarrow \mathcal{R}_{n+1,k+1}$ extend Φ_k, Ψ_k, Θ_k, respectively, so that whenever R is a $(k+1)$-fold section, then:

- $\Phi_{k+1}(R) = \alpha + k - 1$,
- $\Psi_{k+1}(R)$ is a section of $\pi[R]$,
- whenever $m < \omega$, then $\Theta_{k+1}(R, m)$ is a k-fold section of I_m^n and $\Theta_{k+1}(R, m) \cap (\mathbb{N}^n \times I_m) = \varnothing$.

Clearly, there are such Φ_{k+1}, Ψ_{k+1} and Θ_{k+1}. Moreover, there are computable such functions, so we choose them to be computable.

We prove that $\langle \alpha + k, \Phi_{k+1}, \Psi_{k+1}, \Theta_{k+1} \rangle$ is an effective ranking of $\mathcal{R}_{n+1,k+1}$. Consider $R \in \mathcal{R}_{n+1,k+1}$. Conditions (1), (2) and (4) of Definition 2.11 are immediate. To prove (3), suppose that $|R| \geq 2$. We need be concerned only when $R \notin \mathcal{R}_{n+1,k}$. Suppose that $\Psi_{k+1}(R) \cong Q \subseteq I_m^{n+1}$, so that Q is a section of $\pi[Q] \subseteq I_m^n$. Let $Q_m = \Theta_{k+1}(R, m)$, so Q_m is a k-fold section of I_m^n. Since we carefully required that $Q_m \cap (\mathbb{N}^n \times I_m) = \varnothing$, we have that $R' = \pi^{-1}(\pi[Q]) \cap (Q \cup Q_m)$ is a $(k+1)$-fold section of $\pi[Q]$, so that $R' \cong R$. Thus, R is embeddable into $Q \cup Q_m$. ⊣

Corollary 2.12.6. *If $\mathcal{P}_n$ is effectively rankable, then $\mathcal{R}_{n+1}$ is effectively rankable.*

Proof. This is not so much a corollary of Sublemma 2.12.5 as it is of its proof. Let $\langle \alpha, \Phi, \Psi, \Theta \rangle$ be an effective ranking of $\mathcal{P}_n$. We just need to observe that the proof of Sublemma 2.12.5 is uniform in k. That is, there is a computable sequence $\langle\langle\alpha + k - 1, \Phi_k, \Psi_k, \Theta_k\rangle : 1 \le k < \omega\rangle$ such that $\langle\alpha + k - 1, \Phi_k, \Psi_k, \Theta_k\rangle$ is an effective ranking of $\mathcal{R}_{n+1,k}$ and $\Phi_k \subseteq \Phi_{k+1}$, $\Psi_k \subseteq \Psi_{k+1}$ and $\Theta_k \subseteq \Theta_{k+1}$ whenever $1 \le k < \omega$. Then, letting $\Phi = \bigcup_k \Phi_k$, $\Psi = \bigcup_k \Psi_k$ and $\Theta = \bigcup_k \Theta_k$, we easily see that $\langle\alpha + \omega, \Phi, \Psi, \Theta\rangle$ is an effective ranking of $\mathcal{R}_{n+1}$. $\dashv$

Definition 2.12.7. Suppose that $C \in \mathcal{P}_n$. We say that C is a **cylinder** if there are finite sets $P \subseteq \mathbb{N}^{n-1}$ and $H \subseteq \mathbb{N}$ such that $C = P \times H$.

We let $\mathcal{C}_n$ be the set of all cylinders in $\mathcal{P}_n$. Every 1-pattern is a cylinder. Obviously, $\mathcal{C}_n$ is computable.

Every n-pattern is a subset of some cylinder. Thus, we could show that $\mathcal{P}_n$ is effectively rankable if $\mathcal{C}_n$ were effectively rankable. Unfortunately, $\mathcal{C}_n$ is, in general, not effectively rankable. The sticking point are the cylinders of the form $\{x\} \times H$. However, every such cylinder is a k-fold section, where $k = |H|$. This explains why, in the following sublemma, we consider $\mathcal{R}_{n+1} \cup \mathcal{C}_{n+1}$ rather than simply $\mathcal{C}_{n+1}$.

Sublemma 2.12.8. *If $\mathcal{P}_n$ is effectively rankable, then $\mathcal{R}_{n+1} \cup \mathcal{C}_{n+1}$ is effectively rankable.*

Proof. Let $\langle\alpha_n, \Phi_n, \Psi_n, \Theta_n\rangle$ be an effective ranking of $\mathcal{P}_n$. By Corollary 2.12.6, $\mathcal{R}_{n+1}$ is effectively rankable, so we let $\langle\beta, \Phi_0, \Psi_0, \Theta_0\rangle$ be an effective ranking of $\mathcal{R}_{n+1}$. Following the proof of Corollary 2.12.6, we can let $\beta = \alpha_n + \omega$ so that $\beta > \alpha_n$. We will get $\langle\beta + \alpha_n, \Phi, \Psi, \Theta\rangle$ that is an effective ranking of $\mathcal{R}_{n+1} \cup \mathcal{C}_{n+1}$.

Notice that $\mathcal{R}_{n+1} \cap \mathcal{C}_{n+1}$ consists precisely of those $(n + 1)$-patterns $C = P \times H$, where $P \in \mathcal{P}_n$, $H \subseteq \mathbb{N}$, and $|P| \le 1$ or $H = \varnothing$. Let Φ, Ψ, Θ extend Φ_0, Ψ_0, Θ_0, respectively, so that whenever $C \notin \mathcal{R}_{n+1}$ (that is, $C = P \times H \in \mathcal{C}_{n+1}$, where $|P| \ge 2$ and $H \ne \varnothing$), then:

- $\Phi(C) = \beta + \Phi_n(P)$,
- $\Psi(C) = \Psi_n(P) \times H$,
- $\Theta(C, m) = \Theta_n(P, m) \times I_m$.

We prove that $\langle\beta + \alpha_n, \Phi, \Psi, \Theta\rangle$ is an effective ranking of $\mathcal{R}_{n+1} \cup \mathcal{C}_{n+1}$. Clearly, $\beta + \alpha_n$ is a computable ordinal and Φ, Ψ, Θ are computable functions. Consider $C \in \mathcal{R}_{n+1} \cup \mathcal{C}_{n+1}$. We must check that conditions $(1) - (4)$ of Definition 2.11 hold. Clearly, they do so if $C \in \mathcal{R}_{n+1}$, so assume that $C = P \times H$, where $|P| \ge 2$ and $H \ne \varnothing$. Then (1) holds vacuously. For (2), we first have that

$$\Phi(\Psi(C)) = \Phi(\Psi_n(P) \times H) = \beta + \Phi_n(\Psi_n(P)) < \beta + \Phi_n(P) = \Phi(C).$$

Secondly, if $|\Theta_n(P, m)| = 1$, then $\Theta_n(P, m) \times I_m \in \mathcal{R}_{n+1}$, so

$$\Phi(\Theta(C, m)) = \Phi(\Theta_n(P, m) \times I_m) < \beta \le \Phi(C),$$

and if $|\Theta_n(P, m)| > 1$, then

$$\Phi(\Theta(C, m)) = \beta + \Phi_n(\Theta_n(P, m)) < \beta + \Phi_n(P) = \Phi(C).$$

Since $\Theta(C, m) = \Theta_n(P, m) \times I_m$ and $\Theta_n(P, m) \cap I_m^n = \varnothing$, it follows that (4) holds.

To prove (3), we will prove the following statement $*(\gamma)$ for each ordinal γ:

> Suppose that $P \in \mathcal{P}_n$, $|P| \geq 2$, $\varnothing \neq H \subseteq \mathbb{N}$, $\gamma = \Phi_n(P)$ and $C = P \times H \in \mathcal{C}_{n+1}$. Whenever $m < \omega$ and $\Psi(C) \cong Q \subseteq I_m^{n+1}$, then C is embeddable into $Q \cup \Theta(C, m)$.

The proof proceeds by induction on γ. Suppose that α is an ordinal and that $*(\gamma)$ whenever $\gamma < \alpha$. We will prove $*(\alpha)$.

Clearly, $\alpha \geq 2$, else we would have that $C \in \mathcal{R}_{n+1}$. Suppose that $m < \omega$ and $\Psi(C) \cong Q \subseteq I_m^{n+1}$. Then $Q = Q' \times H'$, where $\Psi_n(P) \cong Q' \subseteq I_m^n$ and $H' \subseteq I_m$. Then, P is embeddable into $Q' \cup \Theta_n(P, m)$. Let P' be such that $P \cong P' \subseteq Q' \cup \Theta_n(P, m)$. Then we have that $C = P \times H \cong P' \times H' \subseteq (Q' \cup \Theta_n(P, m)) \times H' = (Q' \times H') \cup (\Theta_n(P, m) \times H') = Q \cup (\Theta_n(P, m) \times H') \subseteq Q \cup \Theta(C, m)$. ⊣

Sublemma 2.12.9. *If $\mathcal{P}_n$ is effectively rankable, then $\mathcal{P}_{n+1}$ is effectively rankable.*

Proof. The proof is straightforward since every $(n + 1)$-pattern is a subset of a pattern in $\mathcal{C}_{n+1}$. Let $\langle \alpha_n, \Phi_n, \Psi_n, \Theta_n \rangle$ be an effective ranking of $\mathcal{P}_n$. By Sublemma 2.12.8, let $\langle \alpha_{n+1}, \Phi, \Psi, \Theta \rangle$ be an effective ranking of $\mathcal{R}_{n+1} \cup \mathcal{C}_{n+1}$. We will get $\langle \alpha_{n+1}, \Phi_{n+1}, \Psi_{n+1}, \Theta_{n+1} \rangle$, that is an effective ranking of $\mathcal{P}_{n+1}$.

Let $\Phi_{n+1}, \Psi_{n+1}, \Theta_{n+1}$ extend Φ, Ψ, Θ, respectively, so that whenever $P \in \mathcal{P}_{n+1} \setminus (\mathcal{R}_{n+1} \cup \mathcal{C}_{n+1})$, there is $Q \in \mathcal{C}_{n+1}$ such that $P \subseteq Q$ and:

- $\Phi_{n+1}(P) = \Phi(Q)$,
- $\Psi_{n+1}(P) = \Psi(Q)$,
- $\Theta_{n+1}(P, m) = \Theta(Q, m)$ for all $m < \omega$.

It is easily checked that $\langle \alpha_{n+1}, \Phi_{n+1}, \Psi_{n+1}, \Theta_{n+1} \rangle$ is an effective ranking of $\mathcal{P}_{n+1}$. ⊣

Lemma 2.12 now follows from Sublemmas 2.12.1 and 2.12.9. This completes the proof of Lemma 2.3 and, hence, of Lemma 2.2 and finally Theorem 2.1. ⊣

For what it's worth, our proof of Lemma 2.10 shows that, for every n-pattern P, $\mathsf{rank}(P) < \omega^2$.

3 Distributive Lattices

This section is devoted to completing the proof of the Theorem.

Let D be a finite distributive lattice and that $|D| \geq 2$. Then we can assume that $1 \leq n < \omega$ and that D is a $0, 1$-sublattice of $\mathbf{B}_n$. Let $\langle \beta_{n,i} : i < \omega \rangle$ be the canonical sequence of representations of $\mathbf{B}_n$ as defined in §2. For each $i < \omega$, let $\alpha_i = \beta_{n,i} \!\restriction\! D$. It is easy to see that $\langle \alpha_i : i < \omega \rangle$ is a computable sequence of finite 0-CPP representations of D. Thus, (C1) – (C3) of Definition 1.1 are satisfied. If we knew that (C4) also were, then the sequence would be precanonical, and then we could easily deduce from Lemma 2.3 that (C5) of Definition 1.1 is satisfied, implying that the sequence is canonical. Then, Theorem 1.3 could be invoked. Thus, to complete the proof of the Theorem, we need to show (C4). We will prove the following theorem, which can be viewed as a generalization of the finite Canonical Ramsey Theorem, which appeared in §2.

Theorem 3.1. *For every $i < \omega$, there is $j < \omega$ such that $\alpha_j \longrightarrow \alpha_i$.*

Proof. Let $e : B_n \longrightarrow D$ be such that for each $I \subseteq n$, $e(I) = \bigcup \{ J \in D : J \subseteq I \}$. We now make the following claim.

Claim 1: For each finite $X \subseteq \mathbb{N}^n$, there is a (necessarily injective) function $f : X \longrightarrow \mathbb{N}^n$ such that whenever $x, y \in X$, then $E(f(x), f(y)) = e(E(x, y))$.

We prove Claim 1 by induction on $|X|$. The claim is vacuously true for $X = \varnothing$. Now, suppose that $X \subseteq \mathbb{N}^n$ is finite and $f : X \longrightarrow \mathbb{N}^n$ is as required. Suppose that $y \in \mathbb{N}^n \backslash X$ with the intent of extending f to an appropriate function on the set $X \cup \{y\}$. It suffices to show that there is $z \in \mathbb{N}^n$ such that whenever $x \in X$, then $E(f(x), z) = e(E(x, y))$. For, given such z, we then have the function $f \cup \{\langle y, z \rangle\}$.

We make a second claim.

Claim 2: For each $k < n$, there is at most one $m \in \mathbb{N}$ for which there is $x \in X$ such that $k \in e(E(x, y))$ and $f(x)_k = m$.

Suppose that $u, v \in X$, $k \in e(E(u, y))$ and $k \in e(E(v, y))$. Then, $k \in e(E(u, y)) \cap e(E(v, y)) \subseteq e(E(u, v)) = E(f(u), f(v))$, so that $f(u)_k = f(v)_k$, proving Claim 2.

To complete the proof of Claim 1, let $z \in \mathbb{N}^n$ be such that for each $k < n$, $z_k = f(x)_k$ if $k \in e(E(x, y))$ (well defined by Claim 2) and $z_k \notin \{f(u)_k : u \in X\}$ if no such x exists. We show that if $x \in X$, then $E(f(x), z) = e(E(x, y))$.

$E(f(x), z) \subseteq e(E(x, y))$: Suppose that $k \in E(f(x), z)$. Then, $z_k = f(x)_k$, so there is $u \in X$ such that (a) $k \in e(E(u, y))$ and (b) $f(u)_k = z_k = f(x)_k$. It follows from (b) that $k \in e(E(x, u))$, and then from (a) that $k \in e(E(x, y))$.

$e(E(x,y)) \subseteq E(f(x),z)$: Suppose that $k \notin E(f(x),z)$. Thus, $z_k \neq f(x)_k$, so $k \notin e(E(x,y))$. This completes the proof of Claim 1.

Fix $i < \omega$. Let $X = I_i^n$, and then obtain $f : I_i^n \longrightarrow \mathbb{N}^n$ as in Claim 1. Without loss, we assume that $k < \omega$ is such that $f : I_i^n \longrightarrow I_k^n$. Let $C = f[I_i^n] \subseteq I_k^n$. Observe that f demonstrates that $\alpha_i \cong \alpha_k | C$. To strengthen this last statement, we need another definition. Let $e' : B_n \longrightarrow D$ be such that for each $I \subseteq n$, $e'(I) = \bigcap \{J \in D : J \supseteq I\}$. Then e and e' are "dual" to each other in the sense that whenever $I, J \subseteq n$, then

$$I \subseteq e(J) \iff e'(I) \subseteq e(J) \iff e'(I) \subseteq J.$$

We then get that if $r \in B_n$ and $x, y \in I_i^n$, then

$$\langle x,y \rangle \in \alpha_i(e'(r)) \Leftrightarrow \langle f(x),f(y) \rangle \in \alpha_k(e'(r)) \Leftrightarrow \langle f(x),f(y) \rangle \in \beta_k(r).$$

We have already seen the first of the above equivalences. For the second one, onsider $x, y \in I_i^n$, so that $f(x), f(y) \in C$. Then,

$$
\begin{aligned}
\langle f(x),f(y) \rangle \in \beta_k(r) &\iff r \subseteq E(f(x),f(y)) \\
&\iff r \subseteq e(E(x,y)) \\
&\iff e'(r) \subseteq E(x,y) \\
&\iff \langle x,y \rangle \in \beta_i(e'(r)) = \alpha_i(e'(r)) \\
&\iff \langle f(x),f(y) \rangle \in \alpha_k(e'(r)).
\end{aligned}
$$

Now let $j < \omega$ be such that $\beta_j \longrightarrow \beta_k$. We will show that $\alpha_j \longrightarrow \alpha_i$. Let $\Theta \in \mathrm{Eq}(I_j^n)$. Since $\beta_j \longrightarrow \beta_k$, we can let $Y \subseteq I_j^n$ and $r \in B_n$ be such that $\Theta \cap Y^2 = \beta_j(r) \cap Y^2$ and $\beta_k \cong \beta_j | Y$, as demonstrated by the bijection $g : I_k^n \longrightarrow Y$. Let $Z = g[C] \subseteq I_j^n$. Then gf demonstrates that $\alpha_i \cong \alpha_j | Z$. Moreover, $\Theta \cap Z^2 = \alpha_j(e'(r)) \cap Z^2$ since, if $x, y \in I_i^n$, then

$$
\begin{aligned}
\langle gf(x),gf(y) \rangle \in \Theta &\iff \langle f(x),f(y) \rangle \in \beta_k(r) \\
&\iff \langle f(x),f(y) \rangle \in \alpha_k(e'(r)) \\
&\iff \langle gf(x),gf(y) \rangle \in \alpha_j(e'(r)). \qquad \dashv
\end{aligned}
$$

References

[1] P. Erdős and R. Rado, A combinatorial theorem, J. London Math. Soc. **25** (1950), 249–255.

[2] Joram Hirschfeld, Models of arithmetic and the semi-ring of recursive functions, in: *Victoria Symposium on Nonstandard Analysis (Univ. Victoria, Victoria, B.C., 1972)*, Lecture Notes in Math. **369**, (Springer, Berlin, 1974), pp. 99–105.

[3] Joram Hirschfeld, Models of arithmetic and recursive functions, Israel J. Math. **20** (1975), 111–126.

[4] Joram Hirschfeld and William H. Wheeler, *Forcing, Arithmetic, Division Rings*, Lecture Notes in Mathematics **454** (Springer-Verlag, 1975).

[5] Roman Kossak and James H. Schmerl, *The Structure of Models of Peano Arithmetic*, Oxford Logic Guides **50**, Oxford Science Publications, Clarendon Press, Oxford, 2006.

[6] Thomas G. McLaughlin, Some extension and rearrangement theorems for Nerode semirings, Z. Math. Logik Grundlag. Math. **35** (1989), 197–209.

[7] Thomas G. McLaughlin, Some observations on the substructure lattice of a Δ_1 ultrapower, Math. Log. Quart. **56** (2010), 323–330.

[8] James H. Schmerl and V. Yu. Shavrukov, (in preparation)

8

Towards an Effective Theory of Absolutely Continuous Measures

HENRY TOWSNER

Abstract: We give a constructive, metastable formulation of a theorem about the exchange of limits for convergent sequence L_1 functions. A crucial tool is a one-dimensional version of Szemeredi's regularity lemma for L_1 functions.

1 Introduction

There has been a great deal of work extracting quantitative results from non-constructive theorems in analysis (see [8], and for some recent examples, [9, 15, 14, 13]), often from fairly new results involving sophisticated techniques. However even very basic results can turn out to be deeply non-constructive, and a library of quantitative versions of such results is a needed resource for extracting bounds from theorems which depend on them.

In this paper we consider the following innocuous looking theorem:

Theorem 1. *Let $(f_n)_n$ and $(g_p)_p$ be sequences of L^1 functions such that*

- *the sequences $(f_n)_n$ and $(g_p)_p$ converge weakly,*

- *all the functions $f_n g_p$ are L^1,*

- *for each fixed n, the sequence $(f_n g_p)_p$ converges weakly, and*

- *for each fixed p, the sequence $(f_n g_p)_n$ converges weakly.*

Studies in Weak Arithmetics, Volume 3.
Patrick Cégielski, Ali Enayat,
Roman Kossak.

Then

$$\lim_n \lim_p \int f_n g_p \, d\mu = \lim_p \lim_n \int f_n g_p \, d\mu.$$

Replacing μ with the measure concentrating on σ, this immediately implies that for all sets σ,

$$\lim_n \lim_p \int_\sigma f_n g_p \, d\mu = \lim_p \lim_n \int_\sigma f_n g_p \, d\mu.$$

This is part of (or at least follows from) the standard development of L^1 functions, as considered in [4] for instance. The proof, however, is surprisingly non-trivial—a crucial step passes through the Radon-Nikodym derivative. Our motivation is that this result is the crucial step in a theorem about Banach spaces ("the non-local unconditionality of the James space") whose proof makes use of an ultraproduct construction.

As a general principle, we expect that quantified statements in ultra-products correspond to uniformly bounded quantifiers in the original models. In particular, since the ultraproduct proof uses the fact that these two limits commute in the ultraproduct, a finitary proof should use the corresponding fact that when the sequences f_n, g_p are sufficiently long, we can find points $n < m$ and $p < q$ so that $\int f_n g_q \, d\mu$ is close to $\int f_m g_p \, d\mu$.

This uniform version of Theorem 1 is the main result we prove below. The application of the results in this paper to producing a constructive version of that theorem is given in [18].

While the proof below is intricate, one feature of our methods is that they are modular: each analytic lemma like Theorem 1 corresponds to a single uniform finitary statement. Thenceforth, we expect that any ultraproduct proof involving this theorem can be turned into a finitary proof using the methods below. More generally, this paper is the first step towards a general library of constructive, uniform versions of the basic theory of L^1 spaces. Given such a library, many proofs involving ultraproducts can be (relatively) straightforwardly converted to finitary proofs.

Relevant to the construction of such a library, a crucial intermediate step is the intermediate result developed in Section 5, a one-dimensional analog of Szemerédi's regularity lemma for L^1 spaces, Theorem 5, which may independently useful. This regularity lemma is the constructive analog of the statement that an L^1 function can be approximated by its level sets; the appearance of a regularity-like statement is a reflection of the general connection between infinitary Π_3 statements and finitary regularity-like statements [7, 16].

The main technique used to obtain the results in this paper is the functional (or "Dialectica") translation [1]; in particular the variant known as the monotone functional interpretation [12]. We do not describe the process of using the functional interpretation to obtain these results here, but see [19] for more about the general method.

In the case of Theorem 1, we are interested in how long it takes for the convergence to occur—that is, how big do n and p have to be for the two sides to be close to each other. More precisely, since the actual rate of convergence may be both non-computable and non-uniform, we are interested in the *metastable convergence* of these limits.

Metastable convergence was introduced in the context of ergodic theory in [2, 17]. Suppose $(r_n)_n$ is a sequence of real numbers with the property that $\lim_n r_n$ exists (for some fixed σ); that is, for each E, there is an n so that for every $m \geq n$, $|r_n - r_m| < 1/E$. It is well known that the function mapping E to the corresponding bound n may be uncomputable, and (worse for our purposes) may be highly non-uniform.

Metastable convergence is a seemingly weaker property which addresses this: we say the sequence $(r_n)_n$ is metastably convergent if for each E and each function $\widehat{m} : \mathbb{N} \to \mathbb{N}$ there exists an n so that

$$|r_n - r_{\widehat{m}(n)}| < 1/E.$$

Metastable convergence is essentially the functional interpretation of the ordinary statement of convergence, and is equivalent to ordinary convergence. More generally, we will speak of the "metastable analog" of a statement, by which we mean the functional interpretation of that statement. In particular, our main results will involve the convergence of double limits—that is, limits of the form $\lim_n \lim_p r_{n,p}$; in this case the appropriate notion of metastable convergence will be more complicated and will *not* be equivalent to ordinary convergence. However the metastable notions always capture all the computable content of the original result: any computation which could be proven to halt using the original convergence result can also be shown to halt using metastable convergence. This is a well-known property of the functional interpretation [8, 1].

Abstract meta-theorems of the sort in [6, 10, 11, 12] say that, even though the proof of Theorem 1 goes through the highly non-constructive Radon-Nikodym theorem, it should be possible to extract from the proof explicit, computable, bounds on the metastable convergence, uniformly in computable bounds on the premises—that from bounds on the L^1 norms of the functions in question and the rates of metastable convergence of the sequences $(f_n)_n, (g_p)_p, (f_n g_p)_n$, and $(f_n g_p)_p$. Because the resulting argument would be unreasonably complicated, we settle for a slightly weaker result where we make some additional uniformity assumptions.

In this case, because we are dealing with a double limit, the right notion of metastable convergence is more complicated. Our main theorem is essentially the following. (It is stated and proved as Theorem 8 in a slightly more general form.)

Theorem 2. *Suppose* $(f_n)_n$ *and* $(g_p)_p$ *are sequences of* L^1 *functions such that*

$(*)_1$: *For each* $\epsilon > 0$ *there is a bound* V *so that for every function* $\widehat{\mathbf{m}}$: $\mathbb{N} \to \mathbb{N}$, *every* p, *every* n_0, *and every measurable set* σ *with* $\mu(\sigma) > 0$, *there is an* $n \in [n_0, \ldots, \widehat{\mathbf{m}}^V(n_0)]$ *so that for each* $m, m' \in [n, \widehat{\mathbf{m}}(n)]$, $|\int_\sigma f_m g_p \, d\mu - \int_\sigma f_{m'} g_p \, d\mu| < \epsilon$,

$(*)_2$: *For each* $\epsilon > 0$ *there is a bound* V *so that for every function* $\widehat{\mathbf{q}} : \mathbb{N} \to \mathbb{N}$, *every* n, *every* p_0, *and every measurable set* σ *with* $\mu(\sigma) > 0$, *there is a* $p \in [p_0, \ldots, \widehat{\mathbf{q}}^V(p_0)]$ *so that for each* $q, q' \in [p, \widehat{\mathbf{q}}(p)]$, $|\int_\sigma f_n g_q \, d\mu - \int_\sigma f_n g_{q'} \, d\mu| < \epsilon$,

$(*)_3$: *There is a* B *so that, for each* n, $\|f_n\|_{L^1} \le B$ *and for each* p, $\|g_p\|_{L^1} \le B$.

Then for every $\epsilon > 0$, *every* p *and* n, *and all functions* $\widehat{\mathbf{k}}$ *and* $\widehat{\mathbf{r}}$, *there exist:*

- *Values* $m \ge n$ *and* $q \ge p$, *and*
- *Functions* $\widehat{\mathbf{l}}$ *and* $\widehat{\mathbf{s}}$,

such that, setting $k = \widehat{\mathbf{k}}(m, q, \widehat{\mathbf{l}}, \widehat{\mathbf{s}})$ *and* $r = \widehat{\mathbf{r}}(m, q, \widehat{\mathbf{l}}, \widehat{\mathbf{s}})$, *we have* $\widehat{\mathbf{l}}(k, r) \ge k$, $\widehat{\mathbf{s}}(k, r) \ge r$, *and*

$$\left| \int f_m g_{\widehat{\mathbf{s}}(k,r)} d\mu - \int f_{\widehat{\mathbf{l}}(k,r)} g_q d\mu \right| < \epsilon.$$

In Section 7, we illustrate the resulting bounds by calculating them explicitly in the simplest interesting case, where $\widehat{\mathbf{l}}(m, q, \widehat{\mathbf{l}}, \widehat{\mathbf{s}}) = q + 1$, $p = n = 0$, and $\widehat{\mathbf{s}}(m, q, \widehat{\mathbf{l}}, \widehat{\mathbf{s}}) = m + 1$. Recall the fast-growing hierarchy

- $f_0(m) = m + 1$,
- $f_{j+1}(m) = f_j^m(m)$.

Theorem 3. *Suppose* $(f_n)_n$ *and* $(g_p)_p$ *are sequences of* L^1 *functions satisfying* $(*)_3$ *and*

$(*)_1^Q$: $(*)_1$ *holds with* $V = 8B^2/\epsilon^2$,

$(*)_2^Q$: $(*)_2$ *holds with* $V = 8B^2/\epsilon^2$,

$(*)_3^Q$: $\|f_n\|_{L^\infty} \le 2^n$ *and* $\|g_p\|_{L^\infty} \le 2^p$.

Then for every $\epsilon > 0$ there exist $s > m$ and $l > q$ so that

$$\left| \int f_m g_s d\mu - \int f_l g_q d\mu \right| < \epsilon$$

and

$$s, l \leq f_{\lceil 2^{21} B^4 / \epsilon^4 \rceil}^{\lceil 2^{29} B^6 / \epsilon^6 \rceil} (6 + \lceil 1/\epsilon \rceil + B).$$

2 Absolutely Continuous Measures

Rather than work with L^1 functions, it turns out to be more natural to work with the corresponding absolutely continuous measures.

2.1 Measures

We fix a Boolean algebra Σ containing a largest element Ω and a smallest element $\emptyset$. Because we are thinking of Σ as an algebra of sets, we write $\cup$ and $\cap$ for the the lattice operations on Σ, and write $\sigma \subseteq \tau$ as an abbreviation for "$\sigma \cup \tau = \tau$".

Definition 1. *If $\nu : \Sigma \to \mathbb{R}$, we say ν is additive if $\nu(\emptyset) = 0$ and whenever $\sigma, \tau \in \Sigma$, $\nu(\sigma \cup \tau) = \nu(\sigma) + \nu(\tau) - \nu(\sigma \cap \tau)$.*

We write $|\nu|$ for the function $|\nu|(\sigma) = |\nu(\sigma)|$. Note that for a general additive ν, $|\nu|$ need not be additive.

A partition in Σ is a finite set $\mathcal{A} \subseteq \Sigma$ such that the elements of $\mathcal{A}$ are pairwise disjoint (We do not assume that $\bigcup \mathcal{A} = \Omega$.) We define $\nu(\mathcal{A}) = \sum_{\sigma \in \mathcal{A}} \nu(\sigma)$. By abuse of notation we will write σ for the partition $\{\sigma\}$.

We write $\mathcal{A} \preceq \mathcal{B}$ ($\mathcal{B}$ refines $\mathcal{A}$) if $\bigcup \mathcal{B} = \bigcup \mathcal{A}$ and for every $\sigma \in \mathcal{B}$ there is a $\sigma_\mathcal{A} \in \mathcal{A}$ with $\sigma \subseteq \sigma_\mathcal{A}$. When $\mathcal{A} \preceq \mathcal{B}$ and $\sigma \in \mathcal{A}$, we write $\mathcal{B}_\sigma = \{\sigma' \in \mathcal{B} \mid \sigma' \subseteq \sigma\}$. Clearly $\sigma \preceq \mathcal{A}_\sigma$. For any $\tau \in \mathcal{B}$ we write $\tau_\mathcal{A}$ for the unique $\sigma \in \mathcal{A}$ so that $\tau \subseteq \sigma$.

We write $[\mathcal{A}, \mathcal{B}]$ for the set of all partitions $\mathcal{C}$ with $\mathcal{A} \preceq \mathcal{C} \preceq \mathcal{B}$.

To help keep the notation straight, note that $\sigma_\mathcal{A}$ is itself a set—the same type as σ—namely the element of $\mathcal{A}$ containing σ, while $\mathcal{B}_\tau$ is a partition—the same type as $\mathcal{B}$—namely a partition refining τ.

Throughout this paper we work with a fixed additive function $\mu : \Sigma \to [0, 1]$ such that $\mu(\Omega) = 1$.

Definition 2. *When $\mu(\mathcal{A}) \neq 0$, we write $\delta_\nu(\mathcal{A})$, the density of ν on $\mathcal{A}$, for $\frac{\nu(\mathcal{A})}{\mu(\mathcal{A})}$.*

We say $\nu : \Sigma \to \mathbb{R}$ is absolutely continuous if for every E there is a D so that whenever $\mathcal{A}$ is a partition with $\mu(\mathcal{A}) < 1/D$, $|\nu|(\mathcal{A}) < 1/E$. A modulus of continuity for ν is a function $\omega_\nu : \mathbb{N} \to \mathbb{N}$ such that for every E and every $\mathcal{A}$ with $\mu(\mathcal{A}) < 1/\omega_\nu(E)$, $|\nu|(\mathcal{A}) < 1/E$.

Here, and throughout the paper, we will prefer to work with bounds given by natural numbers. Thus, we write $1/E$ in place of ϵ and $1/D$

in place of δ.

In general, if ν is absolutely continuous, we write ω_ν for some canonical modulus of continuity (if there is one).

We will use the letters ρ, λ, ν, and μ exclusively to refer to additive functions.

Lemma 1. *If $\mathcal{A} \preceq \mathcal{B}$ then $\delta_{|\nu|}(\mathcal{A}) \leq \delta_{|\nu|}(\mathcal{B})$.*

Proof. Since

$$\delta_{|\nu|}(\mathcal{A}) = \frac{1}{\mu(\mathcal{A})} \sum_{\sigma \in \mathcal{A}} \mu(\sigma) \delta_{|\nu|}(\sigma),$$

and

$$\delta_{|\nu|}(\mathcal{B}) = \frac{1}{\mu(\mathcal{A})} \sum_{\sigma \in \mathcal{A}} \mu(\sigma) \delta_{|\nu|}(\mathcal{B}_\sigma),$$

it suffices to show that $\delta_{|\nu|}(\sigma) \leq \delta_{|\nu|}(\mathcal{B}_\sigma)$.

$$\delta_{|\nu|}(\sigma) = \frac{|\nu(\sigma)|}{\mu(\sigma)} = \frac{1}{\mu(\sigma)} | \sum_{\sigma' \in \mathcal{B}_\sigma} \nu(\sigma')| \leq \frac{1}{\mu(\sigma)} \sum_{\sigma' \in \mathcal{B}_\sigma} |\nu(\sigma')| = \delta_{|\nu|}(\mathcal{B}_\sigma).$$

$\dashv$

Definition 3. *The L^1 norm of ν, $||\nu||_{L^1}$, is $\sup_\mathcal{A} |\nu|(\mathcal{A})$.*

Lemma 2. *If ν is absolutely continuous, $||\nu||_{L^1}$ is finite.*

Proof. Apply absolute continuity with $E = 1$. Then there is a D so that whenever $\mu(\mathcal{A}) < 1/D$, $|\nu|(\mathcal{A}) < 1$. We claim that for any $\mathcal{B}$, $|\nu|(\mathcal{B}) < 2D$. Take any $\mathcal{B}$ and choose $\mathcal{B}_0 \subseteq \mathcal{B}$ so that $\mu(\mathcal{B}_0) < 1/D$ and $\mu(\mathcal{B}_0)$ is maximal among subsets of $\mathcal{B}$ with measure $< 1/D$. (Such a $\mathcal{B}_0$ exists because there are only finitely many subsets of $\mathcal{B}$.) Choose $\mathcal{B}_1 \subseteq \mathcal{B} \setminus \mathcal{B}_0$ similarly, and repeat until we have $\mathcal{B}_0, \ldots, \mathcal{B}_k$. For $i < k$, we must have $1/2D \leq \mu(\mathcal{B}_i) < 1/D$. In particular, $k \leq 2D$. Since $\mu(\mathcal{B}_i) < 1/D$ for all i, $|\nu|(\mathcal{B}_i) < 1$ for all i. Since $|\nu|(\mathcal{B}) = \sum_i |\nu|(\mathcal{B}_i)$, $|\nu|(\mathcal{B}) < 2D$.

This holds for any $\mathcal{B}$, so $||\nu||_{L^1} < 2D$.

$\dashv$

This gives us an easily expressed bound on densities of large partitions:

Lemma 3. *If $\mu(\mathcal{A}) \geq 1/D$ then $\delta_{|\nu|}(\mathcal{A}) \leq D||\nu||_{L^1}$.*

Proof. For any $\mathcal{A}$ we have $|\nu|(\mathcal{A}) \leq ||\nu||_{L^1}$, and therefore $\delta_{|\nu|}(\mathcal{A}) \leq D||\nu||_{L^1}$.

$\dashv$

2.2 Products

When ρ and λ are induced by integrals—that is, $\rho(\sigma) = \int_\sigma f \, d\mu$ and $\lambda(\sigma) = \int_\sigma g \, d\mu$—we can consider a product $(\rho\lambda)(\sigma) = \int_\sigma fg \, d\mu$. Of course, since f and g need only be L^1 functions, the product may be infinite on some sets. As a result, the relationship between the separate measures ρ and λ and the product $\rho\lambda$ is not trivial to compute.

We can define a local version of the product:

Definition 4. *If ρ, λ are functions from Σ to $\mathbb{R}$, we define $\rho * \lambda$ to be the function*

$$(\rho * \lambda)(\sigma) = \frac{\rho(\sigma)\lambda(\sigma)}{\mu(\sigma)}.$$

Note that $\rho * \lambda$ need not be additive or absolutely continuous.

It is not hard to verify that, when the product $(\rho\lambda)(\sigma)$ is defined,

$$(\rho\lambda)(\sigma) = \lim_{\mathcal{A} \succeq \sigma} (\rho * \lambda)(\mathcal{A})$$

in the sense that, for every $\epsilon > 0$, there is a $\mathcal{A} \succeq \sigma$ so that whenever $\mathcal{B} \succeq \mathcal{A}$, $|(\rho\lambda)(\sigma) - (\rho * \lambda)(\mathcal{B})| < \epsilon$. We will not use this fact directly, but of the complexity of the proof will come from our need to approximate $\rho\lambda$ using $\rho * \lambda$.

3 Notation

We will ultimately need a series of techical computational lemmas, which will involve a large number of interrelated numeric bounds. In order to keep the values somewhat organized, we adopt the following notation. Most of our theorems and definitions will have the general form

> For all data E, n, etc., there exist values D, m, etc., such that something happens.

We adopt the convention that the given data in a statement will always use use subscript $\flat$, while the values shown to exist will always have subscript $\sharp$. Thus the statement above would be written:

> For all data $E_\flat$, $n_\flat$, etc., there exist values $D_\sharp$, $m_\sharp$, etc., such that something happens.

We also need to avoid notation conflicts when applying theorems. We adopt the rule that all the data corresponding to a single application of a theorem or definition will share a subscript, which will take the place of the $\flat$ or $\sharp$ which was used in the original statement. Thus, if some later theorem makes use of the statement above, it would say:

> We apply the statement to the case $E_0 = \cdots$ and $n_0 = \cdots$, and the statement guarantees the existence of values D_0 and m_0 such that...

We also adopt the rule that functions are always written in bold with a hat, so a function whose output is $m_\flat$ would be written $\widehat{\mathbf{m}}_\flat$. Functions whose output is itself a function have the same name with a capital letter, so $\widehat{\mathbf{M}}_\flat(\cdots) = \widehat{\mathbf{m}}_\flat$ and $\widehat{\mathbf{m}}_\flat(\cdots) = m_\flat$.

Because most of our lemmas involve a sequence of numeric values, we use the letters n, m, k, l for the indices of such a sequence, with the convention that typically $n \leq m \leq k \leq l$ (these letters will typically have subscripts as well). When we have two distinct sets of indices, we use $p \leq q \leq r \leq s$ for the other indices. When a theorem is stated involving the values n, m, k, l, we will sometimes apply to values of the form p, q, r, s; when we do so, we will be consistent—m in the original theorem will correspond to q in the application, and so on.

The letters E and D are always positive integers; the values $\epsilon = 1/E$ and $\delta = 1/D$ are used, as they often are, to be small error bounds.

We assume throughout that all functions are monotone [12]—that is, if $n \leq m$ then $\widehat{\mathbf{m}}(n) \leq \widehat{\mathbf{m}}(m)$—and that $\widehat{\mathbf{m}}(m) \geq m$. These assumptions are harmless, since we could always specify our theorems to replace $\widehat{\mathbf{m}}$ with $\widehat{\mathbf{m}}'(m) = \max\{m, \max_{n \leq m} \widehat{\mathbf{m}}(n)\}$. We further assume monotonicity in function arguments (if $\widehat{\mathbf{s}}(m) \leq \widehat{\mathbf{t}}(m)$ for all m then $\widehat{\mathbf{n}}(\widehat{\mathbf{s}}) \leq \widehat{\mathbf{n}}(\widehat{\mathbf{t}})$).

4 Sequences

4.1 Convergence

The metastable analog of weak convergence is:

Definition 5. *We say $(\nu_n)_n$ is* metastably weakly convergent *if for every $E_\flat, \widehat{\mathbf{m}}_\flat, n_\flat$, there is an $M_\sharp \geq n_\flat$ so that for every σ, there is an $m_\sharp \leq M_\sharp$ such that whenever $m, m' \in [m_\sharp, \widehat{\mathbf{m}}_\flat(m_\sharp)]$, $|\nu_m(\sigma) - \nu_{m'}(\sigma)| < 1/E_\flat$.*

This is slightly more complicated than the notion for sequences of real numbers because of the uniformity. (We are also following our general notation for the complicated functions produced by the functional interpretation, which creates an excessive number of subscripts on a simple statement like this.) Note that the precise amount of uniformity is important: we find a single bound $M_\sharp$ which suffices for all σ simultaneously. However we cannot, in general, find a single bound $M_\sharp$ which works for all partitions $\mathcal{A}$; indeed, having a bound $M_\sharp$ independently of the size of $\mathcal{A}$ is the analog of L^1-convergence.

If we want to consider partitions, we have the following statement, which is *not* uniform in the size of the partition:

Lemma 4. *If (ν_n) is metastably weakly convergent then for every*

$E_\flat$, $\mathcal{B}_\flat$, $\widehat{\mathbf{m}}_\flat$, $n_\flat$ there is an $m_\sharp \geq n_\flat$ such that whenever $m, m' \in [m_\sharp, \widehat{\mathbf{m}}_\flat(m_\sharp)]$, for each $\sigma_\flat \in \mathcal{B}_\flat$, $|\nu_m - \nu_{m'}|(\sigma_\flat) < 1/E_\flat$.

Proof. By induction on $|\mathcal{B}_\flat|$. When $|\mathcal{B}_\flat| = 1$, this follows immediately from metastable weak convergence applied to $E_\flat, \widehat{\mathbf{m}}_\flat, n_\flat$.

Suppose the claim holds for $\mathcal{B}_\flat$ and we have some $\sigma_0 \notin \mathcal{B}_\flat$. Given any m_0, by metastable weak convergence applied to $E_\flat, \widehat{\mathbf{m}}_\flat, n_\flat$, there is some $m_{m_0} \geq m_0$ so that for all $m, m' \in [m_{m_0}, \widehat{\mathbf{m}}_\flat(m_{m_0})]$, $|\nu_m - \nu_{m'}|(\sigma_0) < 1/E_\flat$. Define $\widehat{\mathbf{m}}_0(m_0) = \widehat{\mathbf{m}}_\flat(m_{m_0})$ and apply the inductive hypothesis to $E_\flat, \mathcal{B}_\flat, \widehat{\mathbf{m}}_0, n_\flat$. We obtain $m_0 \geq n_\flat$ so that for all $m, m' \in [m_0, \widehat{\mathbf{m}}_0(m_0)]$ and all $\sigma_\flat \in \mathcal{B}_\flat$, $|\nu_m - \nu_{m'}|(\sigma_\flat) < 1/E_\flat$.

We set $m_\sharp = m_{m_0} \geq m_0$. Then $[m_{m_0}, \widehat{\mathbf{m}}_\flat(m_{m_0})] \subseteq [m_0, \widehat{\mathbf{m}}_0(m_0)]$, so $m_\sharp$ satisfies the claim. ⊣

There is a natural strengthening of metastable weak convergence:

Definition 6. (ν_n) *has* bounded fluctuations *if for every* $E_\flat$ *there is a natural number* $V_\sharp$ *so that for every* $\widehat{\mathbf{m}}_\flat, n_\flat, \sigma$ *there is an* $m_\sharp \in [n_\flat, \widehat{\mathbf{m}}_\flat^{V_\sharp}(n_\flat)]$ *such that whenever* $m, m' \in [m_\sharp, \widehat{\mathbf{m}}_\flat(m_\sharp)]$,

$$|\nu_m(\sigma) - \nu_{m'}(\sigma)| < 1/E_\flat.$$

Metastable weak convergence corresponds to the statement that a certain tree is well-founded (see [5]); having bounded fluctuations implies that the height of this tree is bounded by ω.

It will be convenient to be able to assume that $m_\sharp = \widehat{\mathbf{m}}_\flat^v(n_\flat)$ exactly for some v:

Lemma 5. *Suppose* (ν_n) *has bounded fluctuations. Then for every* $E_\flat$ *there is a* $V_\sharp$ *so that for every* $\widehat{\mathbf{m}}_\flat, n_\flat, \sigma$ *there is a* $v_\sharp \leq V_\sharp$ *such that whenever* $m, m' \in [\widehat{\mathbf{m}}_\flat^{v_\sharp}(n_\flat), \widehat{\mathbf{m}}_\flat^{v_\sharp+1}(n_\flat)]$, $|\nu_m(\sigma) - \nu_{m'}(\sigma)| < 1/E_\flat$.

Proof. Let V_0 be the bound for the bounded fluctuation of (ν_n); applying this to the function $\widehat{\mathbf{m}}_\flat^2$, for any $\sigma, n_\flat$ there is an $m_\sharp \in [n_\flat, \widehat{\mathbf{m}}^{2V_0}(n_\flat)]$ such that whenever $m, m' \in [m_\sharp, \widehat{\mathbf{m}}_\flat^2(m_\sharp)]$, $|\nu_m(\sigma) - \nu_{m'}(\sigma)| < 1/E_\flat$. Let $v_0 < 2V_0$ be greatest such that $\widehat{\mathbf{m}}_\flat^{v_0}(n_\flat) < m_\sharp$. Then $\widehat{\mathbf{m}}_\flat^{v_0+1}(n_\flat) \geq m_\sharp$, so $m_\sharp \leq \widehat{\mathbf{m}}_\flat^{v_0+1}(n_\flat) \leq \widehat{\mathbf{m}}_\flat(m_\sharp)$ and $\widehat{\mathbf{m}}_\flat^{v_0+2}(n_\flat) \leq \widehat{\mathbf{m}}_\flat^2(m_\sharp)$, so for any $m, m' \in [\widehat{\mathbf{m}}_\flat^{v_0+1}(n_\flat), \widehat{\mathbf{m}}_\flat^{v_0+2}(n_\flat)] \subseteq [m_\sharp, \widehat{\mathbf{m}}_\flat^2(m_\sharp)]$ we have $|\nu_{m_\sharp}(\sigma) - \nu_m(\sigma)| < 1/E_\flat$ as desired. ⊣

In this case we can get also get some uniform bounds on partitions if we are willing to accept a set of defective σ of small measure:

Lemma 6. *If* (ν_n) *has bounded fluctuations then for every* $E_\flat$, $D_\flat$, $\widehat{\mathbf{m}}_\flat$, $n_\flat$, $\mathcal{B}_\flat$ *there is an* $m_\sharp \geq n_\flat$ *such that, taking*

$$\mathcal{B} = \{\sigma \in \mathcal{B}_\flat \mid \text{for every } m, m' \in [m_\sharp, \widehat{\mathbf{m}}_\flat(m_\sharp)], |\nu_m - \nu_{m'}|(\sigma) < 1/E_\flat\},$$

we have $\mu(\mathcal{B}) \geq (1 - 1/D_\flat)\mu(\mathcal{B}_\flat)$.

Proof. Let $V_\sharp$ be the bound on the number of fluctuations for $2E_\flat$. Given $n_\flat, \widehat{\mathbf{m}}_\flat, v$, let

$$\mathcal{E}(v, n_\flat, \widehat{\mathbf{m}}_\flat) = \{\sigma \in \mathcal{B}_\flat \mid \text{for some } m, m' \in [\widehat{\mathbf{m}}_\flat^v(n_\flat), \widehat{\mathbf{m}}_\flat^{v+1}(n_\flat)],$$
$$|\nu_m - \nu_{m'}|(\sigma) \geq 1/E_\flat\},$$

the "exceptional" σ. We will show that $\mu(\mathcal{E}(v, n_\flat, \widehat{\mathbf{m}}_\flat)) < 1 - \mu(\mathcal{B}_\flat)/D_\flat$ for some v.

By induction on k we will show that, for any $\widehat{\mathbf{m}}_\flat$, there is a $v \leq V_\sharp^k$ so that $\mathcal{E}(v, n_\flat, \widehat{\mathbf{m}}_\flat) < (1 - 1/V_\sharp)^k \mu(\mathcal{B}_\flat)$.

When $k = 1$, since (ν_n) has bounded fluctuations, for each $\sigma \in \mathcal{B}_\flat$ there is a $v_\sigma \leq V_\sharp$ so that for each $m, m' \in [\widehat{\mathbf{m}}_\flat^{v_\sigma}(n_\flat), \widehat{\mathbf{m}}_\flat^{v_\sigma+1}(n_\flat)]$, $|\nu_m - \nu_{m'}|(\sigma) < 1/E_\flat$—that is, $\sigma \notin \mathcal{E}(v_\sigma, n_\flat, \widehat{\mathbf{m}}_\flat)$. In particular, there must be some $v \leq V_\sharp$ such that the set of σ with $v_\sigma = v$ has measure $\geq \mu(\mathcal{B}_\flat)/V_\sharp$, so $\mathcal{E}(v, n_\flat, \widehat{\mathbf{m}}_\flat) < (1 - 1/V_\sharp)\mu(\mathcal{B}_\flat)$.

Suppose the claim holds for k. We apply the inductive hypothesis to the function $\widehat{\mathbf{m}}_\flat^{V_\sharp}$, so there is some $v \leq V_\sharp^k$ so that $\mu(\mathcal{E}(v, n_\flat, \widehat{\mathbf{m}}_\flat^{V_\sharp})) < (1 - 1/V_\sharp)^k \mu(\mathcal{B}_\flat)$. Then applying the $k = 1$ case to $\widehat{\mathbf{m}}_\flat^{v \cdot V_\sharp}(n_\flat), \widehat{\mathbf{m}}_\flat$, $\mathcal{E}(v, n_\flat, \widehat{\mathbf{m}}_\flat^{V_\sharp})$, we obtain a v' so that

$$\mu(\mathcal{E}(v', \widehat{\mathbf{m}}_\flat^{v \cdot V_\sharp}(n_\flat), \widehat{\mathbf{m}}_\flat)) < (1 - 1/V_\sharp)(1 - 1/V_\sharp)^k \mu(\mathcal{B}_\flat).$$

Therefore

$$\mu(\mathcal{E}(v \cdot V_\sharp + v', n_\flat, \widehat{\mathbf{m}}_\flat)) < (1 - 1/V_\sharp)^{k+1} \mu(\mathcal{B}_\flat).$$

Therefore $v \cdot V_\sharp + v'$ $v \cdot V_\sharp + v'$ satisfies the claim.

The lemma follows by taking $k = \lceil \frac{\ln(1/D_\flat)}{\ln(1 - 1/V_\sharp)} \rceil$. ⊣

4.2   Uniform Continuity

The Vitali-Hahn-Saks Theorem says roughly that a weakly convergent sequence of additive functions ν_m is actually uniformly continuous— that is, for each $\epsilon > 0$ there is a $\delta > 0$ so that when $\mu(\sigma) < \delta$, $|\nu_m(\sigma)| < \epsilon$ for all m simultaneously. The metastable analog of uniform continuity is:

Definition 7. *We say a sequence of functions* $(\nu_n)_n$ *is metastably uniformly continuous if for every* $E_\flat, \widehat{\mathbf{m}}_\flat, n_\flat$ *there exist* $m_\sharp \geq n_\flat$ *and* $D_\sharp$ *such that whenever* $\mu(\sigma) < 1/D_\sharp$ *and* $m \in [m_\sharp, \widehat{\mathbf{m}}_\flat(D_\sharp, m_\sharp)]$, $|\nu_m(\sigma)| < 1/E_\flat$.

Note that this immediately implies the same statement with uniformity over partitions:

Lemma 7. *Let $(\nu_n)_n$ be metastably uniformly continuous. Then for any $E_\flat, \widehat{\mathbf{m}}_\flat, n_\flat$ there are $m_\sharp \geq n_\flat$ and $D_\sharp$ such that whenever $\mu(\mathcal{A}) < 1/D_\sharp$ and $m \in [m_\sharp, \widehat{\mathbf{m}}_\flat(D_\sharp, m_\sharp)]$, $|\nu_m|(\mathcal{A}) < 1/E_\flat$.*

Proof. Given $E_\flat, \widehat{\mathbf{m}}_\flat, n_\flat$, apply metastable uniform continuity to $2E_\flat$, $\widehat{\mathbf{m}}_\flat, n_\flat$ to obtain $m_\sharp \geq n_\flat$ and $D_\sharp$. Then for any $m \in [m_0, \widehat{\mathbf{m}}(D, m_0)]$ and any $\mathcal{A}$, we may decompose $\mathcal{A} = \mathcal{A}_+ \cup \mathcal{A}_-$ where

$$\mathcal{A}_+ = \{\sigma \in \mathcal{A} \mid \nu_m(\sigma) \geq 0\}, \quad \mathcal{A}_- = \{\sigma \in \mathcal{A} \mid \nu_m(\sigma) < 0\}.$$

Then

$$|\nu_m|(\mathcal{A}) = \nu_m(\mathcal{A}_+) - \nu_m(\mathcal{A}_-) = \nu_m(\bigcup \mathcal{A}_+) - \nu_m(\bigcup \mathcal{A}_-) < 2/2E_\flat.$$

$\dashv$

We now give a quantitative version of Vitali-Hahn-Saks.

Theorem 4. *Let $E_\flat, \widehat{\mathbf{m}}_\flat, n_\flat$ be given and let $(\nu_n)_n$ be a metastably weakly convergent of additive functions with moduli of continuity ω_{ν_n}. Then there are $m_\sharp \geq n_\flat$ and $D_\sharp$ so that, for each $m \in [m_\sharp, \widehat{\mathbf{m}}_\flat(D_\sharp, m_\sharp)]$, whenever $\mu(\sigma) < 1/D_\sharp$, $|\nu_m(\sigma)| < 1/E_\flat$.*

Proof. We assume that the moduli of continuity are rapidly growing, specifically that $\omega_{\nu_{m+1}}(E) \geq 2\omega_{\nu_m}(E)$. (This is without loss of generality, since we can always replace ω_ν with a larger function.)

We define a function

$$\widehat{\mathbf{m}}(m_0) = \widehat{\mathbf{m}}_\flat(2\omega_{\nu_{m_0}}(16E_\flat), m_0).$$

We now define a sequence of values m_i, σ_i. We always have $D_i = 2\omega_{\nu_{m_i}}(16E_\flat)$. We will always have $m_i < m_{i+1}$, and therefore for any $j < i$ we have $m_i \geq m_j + (j - i)$, and so $D_i \geq 2^{i-j}D_j$.

We set $m_0 = n_\flat$ and $\sigma_0 = \emptyset$. Suppose m_i, σ_i are given. We suppose that there is some $m \in [m_i, \widehat{\mathbf{m}}(m_i)]$ and a σ with $\mu(\sigma_i \bigtriangleup \sigma) < 1/D_i$ so that $|\nu_{m_i}(\sigma) - \nu_m(\sigma)| \geq 1/4E_\flat$. (If not, the process stops and we will be able to prove the theorem as described below.) We define m_{i+1} to be this value of m and $\sigma_{i+1} = \sigma$.

Note that for any $j < i$,

$$\mu(\sigma_j \bigtriangleup \sigma_i) \leq \sum_{j' \in [j,i)} \mu(\sigma_{j'} \bigtriangleup \sigma_{j'+1})$$

$$\leq \sum_{j' \in [j,i)} 1/D_{j'}$$

$$\leq \sum_{j' \in [j,i)} 2^{j-j'}/D_j < 2/D_j.$$

Suppose we construct m_i for all i. Now define a function $\widehat{\mathbf{m}}'(m')$ to be m_{i+1} where i is least so $m_i \geq m'$. Let M' be given by metastable weak convergence applied to $8E_\flat, \widehat{\mathbf{m}}', n_\flat$ and let $m' \leq M'$ be such that whenever $k, k' \in [m', \widehat{\mathbf{m}}'(m')]$, $|\nu_k(\sigma_{M'}) - \nu_{k'}(\sigma_{M'})| < 1/8E_\flat$. In particular, since $m_i, m_{i+1} \in [m', \widehat{\mathbf{m}}'(m')]$, $|\nu_{m_i}(\sigma_{M'}) - \nu_{m_{i+1}}(\sigma_{M'})| < 1/8E_\flat$.

As noted above, we have $\mu(\sigma^{i+1} \triangle \sigma^{M'}) < 2/D_{i+1} \leq 2/D_i$. This means

$$\begin{aligned}
|\nu_{m_i}(\sigma_{i+1}) - \nu_{m_{i+1}}(\sigma_{i+1})| &\leq |\nu_{m_i}(\sigma_{M'}) - \nu_{m_{i+1}}(\sigma_{M'})| \\
&\quad + |\nu_{m_i}(\sigma_{i+1} \triangle \sigma_{M'})| \\
&\quad + |\nu_{m_{i+1}}(\sigma_{i+1} \triangle \sigma_{M'})| \\
&< 1/8E_\flat + 1/16E_\flat + 1/16E_\flat \\
&= 1/4E_\flat.
\end{aligned}$$

But this contradicts the choice of σ_{i+1}.

So the process must eventually stop, and we find some m_i, σ_i so that for every $m \in [m_i, \widehat{\mathbf{m}}(m_i)]$ and σ with $\mu(\sigma_i \triangle \sigma) < 1/D_i$ we have $|\nu_{m_i}(\sigma) - \nu_m(\sigma)| < 1/4E_\flat$. We take $D_\sharp = D_i \geq D_0$ and $m_\sharp = m_i$. Then for any $m \in [m_\sharp, \widehat{\mathbf{m}}_\flat(D_\sharp, m_\sharp)] = [m_i, \widehat{\mathbf{m}}(m_i)]$ and any σ with $\mu(\sigma) < 1/D_\sharp$,

$$\begin{aligned}
|\nu_m(\sigma)| &= |\nu_m(\sigma_i \cup \sigma) - \nu_m(\sigma_i \setminus \sigma)| \\
&< |\nu_{m_\sharp}(\sigma_i \cup \sigma) - \nu_{m_\sharp}(\sigma_i \setminus \sigma)| + 1/2E_\flat \\
&\leq |\nu_{m_\sharp}(\sigma_i) - \nu_{m_\sharp}(\sigma_i)| + 1/2E_\flat \\
&\quad + |\nu_{m_\sharp}(\sigma_i) - \nu_{m_\sharp}(\sigma_i \cup \sigma)| \\
&\quad + |\nu_{m_\sharp}(\sigma_i) - \nu_{m_\sharp}(\sigma_i \setminus \sigma)| \\
&< 1/E_\flat.
\end{aligned}$$

$\dashv$

4.3 Double Sequences

We need a similar notion for doubly indexed sequences—that is, given a collection of measures $(\rho_n \lambda_p)_{n,p}$, we need to be able to express uniform continuity.

Definition 8. *We say $(\rho_n \lambda_p)_{n,p}$ is n/p-metastably uniformly continuous ("n over p metastably uniformly continuous") if for every $E_\flat, \widehat{\mathbf{m}}_\flat$, and $\widehat{\mathbf{q}}_\flat$ there are $D_\sharp, m_\sharp, p_\sharp, \widehat{\mathbf{r}}_\sharp$ so that for if $\widehat{\mathbf{m}}_\flat(D_\sharp, m_\sharp, p_\sharp, \widehat{\mathbf{r}}_\sharp) \geq m_\sharp$ and $\widehat{\mathbf{q}}_\flat(D_\sharp, m_\sharp, p_\sharp, \widehat{\mathbf{r}}_\sharp) \geq p_\sharp$ then $\widehat{\mathbf{r}}_\sharp(\widehat{\mathbf{m}}_\flat(D_\sharp, m_\sharp, p_\sharp, \widehat{\mathbf{r}}_\sharp), \widehat{\mathbf{q}}_\flat(D_\sharp, m_\sharp, p_\sharp, \widehat{\mathbf{r}}_\sharp)) \geq \widehat{\mathbf{q}}_\flat(D_\sharp, m_\sharp, p_\sharp, \widehat{\mathbf{r}}_\sharp)$ and for any σ with $\mu(\sigma) < 1/D_\sharp$,*

$$|(\rho_{\widehat{\mathbf{m}}_\flat(D_\sharp, m_\sharp, p_\sharp, \widehat{\mathbf{r}}_\sharp)} \lambda_{\widehat{\mathbf{r}}_\sharp(\widehat{\mathbf{m}}_\flat(D_\sharp, m_\sharp, p_\sharp, \widehat{\mathbf{r}}_\sharp), \widehat{\mathbf{q}}_\flat(D_\sharp, m_\sharp, p_\sharp, \widehat{\mathbf{r}}_\sharp))})(\sigma)| < 1/E_\flat.$$

Of course there is also a dual version, $p_{/n}$-metastable uniform continuity, with the indices flipped.

Note that this is the metastable statement corresponding to the double limit $\lim_n \lim_p (\rho_n \lambda_p)(\sigma)$; the additional complexity is due to the higher quantifier complexity of a double limit.

In general we could prove that that "$n_{/p}$-metastable weak convergence" (which could be defined analogously) implies $n_{/p}$-metastable uniform continuity. For our purpose we only need a special case which lets us avoid this notion. The following lemma is the main step, which includes a stronger inductive hypothesis we need to complete the proof.

Lemma 8. *Suppose that*

- $(\rho_1 \lambda_p)_p$ *has bounded fluctuations, and*
- *for each* m, $(\rho_m \lambda_r)_r$ *is metastably uniformly continuous.*

Then for any $E_\flat, \widehat{\mathbf{m}}_\flat, \widehat{\mathbf{q}}_\flat, n_\flat, p_\flat$, *there are* $D_\sharp, m_\sharp \geq n_\flat, q_\sharp \geq p_\flat, \widehat{\mathbf{r}}_\sharp$ *so that setting*

- $m_\flat = \widehat{\mathbf{m}}_\flat(D_\sharp, m_\sharp, q_\sharp, \widehat{\mathbf{r}}_\sharp)$,
- $q_\flat = \widehat{\mathbf{q}}_\flat(D_\sharp, m_\sharp, q_\sharp, \widehat{\mathbf{r}}_\sharp)$, *and*
- $r_\sharp = \widehat{\mathbf{r}}_\sharp(m_\flat, q_\flat)$,

if $m_\flat \geq m_\sharp$ *and* $q_\flat \geq q_\sharp$ *then*

- $r_\sharp \geq q_\flat$,
- *there is a* σ_0 *such that whenever* $\mu(\sigma_0 \triangle \sigma) < 1/D_\sharp$,

$$|(\rho_{m_\sharp} \lambda_{r_\sharp})(\sigma) - (\rho_{m_\flat} \lambda_{r_\sharp})(\sigma)| < 1/E_\flat, \ and$$

- *whenever* $\mu(\sigma) < 2/D_\sharp$, $|(\rho_{m_\sharp} \lambda_{r_\sharp})(\sigma)| < 1/4E_\flat$.

Proof. We define functions $\widehat{\mathbf{r}}_{i,D,n,p}$ so that for any m, q we have

$$\widehat{\mathbf{r}}_{i,D,n,p}(m, q) \geq \max\{p, q\},$$

and for any σ_0 one of the following holds:

- There exist $D_\sharp, m_\sharp, q_\sharp, \widehat{\mathbf{r}}_\sharp$ satisfying the lemma,
- There is a σ with $\mu(\sigma) < 2/D$ such that

$$|\rho_n \lambda_{\widehat{\mathbf{r}}_{i,D,n,p}(m,q)}(\sigma)| \geq 1/4E_\flat,$$

- Whenever $\mu(\sigma_0 \triangle \sigma) < 1/D$,

$$|(\rho_n \lambda_{\widehat{\mathbf{r}}_{i,D,n,p}(m,q)})(\sigma) - (\rho_m \lambda_{\widehat{\mathbf{r}}_{i,D,n,p}(m,q)})(\sigma)| < 1/E_\flat,$$

- There is a sequence $n = k_0 < \cdots < k_i$ and a σ with

$$\mu(\sigma_0 \triangle \sigma) < 2/D$$

such that for each $j < i$,

$$|(\rho_{k_j} \lambda_{\widehat{\mathbf{r}}_{i,D,n,p}(m,q)})(\sigma) - (\rho_{k_{j+1}} \lambda_{\widehat{\mathbf{r}}_{i,D,n,p}(m,q)})(\sigma)| \geq 1/2E_\flat.$$

For $i = 0$ we take $\widehat{\mathbf{r}}_{0,D,n,p}(m,q) = \max\{p,q\}$ since the final clause is satisfied trivially.

Suppose we have defined $\widehat{\mathbf{r}}_{i,D,n,p}(m,q)$ for all D,n,p,m,q,σ_0. We now define $\widehat{\mathbf{r}}_{i+1,D,n,p}(m,q)$ for some fixed values D,n,p,m,q. We assume $m \geq n$ and $q \geq p$; if not, we replace m with n or q with p as necessary. We define $\widehat{\mathbf{r}}^*(D^*,q^*)$ by

$$\widehat{\mathbf{r}}_0^*(D^*,q^*) = \widehat{\mathbf{r}}_{i,D^*,m,q^*}(\widehat{\mathbf{m}}_b(D^*,m,q^*,\widehat{\mathbf{r}}_{i,D^*,m,q^*}),$$
$$\widehat{\mathbf{q}}_b(D^*,m,q^*,\widehat{\mathbf{r}}_{i,D^*,m,q^*}))$$

and $\widehat{\mathbf{r}}^*(D^*,q^*) = \widehat{\mathbf{r}}_0^*(\max\{D^*,2D\},q^*)$.

By the metastable uniform continuity of $(\rho_m \lambda_r)_r$, we obtain D^*,q^* such that whenever $\mu(\sigma) < 2/D^*$ and

$$q \in [q^*, \widehat{\mathbf{r}}^*(D^*,q^*)],$$

$|\rho_m \lambda_q(\sigma)| < 1/4E_b$. Without loss of generality we may assume $D^* \geq 2D$. Let $m' = \widehat{\mathbf{m}}_b(D^*,m,q^*,\widehat{\mathbf{r}}_{i,D^*,m,q^*})$, $q' = \widehat{\mathbf{q}}_b(D^*,m,q^*,\widehat{\mathbf{r}}_{i,D^*,m,q^*})$, and $r = \widehat{\mathbf{r}}^*(D^*,q^*) = \widehat{\mathbf{r}}_{i,D^*,m,q^*}(m',p')$. We define $\widehat{\mathbf{r}}_{i+1,D,n,p}(m,q) = r$.

We now check that for every σ_0, one of the four properties holds. If there is any σ with $\mu(\sigma) < 2/D$ and $|(\rho_n \lambda_r)(\sigma)| \geq 1/4E_b$ then the second case holds, so assume not. Similarly, if for every σ with $\mu(\sigma_0 \triangle \sigma) < 1/D$ we have $|(\rho_n \lambda_r)(\sigma) - (\rho_m \lambda_r)(\sigma)| < 1/E_b$ then the third case holds, so assume not, and fix a counterexample σ.

We apply the inductive hypothesis to $\widehat{\mathbf{r}}_{i,D^*,m,q^*}(m',q')$ and σ, so one of the four cases above must hold. If the first case holds, we are done, since it resolves the first case for $\widehat{\mathbf{r}}_{i+1,D,n,q}$ as well. We have chosen D^*,q^* to rule out the second case. If the third case holds then $D^*,m,q^*,\widehat{\mathbf{r}}_{i,D^*,m,q^*}$ satisfies the lemma.

The remaining possibility is the fourth case: there is a sequence $m = k_1 < \cdots < k_{i+1}$ and a σ' with $\mu(\sigma \triangle \sigma') < 1/D^*$ such that for each $0 < j < i+1$, $|(\rho_{k_j} \lambda_r)(\sigma) - (\rho_{k_{j+1}} \lambda_r)(\sigma)| \geq 1/2E_b$. We take $n = k_0$. Since $\mu(\sigma \triangle \sigma') < 1/D^* < 2/D^* \leq 1/D$, $|(\rho_m \lambda_r)(\sigma \triangle \sigma')| < 1/4E_b$ and $|(\rho_n \lambda_r)(\sigma \triangle \sigma')| < 1/4E_b$, so

$$1/E_b \leq |(\rho_n \lambda_r)(\sigma) - (\rho_m \lambda_r)(\sigma)|$$
$$\leq |(\rho_n \lambda_r)(\sigma') - (\rho_m \lambda_r)(\sigma')|$$
$$+ |(\rho_n \lambda_r)(\sigma \triangle \sigma')| + |(\rho_m \lambda_r)(\sigma \triangle \sigma')|$$
$$\leq |(\rho_n \lambda_r)(\sigma') - (\rho_m \lambda_r)(\sigma')| + 1/2E_b,$$

and so $|(\rho_n \lambda_r)(\sigma') - (\rho_m \lambda_r)(\sigma')| \geq 1/2E_b$. Since $\mu(\sigma_0 \triangle \sigma') \leq \mu(\sigma_0 \triangle \sigma) + \mu(\sigma \triangle \sigma') \leq 1/D + 1/D^* \leq 2/D$ as needed, we satisfy the fourth case.

This completes the construction of the functions $\widehat{\mathbf{r}}_{i,D,n,p}$ and shows they have the desired properties.

Now fix B large enough by the bounded fluctuations of $(\rho_1 \lambda_p)_p$ and consider the function

$$\widehat{\mathbf{r}}^*(D^*, q^*) = \widehat{\mathbf{r}}_{B,D^*,1,q^*}(\widehat{\mathbf{m}}_\flat(D^*, 1, q^*, \widehat{\mathbf{r}}_{B,D^*,1,q^*}),$$
$$\widehat{\mathbf{q}}_\flat(D^*, 1, q^*, \widehat{\mathbf{r}}_{B,D^*,1,q^*})).$$

By the metastable uniform continuity of $(\rho_1 \lambda_p)_p$ we obtain D^*, q^* such that whenever $\mu(\sigma) < 2/D^*$, $|\rho_1 \lambda_{\widehat{\mathbf{r}}^*(D^*,q^*)}|(\sigma) < 1/4E_\flat$. Let $m' = \widehat{\mathbf{m}}_\flat(D^*, 1, q^*, \widehat{\mathbf{r}}_{B,D^*,1,q^*})$ and $q' = \widehat{\mathbf{q}}_\flat(D^*, 1, q^*, \widehat{\mathbf{r}}_{B,D^*,1,q^*})$ and consider $r = \widehat{\mathbf{r}}^*_{B,D^*,1,q^*}(m', q')$ with $\emptyset$. One of the four cases must hold; if the first holds, we are done. We have ruled out the second by choice of D^*, q^*. If the third holds then $D^*, 1, q^*, \widehat{\mathbf{r}}_{B,D^*,1,q^*}$ satisfies the claim. If the fourth holds then we have a sequence $k_0 < \cdots < k_B$ and a σ so that for each $j < B$, $|(\rho_{k_j} \lambda_r)(\sigma) - (\rho_{k_{j+1}} \lambda_r)(\sigma)| \geq 1/2E_\flat$. But this violates the choice of B. ⊣

Lemma 9. *Suppose that*

- $(\rho_1 \lambda_p)_p$ *has bounded fluctuations, and*
- *for each m, $(\rho_m \lambda_r)_r$ is metastably uniformly continuous.*

Then $(\rho_n \lambda_p)_{n,p}$ is n/p-metastably uniformly continuous.

Proof. Apply the previous lemma to $4E_\flat, \widehat{\mathbf{m}}_\flat, \widehat{\mathbf{p}}_\flat, 0, 0$ to obtain $D_\sharp, m_\sharp, q_\sharp, \widehat{\mathbf{r}}_\sharp$. Choose σ_0 given by the second clause, let $m_\flat = \widehat{\mathbf{m}}_\flat(D_\sharp, m_\sharp, q_\sharp, \widehat{\mathbf{r}}_\sharp)$, $q_\flat = \widehat{\mathbf{q}}_\flat(D_\sharp, m_\sharp, q_\sharp, \widehat{\mathbf{r}}_\sharp)$, and $r_\sharp = \widehat{\mathbf{r}}_\sharp(m_\flat, q_\flat)$. Then if $\mu(\sigma) < 1/D_\sharp$,

$$\begin{aligned}
|(\rho_{m_\flat} \lambda_{r_\sharp})(\sigma)| &= |(\rho_{m_\flat} \lambda_{r_\sharp})(\sigma_0 \cup \sigma) - (\rho_{m_\flat} \lambda_{r_\sharp})(\sigma_0 \setminus \sigma)| \\
&< |(\rho_{m_\sharp} \lambda_{r_\sharp})(\sigma_0 \cup \sigma) - (\rho_{m_\sharp} \lambda_{r_\sharp})(\sigma_0 \setminus \sigma)| + 1/2E_\flat \\
&= |(\rho_{m_\sharp} \lambda_{r_\sharp})(\sigma)| + 1/2E_\flat \\
&< 1/E_\flat.
\end{aligned}$$

⊣

5 Regularity Lemma

The usual proof of our main theorem, involving actual L^1 functions, would use level sets. In order to obtain an analog for absolutely continuous measures, we need approximate level sets. These are given by a "one-dimensional" L^1 analog of the Szemerédi regularity lemma. (One dimensional regularity lemmas show up in some expositions [3] of the usual regularity lemma.) Roughly, this will say that we can find pairs of partitions $\mathcal{B} \succeq \mathcal{A}$ such that for most $\sigma \in \mathcal{A}$ and most $\sigma' \in \mathcal{B}_\sigma$, $\delta_\nu(\sigma)$ is close to $\delta_\nu(\sigma')$, even though $\mathcal{B}$ is "much finer" than $\mathcal{A}$. To make this precise we will need a number of definitions.

If we were working with L^2 bounded functions, the argument would be much simpler. In order to deal with L^1 functions—equivalently, the absolutely continuous measures were are considering—we need to be able to "cut-off" sets of sufficiently high density.

Definition 9. *Given a partition* $\mathcal{B}$, *we define* $\mathcal{B}_{\nu > K} = \{\sigma \in \mathcal{B} \mid |\delta_\nu(\sigma)| > K\}$ *and* $\mathcal{B}_{\nu \le K} = \{\sigma \in \mathcal{B} \mid |\delta_\nu(\sigma)| \le K\}$.

Then $\mathcal{B} = \mathcal{B}_{\nu > K} \cup \mathcal{B}_{\nu \le K}$, and when K is large relative to $\|\nu\|_{L^1}$, we can be sure that $\mu(\mathcal{B}_{\nu > K})$ is small.

Definition 10. *By a* function on partitions *we mean a function* $\widehat{\mathbf{B}}$ *such that for any* $\mathcal{A}$, $\mathcal{A} \preceq \widehat{\mathbf{B}}(\mathcal{A})$.

Definition 11. *Let* $\mathcal{B}_0 \preceq \mathcal{B}$ *be given. We define*

$$\mathfrak{D}_{E,\mathcal{B}_0,\nu}(\mathcal{B}) = \{\sigma \in \mathcal{B} \mid |\delta_\nu(\sigma) - \delta_\nu(\sigma_{\mathcal{B}_0})| \ge 1/E.\}$$

$\mathfrak{D}$ stands for "difference", since it is those elements of $\mathcal{B}$ on which the density δ_ν has changed significantly.

Our goal is to prove the following theorem:

Theorem 5 (One-dimensional L^1 Regularity). *Let* ν, $\mathcal{A}_\flat$, $E_\flat$, $D_\flat$, *and a function on partitions* $\widehat{\mathbf{B}}_\flat$ *be given. Then there exists a* $\mathcal{B}_\sharp \succeq \mathcal{A}_\flat$ *such that for every* $\mathcal{B} \in [\mathcal{B}_\sharp, \widehat{\mathbf{B}}_\flat(\mathcal{B}_\sharp)]$,

$$\mu(\mathfrak{D}_{E_\flat,\mathcal{B}_\sharp,\nu}(\mathcal{B})) < 1/D_\flat.$$

By analogy with Szemerédi regularity, we expect the proof to proceed as follows: we define a notion of density $\theta(\mathcal{C})$ such that:

- For all partitions $\mathcal{C}$, $\theta(\mathcal{C})$ is non-negative and bounded by some fixed value C,
- If $\mathcal{C} \preceq \mathcal{D}$ then $\theta(\mathcal{C}) \le \theta(\mathcal{D})$,
- If $\mathcal{C}$ is not the desired $\mathcal{B}_\sharp$ then there exists a $\mathcal{C}' \succeq \mathcal{C}$ such that $\theta(\mathcal{C}') \ge \theta(\mathcal{C}) + c$ where c is a fixed constant.

Then failure to witness the theorem means we can increment θ, and so within roughly $1/c$ steps we must find the desired witness. (This method is known as the *density* or *energy increment method*, and is characteristic of finitary analogs of the proofs of Π_3 statements.)

If ν has bounded L^2 norm, the choice of density notion is standard:

$$\theta_{L^2}(\mathcal{C}) = \sum_{\sigma \in \mathcal{C}} \mu(\sigma)\delta_\nu^2(\sigma).$$

It is easy to see that θ_{L^2} is bounded by the square of the L^2 norm of ν.

However since ν need not have bounded L^2 norm, we have to "cut-off" this norm, making it linear when $\delta_{|\nu|}$ gets large enough. By choosing the cut-off large enough, we can ensure that the portion where the cut-off occurs has small measure—say, measure at most $1/2D_\flat$—and is therefore negligible. We choose

$$\theta_{L^1}(\mathcal{C}) = \sum_{\sigma \in \mathcal{C}_{\nu \leq K}} \mu(\sigma)\delta_\nu^2(\sigma) + 2K \sum_{\sigma \in \mathcal{C}_{\nu > K}} \mu(\sigma)\,|\delta_\nu(\sigma)|$$

where $K = \max\{2D_\flat\|\nu\|_{L^1}, 1\}$.

Unfortunately, we have now violated monotonicity under a minor but unavoidable circumstance: if $\mathcal{C} \preceq \mathcal{D}$ and $\sigma \in \mathcal{C}_{\nu > K}$, it could nonetheless be that $\mathcal{D}_\sigma \not\subseteq \mathcal{D}_{\nu > K}$. The second, linear term in θ_{L^1} has some (necessary) leeway built into it—we multiply by $2K$, not just K—and that interferes with monotonicity.

We solve this by weakening the monotonicity requirement to only consider pairs $\mathcal{C} \preceq \mathcal{D}$ where $\mathcal{C}_{\nu > K} \subseteq \mathcal{D}$. Given a $\mathcal{D} \succeq \mathcal{C}$ violating this condition, we can modify $\mathcal{D}$ on a set of small measure to satisfy this condition.

Given a function on partitions $\widehat{\mathbf{B}}$, we can think of $\widehat{\mathbf{B}}(\mathcal{A})$ as specifying, for each $\sigma \in \mathcal{A}$, a partition of σ, $\sigma = \bigcup\{\sigma \in \widehat{\mathbf{B}}(\mathcal{A}) \mid \sigma_\mathcal{A} = \sigma\}$. We modify $\widehat{\mathbf{B}}$ so that we only apply $\widehat{\mathbf{B}}$ to elements of $\mathcal{A}_{\nu \leq K}$.

Definition 12. *Let $\widehat{\mathbf{B}}$ be a function on partitions. We define $\widehat{\mathbf{B}}^K$ to be the function on partitions given by*

$$\widehat{\mathbf{B}}^K(\mathcal{A}) = \mathcal{A}_{\nu > K} \cup \{\sigma \in \widehat{\mathbf{B}}(\mathcal{A}) \mid \sigma_\mathcal{A} \in \mathcal{A}_{\nu \leq K}\}.$$

It is convenient that for any $\mathcal{A}$, $|\widehat{\mathbf{B}}^K(\mathcal{A})| \leq |\widehat{\mathbf{B}}(\mathcal{A})|$.

We now prove some basic properties about θ_{L^1}.

Lemma 10. *For any $\mathcal{C}$, $0 \leq \theta_{L^1}(\mathcal{C}) \leq K^2 + 2K\|\nu\|_{L^1}$*

Proof. The lower bound is obvious. For the upper bound,

$$\theta_{L^1}(\mathcal{C}) = \sum_{\sigma \in \mathcal{C}_{\nu \leq K}} \mu(\sigma)\delta_\nu^2(\sigma) + 2K \sum_{\sigma \in \mathcal{C}_{\nu > K}} \mu(\sigma)\,|\delta_\nu(\sigma)|$$

$$\leq \sum_{\sigma \in \mathcal{C}_{\nu \leq K}} \mu(\sigma)K^2 + 2K \sum_{\sigma \in \mathcal{C}} |\nu|(\sigma)$$

$$\leq K^2 + 2K\|\nu\|_{L^1}.$$

⊣

Lemma 11. *If $\mathcal{C} \preceq \mathcal{D}$ and $\mathcal{C}_{\nu > K} \subseteq \mathcal{D}$ then $\theta_{L^1}(\mathcal{C}) \leq \theta_{L^1}(\mathcal{D})$.*

Proof. It suffices to show that for each $\sigma \in \mathcal{C}_{\nu \leq K}$, $\theta_{L^1}(\sigma) \leq \theta_{L^1}(\mathcal{D}_\sigma)$.

Let us write $\mathcal{D}_{\leq}$ for $(\mathcal{D}_\sigma)_{\nu \leq K}$ and $\mathcal{D}_>$ for $\mathcal{D}_\sigma \setminus \mathcal{D}_{\leq}$. Then

$$\theta_{L^1}(\sigma) = \mu(\sigma)\delta_\nu^2(\sigma)$$

$$\leq \mu(\sigma) \left(\frac{\sum_{\tau \in \mathcal{D}_\sigma} \mu(\tau)\delta_\nu(\tau)}{\mu(\sigma)} \right)^2$$

$$= \frac{\left(\sum_{\tau \in \mathcal{D}_\sigma} \mu(\tau)\delta_\nu(\tau)\right)^2}{\mu(\sigma)}$$

$$= \frac{\left(\sum_{\tau \in \mathcal{D}_{\leq}} \mu(\tau)\delta_\nu(\tau)\right)^2}{\mu(\sigma)}$$

$$+ \frac{\left(\sum_{\tau \in \mathcal{D}_>} \mu(\tau)\delta_\nu(\tau)\right)\left(2\sum_{\tau \in \mathcal{D}_{\leq}} \mu(\tau)\delta_\nu(\tau) + \sum_{\tau' \in \mathcal{D}_>} \mu(\tau')\delta_\nu(\tau')\right)}{\mu(\sigma)}$$

$$= \frac{\left(\sum_{\tau \in \mathcal{D}_{\leq}} \sqrt{\mu(\tau)}\left[\sqrt{\mu(\tau)}\delta_\nu(\tau)\right]\right)^2}{\mu(\sigma)}$$

$$+ \left(\sum_{\tau \in \mathcal{D}_>} \mu(\tau)\delta_\nu(\tau) \right) \frac{\sum_{\tau \in \mathcal{D}_{\leq}} \mu(\tau)\delta_\nu(\tau)}{\mu(\sigma)}$$

$$+ \left(\sum_{\tau \in \mathcal{D}_>} \mu(\tau)\delta_\nu(\tau) \right) \frac{\sum_{\tau \in \mathcal{D}_{\leq}} \mu(\tau)\delta_\nu(\tau) + \sum_{\tau' \in \mathcal{D}_>} \mu(\tau')\delta_\nu(\tau')}{\mu(\sigma)}$$

$$\leq \frac{\sum_{\tau \in \mathcal{D}_{\leq}} \mu(\tau)}{\mu(\sigma)} \sum_{\tau \in \mathcal{D}_{\leq}} \mu(\tau)\delta_\nu^2(\tau)$$

$$+ \left(\sum_{\tau \in \mathcal{D}_>} \mu(\tau)\delta_\nu(\tau) \right) \frac{K \sum_{\tau \in \mathcal{D}_{\leq}} \mu(\tau)}{\mu(\sigma)}$$

$$+ \left(\sum_{\tau \in \mathcal{D}_>} \mu(\tau)\delta_\nu(\tau) \right) \delta_\nu(\sigma)$$

$$\leq \sum_{\tau \in \mathcal{D}_{\leq}} \mu(\tau)\delta_\nu^2(\tau) + 2K \sum_{\tau \in \mathcal{D}_>} \mu(\tau)|\delta_\nu(\tau)|$$

$$= \theta_{L^1}(\mathcal{D}_\sigma).$$

$\dashv$

Lemma 12. *Suppose* $|\delta_\nu(\sigma)| \leq K$ *and* $\sigma \preceq \mathcal{D}$. *Let* $\mathcal{D}^* = \mathfrak{D}_{E_\flat,\sigma,\nu}(\mathcal{D}) \cap \mathcal{D}_{\nu \leq K}$. *Then*

$$\theta_{L^1}(\mathcal{D}) \geq \theta_{L^1}(\sigma) + \mu(\mathcal{D}^*)/E_\flat^2.$$

Proof. For notational simplicity, we consider the case where $\mathcal{D} \setminus \mathcal{D}^*$ is a singleton (possibly a singleton of measure 0); let us write ζ for this element. (The general case follows from combining the two cases below.)

For each $\tau \in \mathcal{D}$, write $\gamma_\tau = \delta_\nu(\tau) - \delta_\nu(\sigma)$. Then since

$$\mu(\sigma)\delta_\nu(\sigma) = \sum_{\tau \in \mathcal{D}} \mu(\tau)\delta_\nu(\tau),$$

we have

$$\sum_\tau \mu(\tau)\gamma_\tau = \sum_\tau \mu(\tau)\delta_\nu(\tau) - \mu(\tau)\delta_\nu(\sigma) = 0.$$

First, suppose $|\delta_\nu(\zeta)| \leq K$, so $\mathcal{D}_{\nu>K} = \emptyset$.

$$\begin{aligned}
\theta_{L^1}(\mathcal{D}) &= \sum_{\tau \in \mathcal{D}} \mu(\tau)\delta_\nu^2(\tau) \\
&= \sum_{\tau \in \mathcal{D}} \mu(\tau)(\delta_\nu(\sigma) + \gamma_\tau)^2 \\
&= \sum_{\tau \in \mathcal{D}} \mu(\tau)\delta_\nu^2(\sigma) + 2\gamma_\tau\mu(\tau)\delta_\nu(\sigma) + \mu(\tau)\gamma_\tau^2 \\
&= \theta_{L^1}(\sigma) + \sum_{\tau \in \mathcal{D}} \mu(\tau)\gamma_\tau^2 \\
&\geq \theta_{L^1}(\sigma) + \mu(\mathcal{D}^*)/E_\flat^2.
\end{aligned}$$

On the other hand, suppose $|\delta_\nu(\zeta)| > K$. Since $|\delta_\nu(\tau)| < K$ for $\tau \in \mathcal{D}$, $\delta_\nu(\zeta)$ has the same sign as $\delta_\nu(\sigma)$, so γ_ζ also has the same sign. In particular, $|\delta_\nu(\sigma) + \gamma_\zeta| = |\delta_\nu(\sigma)| + |\gamma_\zeta|$. Then we have

$$\begin{aligned}
\theta_{L^1}(\mathcal{D}) &= \sum_{\tau \in \mathcal{D}} \mu(\tau)\delta_\nu^2(\tau) + 2K\mu(\zeta)\,|\delta_\nu(\zeta)| \\
&= \sum_{\tau \in \mathcal{D}^*} \mu(\tau)(\delta_\nu(\sigma) + \gamma_\tau)^2 + 2K\mu(\zeta)\,|\delta_\nu(\sigma) + \gamma_\zeta| \\
&\geq \sum_{\tau \in \mathcal{D}^*} (\mu(\tau)\delta_\nu^2(\sigma) + 2\mu(\tau)\delta_\nu(\sigma)\gamma_\tau + \mu(\tau)\gamma_\tau^2) \\
&\quad + 2K\mu(\zeta)\,|\delta_\nu(\sigma)| + 2K\mu(\zeta)\,|\gamma_\zeta| \\
&\geq \sum_{\tau \in \mathcal{D}^*} (\mu(\tau)\delta_\nu^2(\sigma) + \mu(\tau)\gamma_\tau^2) + 2K\mu(\zeta)\,|\delta_\nu(\sigma)| \\
&\quad + 2\mu(\zeta)\,(K\,|\gamma_\zeta| - \gamma_\zeta\delta_\nu(\sigma)) \\
&\geq \sum_{\tau \in \mathcal{D}^*} (\mu(\tau)\delta_\nu^2(\sigma) + \mu(\tau)\gamma_\tau^2) + \mu(\zeta)\delta_\nu^2(\sigma) \\
&\geq \theta_{L^1}(\sigma) + \mu(\mathcal{D}^*)/E_\flat^2.
\end{aligned}$$

$\dashv$

Corollary 1. If $\mathcal{C} \preceq \mathcal{D}$ and $\mathcal{C}_{\nu>K} \subseteq \mathcal{D}$ then

$$\theta_{L^1}(\mathcal{D}) \geq \theta_{L^1}(\mathcal{C}) + \frac{1}{E_\flat^2}\mu(\mathcal{D}_{\nu \leq K} \cap \mathfrak{D}_{E_\flat,\mathcal{C},\nu}(\mathcal{D})).$$

Corollary 2. If $\mathcal{C} \preceq \mathcal{D}$, $\mathcal{C}_{\nu>K} \subseteq \mathcal{D}$ and $\mu(\mathfrak{D}_{E_\flat,\mathcal{C},\nu}(\mathcal{D})) \geq 1/D_\flat$ then $\theta_{L^1}(\mathcal{D}) \geq \theta_{L^1}(\mathcal{C}) + \frac{1}{2D_\flat E_\flat^2}$.

Proof. Follows from the previous corollary using the fact that, by Lemma 3, $\mu(\mathcal{D}_{\nu>K}) < 1/2D_\flat$. ⊣

We can now prove the regularity lemma:

Proof of Theorem 5. We assume $K = 2D_\flat||\nu||_{L^1}$. (In the case where $2D_\flat||\nu||_{L^1} < 1$, we obtain slightly different bounds, but the argument is unchanged.)

Let $\mathcal{A}_0 = \mathcal{A}_\flat$. Given $\mathcal{A}_i$, if there is any $\mathcal{B} \in [\mathcal{A}_i, \widehat{\mathbf{B}}_\flat^K(\mathcal{A}_i)]$ such that $\mu(\mathfrak{D}_{E_\flat,\mathcal{A}_i,\nu}(\mathcal{B})) \geq 1/2D_\flat$, take $\mathcal{A}_{i+1}$ to be such a $\mathcal{B}$.

By Corollary 2, $\theta_{L^1}(\mathcal{A}_{i+1}) \geq \theta_{L^1}(\mathcal{A}_i) + \frac{1}{4D_\flat E_\flat^2}$. Since $\theta_{L^1}(\mathcal{A}_i) \leq K^2 + 2K||\nu||_{L^1} = 4D_\flat^2||\nu||_{L^1}^2 + 4D_\flat||\nu||_{L^1}^2$, there must be some $i \leq 16D_\flat^3 E_\flat^2||\nu||_{L^1}^2 + 16D_\flat^2 E_\flat^2||\nu||_{L^1}^2$ so that for every $\mathcal{B} \in [\mathcal{A}_i, \widehat{\mathbf{B}}_\flat^K(\mathcal{A}_i)]$, $\mu(\mathfrak{D}_{E_\flat,\mathcal{A}_i,\nu}(\mathcal{B})) < 1/2D_\flat$.

Suppose $\mathcal{B} \in [\mathcal{A}_i, \widehat{\mathbf{B}}_\flat(\mathcal{A}_i)]$, and let $\mathcal{B}' = (\mathcal{A}_i)_{\nu>K} \cup \{\sigma \in \mathcal{B} \mid \sigma_{\mathcal{A}_i} \in (\mathcal{A}_i)_{\nu \leq K}\}$. Then $\mathcal{B}' \preceq \widehat{\mathbf{B}}_\flat^K(\mathcal{A}_i)$, so $\mu(\mathfrak{D}_{E_\flat,\mathcal{A}_i,\nu}(\mathcal{B}')) < 1/2D_\flat$. If $\sigma \in \mathfrak{D}_{E_\flat,\mathcal{A}_i,\nu}(\mathcal{B})$ then either $\sigma \in \mathcal{B}'$, and therefore $\sigma \in \mathfrak{D}_{E_\flat,\mathcal{A}_i,\nu}(\mathcal{B}')$, or $\sigma_\mathcal{A} \in (\mathcal{A}_i)_{\nu>K}$. Therefore

$$\mu(\mathfrak{D}_{E_\flat,\mathcal{A}_i,\nu}(\mathcal{B})) \leq \mu(\mathfrak{D}_{E_\flat,\mathcal{A}_i,\nu}(\mathcal{B}')) + \mu(\mathcal{A}_{\nu>K}) < 1/2D_\flat + 1/2D_\flat.$$

⊣

We need to strengthen this theorem to sequences of functions. Since we are no longer able to fix $\widehat{\mathbf{B}}^K$ in advance (we don't know what ν to use), we need a modification.

Theorem 6. Let $(\nu_n)_n$ be a sequence with $||\nu_n||_{L^1} \leq B$ for all n. Let $\mathcal{A}_\flat, E_\flat, D_\flat, \widehat{\mathbf{m}}_\flat$, and $\widehat{\mathbf{B}}_\flat$ be given. Then there exists a $\mathcal{B}_\sharp \succeq \mathcal{A}_\flat$, an $n_\sharp$, and a $\widehat{\mathbf{k}}_\sharp$ so that whenever $\mathcal{B}_\sharp \preceq \mathcal{B} \preceq \mathcal{B}' \preceq \widehat{\mathbf{B}}_\flat(n_\sharp, \widehat{\mathbf{k}}_\sharp, \mathcal{B}_\sharp)$, setting

$$m_\flat = \widehat{\mathbf{m}}_\flat(n_\sharp, \widehat{\mathbf{k}}_\sharp, \mathcal{B}_\sharp, \mathcal{B}, \mathcal{B}')$$

and $k_\sharp = \widehat{\mathbf{k}}_\sharp(m_\flat, \mathcal{B}')$, if $m_\flat \geq n_\sharp$ then we have $k_\sharp \geq m_\flat$ and

$$\mu(\mathfrak{D}_{E_\flat,\mathcal{B},k_\sharp}(\mathcal{B}')) < 1/D_\flat.$$

That $k_\sharp$ is independent of $\mathcal{B}$ is an incidental simplification because of the actual calculations involved.

Proof. As above, we set $K = 4D_\flat B$ and define

$$\theta_k(\mathcal{C}) = \sum_{\sigma \in \mathcal{C}_{\nu_k \leq K}} \mu(\sigma)\delta^2_{\nu_k}(\sigma) + 2K \sum_{\sigma \in \mathcal{C}_{\nu_k > K}} \mu(\sigma)\,|\delta_{\nu_k}(\sigma)|.$$

The main step is the following claim:

Claim 2. *Let m_0, $\mathcal{A}_0 \preceq \ldots \preceq \mathcal{A}_d$ and i be given. Then either:*

- *There are $n_\sharp$, $\widehat{\mathbf{k}}_\sharp$, and $\mathcal{B}_\sharp$ satisfying the theorem or*
- *There is a sequence of extensions $\mathcal{A}_d \preceq \mathcal{A}_{d+1} \preceq \cdots \preceq \mathcal{A}_{d+i}$ and an $m \geq m_0$ so that for all $j < i$ we have $\theta_m(\mathcal{A}_{d+j+1}) \geq \theta_m(\mathcal{A}_{d+j}) + 1/2^5 D_\flat E_\flat^2$.*

Proof of claim. By induction on i. When $i = 0$, this is trivial.

Suppose the claim holds for i; we show it for $i + 1$. Suppose we are given $\mathcal{A}_0, \ldots, \mathcal{A}_d$ and m_0. Consider the function $\widehat{\mathbf{k}}_i(m, \mathcal{B}')$ given by applying the inductive hypothesis to m and $\mathcal{A}_0, \ldots, \mathcal{A}_d, \mathcal{B}'$. If m_0, $\mathcal{A}_d$, $\widehat{\mathbf{k}}_i$ satisfy the theorem (or we ever end up in the first case), we are done, so suppose not. Then there are $\mathcal{B}, \mathcal{B}' \in [\mathcal{A}_d, \widehat{\mathbf{B}}_\flat(m_0, \widehat{\mathbf{k}}_i, \mathcal{A}_d)]$ so that if $m = \widehat{\mathbf{m}}_\flat(m_0, \widehat{\mathbf{k}}_i, \mathcal{A}_d, \mathcal{B}, \mathcal{B}') \geq m_0$ and $k = \widehat{\mathbf{k}}_i(m, \mathcal{B}')$, we have $\mu(\mathfrak{D}_{E_\flat, \mathcal{B}, k}(\mathcal{B}')) \geq 1/D_\flat$. This means that either $\mu(\mathfrak{D}_{2E_\flat, \mathcal{A}_d, k}(\mathcal{B})) \geq 1/2D_\flat$ or $\mu(\mathfrak{D}_{2E_\flat, \mathcal{A}_d, k}(\mathcal{B}')) \geq 1/2D_\flat$.

Let $\mathcal{B}_\star = (\mathcal{A}_d)_{\nu_k > K} \cup \{\sigma \in \mathcal{B} \mid |\delta_{\nu_k}(\sigma_{\mathcal{A}_d})| \leq K\}$ and $\mathcal{B}'_\star = (\mathcal{B}_\star)_{\nu_k > K} \cup \{\sigma \in \mathcal{B}' \mid |\delta_{\nu_k}(\sigma_{\mathcal{B}_\star})| \leq K\}$. Then either $\mu(\mathfrak{D}_{2E_\flat, \mathcal{A}_d, k}(\mathcal{B})) \geq 1/4D_\flat$ or $\mu(\mathfrak{D}_{2E_\flat, \mathcal{A}_d, k}(\mathcal{B}')) \geq 1/4D_\flat$, so $\theta_m(\mathcal{B}'_\star) \geq \theta_m(\mathcal{A}_d) + 1/2^5 D_\flat E_\flat^2$. We may take $\mathcal{A}_{d+1}$ to be $\mathcal{B}'_\star$.

By the definition of $\widehat{\mathbf{k}}_i$, either we obtain $\widehat{\mathbf{k}}_\sharp$ and $\mathcal{B}_\sharp$ satisfying the theorem, in which case we are done, or we find an extension $\mathcal{A}_{d+1} \preceq \mathcal{A}_{(d+1)+1} \preceq \cdots \preceq \mathcal{A}_{(d+1)+i}$ so that

$$\theta_k(\mathcal{A}_{(d+1)+j+1}) \geq \theta_k(\mathcal{A}_{(d+1)+j}) + 1/2^5 D_\flat E_\flat^2$$

for all $j < i$. So $\mathcal{A}_{d+1} \preceq \cdots \preceq \mathcal{A}_{d+i+1}$ is the desired extension. $\dashv$

Since $\theta_m(\mathcal{B}) \leq K^2 + 2KB = 2^4 D_\flat^2 B^2 + 2^3 D_\flat B^2$, applying the claim with $i = 2^{10} D_\flat^3 E_\flat^2 B^2 + 2^9 D_\flat^2 E_\flat^2 B^2$ and $m_0 = 0$ means the second case is impossible, so we must be in the first case, satisfying the theorem. $\dashv$

Before going on, we need the following observation:

Lemma 13. *Suppose $|\nu - \nu'|(\mathcal{B}) < 1/DE$. Then*

$$\mu(\{\sigma \in \mathcal{B} \mid |\delta_\nu(\sigma) - \delta_{\nu'}(\sigma)| \geq 1/E\}) < 1/D.$$

Proof. Suppose not, so setting $\mathcal{B}^* = \{\sigma \in \mathcal{B} \mid |\delta_\nu(\sigma) - \delta_{\nu'}(\sigma)| \geq 1/E\}$. Then

$$
\begin{aligned}
|\nu - \nu'|(\mathcal{B}) &= \sum_{\sigma \in \mathcal{B}} |\delta_\nu(\sigma) - \delta_{\nu'}(\sigma)|\mu(\sigma) \\
&\geq \sum_{\sigma \in \mathcal{B}^*} |\delta_\nu(\sigma) - \delta_{\nu'}(\sigma)|\mu(\sigma) \\
&\geq 1/DE.
\end{aligned}
$$

⊣

We will need a stronger form of Theorem 6 in which we achieve regularity, not for a single ν_n, but for all ν_n with n in an interval.

5.1 An Aside about Notation

The following lemma is the first instance of a pattern we will need many times: we wish to apply several lemmas in a nested fashion, and to do this, we need to define a nested series of functions satisfying the premises of those theorems. We will distinguish variables belonging to a given application of a theorem by subscripts. The outermost theorem will be subscripted with 0 (to indicate no dependencies). For instance, suppose we have two theorems: Theorem A says

For all $\widehat{m}_\flat$ there are $n_\sharp$ and $p_\sharp$ such that $\widehat{m}_\flat(n_\sharp, p_\sharp)$ has a convenient property.

and Theorem B says

For all $\widehat{m}_\flat$ there is an $n_\sharp$ such that $\widehat{m}_\flat(n_\sharp)$ has a desirable property.

We wish to prove a theorem in which the function $\widehat{m}_\flat$ to which we apply Theorem A is defined using Theorem B. We would write this as follows:

Proof of a hypothetical theorem. We define a function $\widehat{m}_0$ so that we can apply Theorem A to it.

Suppose n_0 and p_0 are given. We write † to abbreviate n_0, p_0. We now define a function $\widehat{m}_\dagger$ so that we can apply Theorem B to it.

Suppose $n_\dagger$ is given. Define $\widehat{m}_\dagger(n_\dagger) = f(n_\dagger, n_0, p_0)$.

By Theorem B applied to $\widehat{m}_\dagger$, we obtain a value $n_\dagger$ so that $\widehat{m}_\dagger(n_\dagger)$ has a desirable property. We now define $\widehat{m}_0(n_0, p_0) = g(n_\dagger)$.

By Theorem A applied to $\widehat{m}_0$, we obtain n_0, p_0 so that $\widehat{m}_0(n_0, p_0)$ has a convenient property. This allows us to complete the proof. ⊣

In particular, note that the subscripts distinguish the variables relevant to Theorem A from those relevant to Theorem B, and indicate the dependencies ($n_\dagger$ depends on n_0, p_0, for instance), and the use of the bars on the left of the text to indicate the scope of the variables.

5.2   The Strong Form of Regularity

Theorem 7. *Let $(\nu_n)_n$ be a metastably weakly convergent sequence of functions with $\|\nu_n\|_{L^1} \leq B$ for all n. Let $\mathcal{A}_\flat, E_\flat, D_\flat, \widehat{\mathbf{B}}_\flat, \widehat{\mathbf{m}}_\flat, \widehat{\mathbf{L}}_\flat$ be given. Then there exists a $\mathcal{B}_\sharp \succeq \mathcal{A}_\flat$, an $n_\sharp$, and an $\widehat{\mathbf{k}}_\sharp$ so that, setting $m_\flat = \widehat{\mathbf{m}}_\flat(n_\sharp, \widehat{\mathbf{k}}_\sharp, \mathcal{B}_\sharp)$ and $\widehat{\mathbf{l}}_\flat = \widehat{\mathbf{L}}_\flat(n_\sharp, \widehat{\mathbf{k}}_\sharp, \mathcal{B}_\sharp)$, if $m_\flat \geq n_\sharp$ then whenever $\mathcal{B}_\sharp \preceq \mathcal{B} \preceq \mathcal{B}' \preceq \widehat{\mathbf{B}}_\flat(n_\sharp, \widehat{\mathbf{k}}_\sharp, \mathcal{B}_\sharp)$, we may set*

$$k_\sharp = \widehat{\mathbf{k}}_\sharp(m_\flat, \widehat{\mathbf{l}}_\flat, \mathcal{B}, \mathcal{B}')$$

and then $k_\sharp \geq m_\flat$ and for every $l \in [k_\sharp, \widehat{\mathbf{l}}_\flat(k_\sharp)]$, $\mu(\mathfrak{D}_{E_\flat, \mathcal{B}, l}(\mathcal{B}')) < 1/D_\flat$.

Proof. In order to apply Theorem 6, we prepare to define functions $\widehat{\mathbf{m}}_0(n_0, \widehat{\mathbf{k}}_0, \mathcal{B}_0, \mathcal{B}, \mathcal{B}')$ and $\widehat{\mathbf{B}}_0(n_0, \widehat{\mathbf{k}}_0, \mathcal{B}_0)$.

Suppose $n_0, \widehat{\mathbf{k}}_0$ and $\mathcal{B}_0$ are given. We abbreviate $n_0, \widehat{\mathbf{k}}_0, \mathcal{B}_0$ by $\dagger$.

In order to apply $\widehat{\mathbf{L}}_\flat$, we need to define a function $\widehat{\mathbf{k}}_\dagger(m, \widehat{\mathbf{l}}_\flat, \mathcal{B}, \mathcal{B}')$. On input $m, \widehat{\mathbf{l}}_\flat, \mathcal{B} \preceq \mathcal{B}'$ we may apply the metastable weak convergence of $(\nu_n)_n$ twice to find an $m_{\dagger, m, \widehat{\mathbf{l}}_\flat, \mathcal{B}, \mathcal{B}'} \geq m$ so that for each

$$k, k' \in [m_{\dagger, m, \widehat{\mathbf{l}}_\flat, \mathcal{B}, \mathcal{B}'}, \widehat{\mathbf{l}}_\flat(\widehat{\mathbf{k}}_0(m_{\dagger, n, \widehat{\mathbf{l}}_\flat, \mathcal{B}, \mathcal{B}'}, \mathcal{B}'))],$$

the set of $\sigma \in \mathcal{B}$ such that $|\nu_k - \nu_{k'}|(\sigma) \geq 1/3E_\flat$ has measure $< 1/3D_\flat$ and the set of $\sigma \in \mathcal{B}'$ such that $|\nu_k - \nu_{k'}|(\sigma) \geq 1/3E_\flat$ has measure $< 1/3D_\flat$.

We define

- $\widehat{\mathbf{k}}_\dagger(m, \widehat{\mathbf{l}}_\flat, \mathcal{B}, \mathcal{B}') = \widehat{\mathbf{k}}_0(m_{\dagger, m, \widehat{\mathbf{l}}_\flat, \mathcal{B}, \mathcal{B}'}, \mathcal{B}')$.

We now define

- $\widehat{\mathbf{m}}_0(n_0, \widehat{\mathbf{k}}_0, \mathcal{B}_0, \mathcal{B}, \mathcal{B}') = m_{\dagger, \widehat{\mathbf{m}}_\flat(n_0, \widehat{\mathbf{k}}_\dagger, \mathcal{B}_0), \widehat{\mathbf{L}}_\flat(n_0, \widehat{\mathbf{k}}_\dagger, \mathcal{B}_0), \mathcal{B}, \mathcal{B}'}$,
- $\widehat{\mathbf{B}}_0(n_0, \widehat{\mathbf{k}}_0, \mathcal{B}_0) = \widehat{\mathbf{B}}_\flat(n_0, \widehat{\mathbf{m}}_\dagger, \mathcal{B}_0)$.

By Lemma 6 applied to $\mathcal{A}_\flat, 3E_\flat, 3D_\flat, \widehat{\mathbf{m}}_0, \widehat{\mathbf{B}}_0$, we obtain $n_0, \widehat{\mathbf{k}}_0$ and $\mathcal{B}_0$. We set $n_\sharp = n_0$, $\widehat{\mathbf{k}}_\sharp = \widehat{\mathbf{k}}_\dagger$ and $\mathcal{B}_\sharp = \mathcal{B}_0$. Let $m_\flat = \widehat{\mathbf{m}}_\flat(n_\sharp, \widehat{\mathbf{k}}_\sharp, \mathcal{B}_\sharp)$ and $\widehat{\mathbf{l}}_\flat = \widehat{\mathbf{L}}_\flat(n_\sharp, \widehat{\mathbf{k}}_\sharp, \mathcal{B}_\sharp)$ and consider some $\mathcal{B} \preceq \mathcal{B}' \in [\mathcal{B}_\sharp, \widehat{\mathbf{B}}_\flat(n_\sharp, \widehat{\mathbf{k}}_\sharp, \mathcal{B}_\sharp)] = [\mathcal{B}_0, \widehat{\mathbf{B}}_0(n_0, \widehat{\mathbf{k}}_0, \mathcal{B}_0)]$.

Set $k_\sharp = \widehat{\mathbf{k}}_\sharp(m_\flat, \widehat{\mathbf{l}}_\flat, \mathcal{B}, \mathcal{B}')$. By choice of $\widehat{\mathbf{k}}_0$ and $\mathcal{B}_0$,

$$k_\sharp = \widehat{\mathbf{k}}_0(m_{\dagger, m_\flat, \widehat{\mathbf{l}}_\flat, \mathcal{B}, \mathcal{B}'}, \mathcal{B}') \geq m_{\dagger, m_\flat, \widehat{\mathbf{l}}_\flat, \mathcal{B}, \mathcal{B}'} \geq m_\flat$$

and $\mu(\mathfrak{D}_{3E_\flat, \mathcal{B}, l_0}(\mathcal{B}')) < 1/3D_\flat$.

Consider any

$$k \in [k_\sharp, \widehat{l}_\flat(k_\sharp)] \subseteq [m_{\dagger,m_\flat,\widehat{l}_\flat,\mathcal{B},\mathcal{B}'}, \widehat{l}_\flat(\widehat{k}_0(m_{\dagger,m_\flat,\widehat{l}_\flat,\mathcal{B},\mathcal{B}'}, \mathcal{B}'))].$$

Suppose $\sigma \notin \mathfrak{D}_{3E_\flat,\mathcal{B},k_\sharp}(\mathcal{B}')$. Then

$$|\nu_k(\sigma) - \nu_k(\sigma_\mathcal{B})| \le |\nu_{k_\sharp} - \nu_k|(\sigma) + |\nu_{k_\sharp} - \nu_k|(\sigma_\mathcal{B}) + 1/3E_\flat < 1/E_\flat.$$

Except for a set of σ of measure less than $2/3D_\flat$, the first two values are each bounded by $1/3E_\flat$, so except on a set of measure less than $1/D_\flat$, $|\nu_k(\sigma) - \nu_k(\sigma_\mathcal{B})| < 1/E_\flat$. That is, $\mu(\mathfrak{D}_{E_\flat,\mathcal{B},k}(\mathcal{B}')) < 1/D_\flat$. ⊣

6 Exchanges

6.1 E-Constant Partitions

When $\rho(A) = \int_A f \, d\mu$, a natural and useful operation is to decompose Ω into approximate level sets—to fix E and define $A_c = \{\omega \mid -1/2E \le f(\omega) - c < 1/2E\}$. We could pick a collection of values c so that these sets are pairwise disjoint. This is an infinite partition, but if we know $||f||_{L^1} \le B$ then we could choose a large number of values c so that $\int_{\Omega \setminus \bigcup_{c \in S} A_c} |f| d\mu < 1/E$ by taking $S = \{-K, -K+1/E, -K+2/E, \ldots, -K+2KE/E$ for big enough K.

In our setting, the analog of a partition into sets A_c is the notion of an E-constant set:

Definition 13. *We say ρ is E-constant on $\mathcal{B}$ if for every $\sigma \in \mathcal{B}$ and every λ, $|(\rho\lambda)(\sigma) - (\rho * \lambda)(\sigma)| < |\lambda(\sigma)|/E$.*

The useful situation is to have a partition $\mathcal{B}'$ and a $\mathcal{B}^- \subseteq \mathcal{B}'$ so that some ρ is E-constant on $\mathcal{B}' \setminus \mathcal{B}^-$—we may think of $\mathcal{B}'$ as having the form $\{A_c \mid c \in S\}$ for some large finite S, and $\mathcal{B}^-$ as being some further partition of $\{\omega \mid |f(\omega)| > K\}$.

Lemma 14. *Suppose ρ is E-constant on $\mathcal{B}' \setminus \mathcal{B}^-$ and that $||\rho||_{L^1} \le B$ and $||\lambda||_{L^1} \le B$. Then for any $\mathcal{B} \preceq \mathcal{B}'$ and any C,*

$$\sum_{\sigma \in \mathcal{B}_{\lambda \le C}} |(\rho\lambda)(\sigma) - (\rho * \lambda)(\sigma)| < \frac{2B}{E} + C|\rho|(\mathfrak{D}_{E,\mathcal{B},\lambda}(\mathcal{B}') \cup \mathcal{B}^-)$$

$$+ |\rho\lambda|(\mathfrak{D}_{E,\mathcal{B},\lambda}(\mathcal{B}') \cup \mathcal{B}^-).$$

Proof. Since ρ is E-constant on $\mathcal{B}' \setminus \mathcal{B}^-$, for each $\tau \in \mathcal{B}' \setminus \mathcal{B}^-$ we have

$$|(\rho\lambda)(\tau) - (\rho * \lambda)(\tau)| < \frac{1}{E}|\lambda(\tau)| = \frac{1}{E}|\lambda|(\tau).$$

Write $\mathcal{B}^+ = \mathcal{B}' \setminus (\mathcal{B}^- \cup \mathfrak{D}_{E,\mathcal{B},\lambda}(\mathcal{B}'))$. For any $\sigma \in \mathcal{B}$, let $\upsilon_\sigma = \sigma \setminus \bigcup \mathcal{B}^+$. (We may think of υ_σ as the "bad" subset of σ.) Note that if $\tau \in \mathcal{B}^+$ then $\tau \notin \mathfrak{D}_{E,\mathcal{B},\lambda}(\mathcal{B}')$, so $|\delta_\lambda(\tau) - \delta_\lambda(\sigma)| < 1/E$.

Then for any $\sigma \in \mathcal{B}_{\leq C}$ we have

$$(\rho\lambda)(\sigma) = \sum_{\tau \in \mathcal{B}'_\sigma} (\rho\lambda)(\tau)$$

$$= (\rho\lambda)(v_\sigma) + \sum_{\tau \in \mathcal{B}^+_\sigma} (\rho\lambda)(\tau)$$

$$= (\rho\lambda)(v_\sigma) + \sum_{\tau \in \mathcal{B}^+_\sigma} (\rho * \lambda)(\tau) + \gamma_\tau \qquad |\gamma_\tau| < \frac{1}{E}|\lambda|(\tau)$$

$$= (\rho\lambda)(v_\sigma) + \sum_{\tau \in \mathcal{B}^+_\sigma} \rho(\tau)\delta_\lambda(\tau) + \gamma_\tau \qquad |\gamma_\tau| < \frac{1}{E}|\lambda|(\tau)$$

$$= (\rho\lambda)(v_\sigma) + \sum_{\tau \in \mathcal{B}^+_\sigma} \rho(\tau)\delta_\lambda(\sigma) + \gamma'_\tau \quad |\gamma'_\tau| < \frac{1}{E}|\rho|(\tau) + \frac{1}{E}|\lambda|(\tau)$$

$$= (\rho\lambda)(v_\sigma) + \rho(\sigma \setminus v_\sigma)\delta_\lambda(\sigma) + \gamma'_\sigma \qquad |\gamma'_\sigma| < \sum_{\tau \in \mathcal{B}'_\sigma} |\gamma'_\tau|$$

$$= (\rho\lambda)(v_\sigma) + \rho(\sigma)\delta_\lambda(\sigma) + \gamma''_\sigma \qquad |\gamma''_\sigma| < C|\rho|(v_\sigma) + |\gamma'_\sigma|$$

$$= (\rho\lambda)(v_\sigma) + (\rho * \lambda)(\sigma) + \gamma''_\sigma \qquad |\gamma''_\sigma| < C|\rho|(v_\sigma) + |\gamma'_\sigma|$$

Summing over all $\sigma \in \mathcal{B}_{\lambda \leq C}$, we obtain

$$\sum_{\sigma \in \mathcal{B}_{\lambda \leq C}} |(\rho\lambda)(\sigma) - (\rho * \lambda)(\sigma)| < \frac{2B}{E} + C|\rho|(\mathfrak{D}_{E,\mathcal{B},\lambda}(\mathcal{B}') \cup \mathcal{B}^-)$$

$$+ |\rho\lambda|(\mathfrak{D}_{E,\mathcal{B},\lambda}(\mathcal{B}') \cup \mathcal{B}^-).$$

$\dashv$

6.2 An Exchange of Limits

We now come to a series of lemma constituting the main part of our argument. We make the following assumptions:

$(*)_1$ For each n, $(\rho_n\lambda_p)_p$ has bounded fluctuations with bound independent of n,

$(*)_2$ For each p, $(\rho_n\lambda_p)_n$ has bounded fluctuations with bound independent of p,

$(*)_3$ There is a fixed bound B such that for each n, $\|\rho_n\|_{L^1} \leq B$ and for each p, $\|\lambda_p\|_{L^1} \leq B$,

$(*)_4$ For any E, D, ρ_n (resp. λ_p) and $\mathcal{B}$, there is a $\mathcal{B}' \succeq \mathcal{B}$ and a $\mathcal{B}'' \subseteq \mathcal{B}'$ so that ρ_n (resp. λ_p) is E-constant on $\mathcal{B}' \setminus \mathcal{B}''$ and $\mu(\mathcal{B}'') < 1/D$.

This last assumption is of course true if ρ_n is given by an L^1 function—intersect the elements of $\mathcal{B}$ with the level sets of the function. We could drop this assumption, replacing it by uses of the regularity lemma above, at the cost of further complicating the proof.

We refer to these assumptions collectively as $(*)$.

For technical reasons, we need a variant of Lemma 7 which is essentially the result of combining two applications of it:

Lemma 15. *Suppose* $(*)$ *holds. Let* $E_\flat, D_\flat, \widehat{\mathbf{B}}_\flat^0, \widehat{\mathbf{B}}_\flat^1, \widehat{\mathbf{m}}_\flat, \widehat{\mathbf{L}}_\flat, \widehat{\mathbf{q}}_\flat, \widehat{\mathbf{S}}_\flat$ *be given. There exists a* $\mathcal{B}_\sharp \succeq \{\Omega\}$, $n_\sharp$ *and* $p_\sharp$, *a* $\widehat{\mathbf{k}}_\sharp$, *and an* $\widehat{\mathbf{r}}_\sharp$ *so that, setting*

- $\mathcal{B}_\flat^0 = \widehat{\mathbf{B}}_\flat^0(n_\sharp, p_\sharp, \widehat{\mathbf{k}}_\sharp, \widehat{\mathbf{r}}_\sharp, \mathcal{B}_\sharp)$,
- $\mathcal{B}_\flat^1 = \widehat{\mathbf{B}}_\flat^1(n_\sharp, p_\sharp, \widehat{\mathbf{k}}_\sharp, \widehat{\mathbf{r}}_\sharp, \mathcal{B}_\sharp)$,
- $m_\flat = \widehat{\mathbf{m}}_\flat(n_\sharp, p_\sharp, \widehat{\mathbf{k}}_\sharp, \widehat{\mathbf{r}}_\sharp, \mathcal{B}_\sharp)$,
- $q_\flat = \widehat{\mathbf{q}}_\flat(n_\sharp, p_\sharp, \widehat{\mathbf{k}}_\sharp, \widehat{\mathbf{r}}_\sharp, \mathcal{B}_\sharp)$,
- $\widehat{\mathbf{l}}_\flat = \widehat{\mathbf{L}}_\flat(n_\sharp, p_\sharp, \widehat{\mathbf{k}}_\sharp, \widehat{\mathbf{r}}_\sharp, \mathcal{B}_\sharp)$,
- $\widehat{\mathbf{s}}_\flat = \widehat{\mathbf{S}}_\flat(n_\sharp, p_\sharp, \widehat{\mathbf{k}}_\sharp, \widehat{\mathbf{r}}_\sharp, \mathcal{B}_\sharp)$,
- $k_\sharp = \widehat{\mathbf{k}}_\sharp(m_\flat, q_\flat, \widehat{\mathbf{l}}_\flat, \widehat{\mathbf{s}}_\flat, \mathcal{B}_\flat^0, \mathcal{B}_\flat^1)$,
- $r_\sharp = \widehat{\mathbf{r}}_\sharp(m_\flat, q_\flat, \widehat{\mathbf{l}}_\flat, \widehat{\mathbf{s}}_\flat, \mathcal{B}_\flat^0, \mathcal{B}_\flat^1)$,

if $m_\flat \geq n_\sharp$ *and* $q_\flat \geq p_\sharp$, *we have* $k_\sharp \geq m_\flat$, $r_\sharp \geq q_\flat$, *for every* $l \in [k_\sharp, \widehat{\mathbf{l}}_\flat(k_\sharp, r_\sharp)]$ $\mu(\mathfrak{D}_{E_\flat, \mathcal{B}_\sharp, l}(\mathcal{B}_\flat^0)) < 1/D_\flat$ *and for every* $s \in [r_\sharp, \widehat{\mathbf{s}}_\flat(k_\sharp, r_\sharp)]$ $\mu(\mathfrak{D}_{E_\flat, \mathcal{B}_\sharp, s}(\mathcal{B}_\flat^1)) < 1/D_\flat$.

Proof. We prepare for the first application of Lemma 7.

Let $\mathcal{B}_0, n_0, \widehat{\mathbf{k}}_0$ be given. Write $\dagger = \mathcal{B}_0, n_0, \widehat{\mathbf{k}}_0$. We now prepare for a second application of Lemma 7.

Let $\mathcal{B}_\dagger, p_\dagger, \widehat{\mathbf{r}}_\dagger$ be given. Write $\star$ for $\dagger, \mathcal{B}_\dagger, p_\dagger, \widehat{\mathbf{r}}_\dagger$.

We first define some helper functions:

- $\widehat{\mathbf{l}}_{\widehat{\mathbf{l}},r}(k) = \widehat{\mathbf{l}}(k, r)$,

- $\widehat{\mathbf{k}}_{\star,r}(m, q, \widehat{\mathbf{l}}, \widehat{\mathbf{s}}, \mathcal{B}^0, \mathcal{B}^1) = \widehat{\mathbf{k}}_0(m, \widehat{\mathbf{l}}_{\widehat{\mathbf{l}},r}, \mathcal{B}_\dagger, \mathcal{B}^0)$,

- $\widehat{\mathbf{s}}_{\star, \mathcal{B}^0, \mathcal{B}^1, m, q, \widehat{\mathbf{l}}, \widehat{\mathbf{s}}}(r) = \widehat{\mathbf{s}}(\widehat{\mathbf{k}}_{\star,r}(m, q, \widehat{\mathbf{l}}, \widehat{\mathbf{s}}, \mathcal{B}^0, \mathcal{B}^1), r)$,

- $\widehat{\mathbf{r}}_\star(m, q, \widehat{\mathbf{l}}, \widehat{\mathbf{s}}, \mathcal{B}^0, \mathcal{B}^1) = \widehat{\mathbf{r}}_\dagger(q, \widehat{\mathbf{s}}_{\star, \mathcal{B}^0, \mathcal{B}^1, m, q, \widehat{\mathbf{l}}, \widehat{\mathbf{s}}}, \mathcal{B}_\dagger, \mathcal{B}^1)$,

- $\widehat{\mathbf{k}}_\star(m, q, \widehat{\mathbf{l}}, \widehat{\mathbf{s}}, \mathcal{B}^0, \mathcal{B}^1) = \widehat{\mathbf{k}}_{\star, \widehat{\mathbf{r}}_\star(m, q, \widehat{\mathbf{l}}, \widehat{\mathbf{s}}, \mathcal{B}^0, \mathcal{B}^1)}(m, q, \widehat{\mathbf{l}}, \widehat{\mathbf{s}}, \mathcal{B}^0, \mathcal{B}^1)$,

- $\widehat{\mathbf{l}}_\star = \widehat{\mathbf{L}}_\flat(n_0, p_\dagger, \widehat{\mathbf{k}}_\star, \widehat{\mathbf{r}}_\star, \mathcal{B}_\dagger)$,

- $\widehat{\mathbf{s}}_\star = \widehat{\mathbf{S}}_\flat(n_0, p_\dagger, \widehat{\mathbf{k}}_\star, \widehat{\mathbf{r}}_\star, \mathcal{B}_\dagger)$.

- $m_\star = \widehat{\mathbf{m}}_\flat(n_0, p_\dagger, \widehat{\mathbf{k}}_\star, \widehat{\mathbf{r}}_\star, \mathcal{B}_\dagger)$.

- $q_\star = \widehat{\mathbf{q}}_\flat(n_0, p_\dagger, \widehat{\mathbf{k}}_\star, \widehat{\mathbf{r}}_\star, \mathcal{B}_\dagger)$.

We can now define the functions needed for an application of Lemma 7.

- $\widehat{\mathbf{B}}_\dagger(p_\dagger, \widehat{\mathbf{r}}_\dagger, \mathcal{B}_\dagger) = \widehat{\mathbf{B}}_\flat^1(n_0, p_\dagger, \widehat{\mathbf{k}}_\star, \widehat{\mathbf{r}}_\star, \mathcal{B}_\dagger)$,

- $\widehat{\mathbf{q}}_\dagger(p_\dagger, \widehat{\mathbf{r}}_\dagger, \mathcal{B}_\dagger) = \widehat{\mathbf{q}}_\flat(n_0, p_\dagger, \widehat{\mathbf{k}}_\star, \widehat{\mathbf{r}}_\star, \mathcal{B}_\dagger)$,

- $\widehat{\mathbf{S}}_\dagger(p_\dagger, \widehat{\mathbf{q}}_\dagger, \mathcal{B}_\dagger) = \widehat{\mathbf{s}}_{\star, \mathcal{B}^0_\star, \mathcal{B}^1_\star, m_\star, q_\star, \widehat{\mathbf{l}}_\star, \widehat{\mathbf{s}}_\star}$.

By Lemma 7 applied to $\mathcal{B}_0, E_\flat, D_\flat, \widehat{\mathbf{B}}_\dagger, \widehat{\mathbf{q}}_\dagger, \widehat{\mathbf{S}}_\dagger$ we find $\mathcal{B}_\dagger, p_\dagger, \widehat{\mathbf{r}}_\dagger$ such that, setting:

- $q_\dagger = \widehat{\mathbf{q}}_\dagger(p_\dagger, \widehat{\mathbf{r}}_\dagger, \mathcal{B}_\dagger)$, and

- $\widehat{\mathbf{s}}_\dagger = \widehat{\mathbf{S}}_\dagger(p_\dagger, \widehat{\mathbf{r}}_\dagger, \mathcal{B}_\dagger)$,

if $q_\dagger \geq p_\dagger$ then whenever $\mathcal{B}_\dagger \preceq \mathcal{B} \preceq \mathcal{B}' \preceq \widehat{\mathbf{B}}_\dagger(p_\dagger, \widehat{\mathbf{r}}_\dagger, \mathcal{B}_\dagger)$, we may set $r_\dagger = \widehat{\mathbf{r}}_\dagger(q_\dagger, \widehat{\mathbf{s}}_\dagger, \mathcal{B}, \mathcal{B}')$ and have $r_\dagger \geq q_\dagger$ and for every $s \in [r_\dagger, \widehat{\mathbf{s}}_\dagger(r_\dagger)]$, we have $\mu(\mathfrak{D}_{E_\flat, \mathcal{B}, s}(\mathcal{B}')) < 1/D_\flat$.

Note that we have now defined values $\mathcal{B}_\dagger, p_\dagger, q_\dagger, \widehat{\mathbf{r}}_\dagger, \widehat{\mathbf{s}}_\dagger$, all depending on $\dagger$—that is, as functions of $n_0, \widehat{\mathbf{k}}_0, \mathcal{B}_0$—as well as functions $\widehat{\mathbf{r}}_\star, \widehat{\mathbf{k}}_\star, \widehat{\mathbf{l}}_\star, \widehat{\mathbf{s}}_\star$ which can be derived from these by the definitions above. We now set:

- $\widehat{\mathbf{B}}_0(n_0, \widehat{\mathbf{k}}_0, \mathcal{B}_0) = \widehat{\mathbf{B}}_\flat^0(n_0, p_\dagger, \widehat{\mathbf{k}}_\star, \widehat{\mathbf{r}}_\star, \mathcal{B}_\dagger)$,

- $\widehat{\mathbf{m}}_0(n_0, \widehat{\mathbf{k}}_0, \mathcal{B}_0) = \widehat{\mathbf{m}}_\flat(n_0, p_\dagger, \widehat{\mathbf{k}}_\star, \widehat{\mathbf{r}}_\star, \mathcal{B}_\dagger)$,

- $\widehat{\mathbf{L}}_0(n_0, \widehat{\mathbf{k}}_0, \mathcal{B}_0) = \widehat{\mathbf{l}}_{\widehat{\mathbf{L}}_\flat(n_0, p_\dagger, \widehat{\mathbf{k}}_\star, \widehat{\mathbf{r}}_\star, \mathcal{B}_\dagger), \widehat{\mathbf{r}}_\dagger(q_\dagger, \widehat{\mathbf{s}}_\dagger, \mathcal{B}_\dagger, \mathcal{B}^1_\star)}$.

By Lemma 7 applied to $\{\Omega\}, E_\flat, D_\flat, \widehat{\mathbf{B}}_0, \widehat{\mathbf{m}}_0, \widehat{\mathbf{L}}_0$, we find $\mathcal{B}_0, n_0, \widehat{\mathbf{k}}_0$ such that, setting

- $m_0 = \widehat{\mathbf{m}}_0(n_0, \widehat{\mathbf{k}}_0, \mathcal{B}_0)$, and

- $\widehat{l}_0 = \widehat{L}_0(n_0, \widehat{k}_0, \mathcal{B}_0)$,

if $m_0 \geq n_0$ then whenever $\mathcal{B}_0 \preceq \mathcal{B} \preceq \widehat{B}_0(n_0, \widehat{k}_0, \mathcal{B}_0)$ we may set $k_0 = \widehat{k}_0(m_0, \widehat{l}_0, \mathcal{B}, \mathcal{B}')$ and then $k_0 \geq m_0$ and for every $l \in [k_0, \widehat{l}_0(k_0)]$, $\mu(\mathfrak{D}_{E_\flat, \mathcal{B}, l}(\mathcal{B}')) < 1/D_\flat$.

We may now set:

- $\mathcal{B}_\sharp = \mathcal{B}_\dagger$,
- $n_\sharp = n_0$,
- $p_\sharp = p_\dagger$,
- $\widehat{k}_\sharp = \widehat{k}_\star$, and
- $\widehat{r}_\sharp = \widehat{r}_\star$.

We must check that these satisfy the claim. We define the following values as specified in the statement of this lemma:

- $\mathcal{B}_\flat^0 = \widehat{B}_\flat^0(n_\sharp, p_\sharp, \widehat{k}_\sharp, \widehat{r}_\sharp, \mathcal{B}_\sharp) = \widehat{B}_0(n_0, \widehat{k}_0, \mathcal{B}_0)$,
- $\mathcal{B}_\flat^1 = \widehat{B}_\flat^1(n_\sharp, p_\sharp, \widehat{k}_\sharp, \widehat{r}_\sharp, \mathcal{B}_\sharp) = \widehat{B}_\dagger(p_\dagger, \widehat{r}_\dagger, \mathcal{B}_\dagger)$,
- $m_\flat = \widehat{m}_\flat(n_\sharp, p_\sharp, \widehat{k}_\sharp, \widehat{r}_\sharp, \mathcal{B}_\sharp) = \widehat{m}_0(m_0, \widehat{k}_0, \mathcal{B}_0) = m_0$,
- $q_\flat = \widehat{q}_\flat(n_\sharp, p_\sharp, \widehat{k}_\sharp, \widehat{r}_\sharp, \mathcal{B}_\sharp) = \widehat{q}_\dagger(p_\dagger, \widehat{r}_\dagger, \mathcal{B}_\dagger) = q_\dagger$,
- $\widehat{l}_\flat = \widehat{L}_\flat(n_\sharp, p_\sharp, \widehat{k}_\sharp, \widehat{r}_\sharp, \mathcal{B}_\sharp) = \widehat{l}_\star$,
- $\widehat{s}_\flat = \widehat{S}_\flat(n_\sharp, p_\sharp, \widehat{k}_\sharp, \widehat{r}_\sharp, \mathcal{B}_\sharp) = \widehat{s}_\star$,
- $k_\sharp = \widehat{k}_\sharp(m_\flat, q_\flat, \widehat{l}_\flat, \widehat{s}_\flat, \mathcal{B}_\flat^0, \mathcal{B}_\flat^1)$,
- $r_\sharp = \widehat{r}_\sharp(m_\flat, q_\flat, \widehat{l}_\flat, \widehat{s}_\flat, \mathcal{B}_\flat^0, \mathcal{B}_\flat^1) = \widehat{r}_\dagger(q_\flat, \widehat{s}_{\star, \mathcal{B}_\flat^0, \mathcal{B}_\flat^1, m_\flat, q_\flat, \widehat{l}_\flat, \widehat{s}_\flat}, \mathcal{B}_\sharp, \mathcal{B}_\flat^1)$.

Many of the other quantities we defined above are equal to these values:

- $\widehat{s}_{\star, \mathcal{B}_\flat^0, \mathcal{B}_\flat^1, m_\flat, q_\flat, \widehat{l}_\flat, \widehat{s}_\flat} = \widehat{S}_\dagger(p_\dagger, \widehat{q}_\dagger, \mathcal{B}_\dagger) = \widehat{s}_\dagger$,
- $r_\dagger = \widehat{r}_\dagger(q_\dagger, \widehat{s}_\dagger, \mathcal{B}_\dagger, \mathcal{B}_\star^1) = r_\sharp$,
- $\widehat{l}_0 = \widehat{L}_0(n_0, \widehat{k}_0, \mathcal{B}_0) = \widehat{l}_{\flat, r_\dagger}$,
- $k_\sharp = \widehat{k}_{\star, r_\sharp}(m_\flat, q_\flat, \widehat{l}_\flat, \widehat{s}_\flat, \mathcal{B}_\flat^0, \mathcal{B}_\flat^1) = \widehat{k}_0(m_\flat, \widehat{l}_{\flat, r_\sharp}, \mathcal{B}_\sharp, \mathcal{B}_\flat^1) = k_0$,
- $\widehat{l}_0(k_\sharp) = \widehat{l}_\flat(k_\sharp, r_\sharp)$,
- $\widehat{s}_\dagger(r_\sharp) = \widehat{s}_\flat(k_\sharp, r_\sharp)$.

Suppose $m_\flat \geq n_\sharp$ and $q_\flat \geq p_\sharp$. Then we have $\mathcal{B}_0 \preceq \mathcal{B}_\sharp \preceq \mathcal{B}_\flat^0 = \widehat{B}_0(n_0, \widehat{k}_0, \mathcal{B}_0)$, and so $k_\sharp = k_0 \geq m_0 = m_\flat$ and for every $l \in [k_\sharp, \widehat{l}_0(k_\sharp)] = [k_\sharp, \widehat{l}_\flat(k_\sharp, r_\sharp)]$, $\mu(\mathfrak{D}_{E_\flat, \mathcal{B}_\sharp, l}(\mathcal{B}_\flat^1)) < 1/D_\flat$.

We also have $\mathcal{B}_\dagger = \mathcal{B}_\sharp \preceq \mathcal{B}_\flat^1 = \widehat{B}_\dagger(p_\dagger, \widehat{r}_\dagger, \mathcal{B}_\dagger)$, so $r_\sharp = r_\dagger \geq q_\dagger = q_\flat$ and for every $s \in [r_\sharp, \widehat{s}_\dagger(r_\sharp)] = [r_\sharp, \widehat{s}_\flat(k_\sharp, r_\sharp)]$ we have $\mu(\mathfrak{D}_{E_\flat, \mathcal{B}_\sharp, s}(\mathcal{B}_\flat^1)) < 1/D_\flat$. $\dashv$

The following lemma is our first approximation to the final result; it shows that we can attain some sort of bound on

$$|(\rho_m \lambda_s)(\Omega) - (\rho_l \lambda_q)(\Omega)|$$

when s, l are suitably chosen and much larger than m, q. The remainder of the argument will amount to refining the right side of the inequality to depend only on $E_\flat$.

Lemma 16. *Suppose* $(*)$ *holds. Then for every* $E_\flat$, $D_\flat^0 \leq D_\flat^1$, $p_\flat$, $n_\flat$, $\widehat{\mathbf{L}}_\flat$, $\widehat{\mathbf{S}}_\flat$ *there are* $m_\sharp \geq n_\flat$, $q_\sharp \geq p_\flat$, $\widehat{\mathbf{k}}_\sharp$, $\widehat{\mathbf{r}}_\sharp$ *so that, setting*

- $\widehat{\mathbf{l}}_\flat = \widehat{\mathbf{L}}_\flat(m_\sharp, q_\sharp, \widehat{\mathbf{k}}_\sharp, \widehat{\mathbf{r}}_\sharp)$,
- $\widehat{\mathbf{s}}_\flat = \widehat{\mathbf{S}}_\flat(m_\sharp, q_\sharp, \widehat{\mathbf{k}}_\sharp, \widehat{\mathbf{r}}_\sharp)$,
- $k_\sharp = \widehat{\mathbf{k}}_\sharp(\widehat{\mathbf{l}}_\flat, \widehat{\mathbf{s}}_\flat)$,
- $r_\sharp = \widehat{\mathbf{r}}_\sharp(\widehat{\mathbf{l}}_\flat, \widehat{\mathbf{s}}_\flat)$,
- $l_\flat = \widehat{\mathbf{l}}_\flat(k_\sharp, r_\sharp)$,
- $s_\flat = \widehat{\mathbf{s}}_\flat(k_\sharp, r_\sharp)$,

we have $k_\sharp \geq m_\sharp$, $r_\sharp \geq q_\sharp$, *and if* $l_\flat \geq k_\sharp$ *and* $s_\flat \geq r_\sharp$ *then for any* $s \in [r_\sharp, s_\flat]$ *and* $l \in [k_\sharp, l_\flat]$ *there are sets* $\mathcal{B}^-, \mathcal{B}^{0,-}$, *and* $\mathcal{B}^{1,-}$ *with* $\mu(\mathcal{B}^-) < 4/D_\flat^0$, $\mu(\mathcal{B}^{0,-}) < 2/D_\flat^1$, *and* $\mu(\mathcal{B}^{1,-}) < 2/D_\flat^1$ *so that*

$$\begin{aligned}
\left|(\rho_{m_\sharp} \lambda_s)(\Omega) - (\rho_l \lambda_{q_\sharp})(\Omega)\right| &\leq |\rho_{m_\sharp} \lambda_s|(\mathcal{B}^-) + |\rho_l \lambda_{q_\sharp}|(\mathcal{B}^-) \\
&\quad + BD_\flat^0 |\rho_{m_\sharp}|(\mathcal{B}^{0,-}) + |\rho_{m_\sharp} \lambda_s|(\mathcal{B}^{0,-}) \\
&\quad + BD_\flat^0 |\lambda_{q_\sharp}|(\mathcal{B}^{1,-}) + |\rho_l \lambda_{q_\sharp}|(\mathcal{B}^{1,-}) \\
&\quad + \frac{6}{E_\flat}.
\end{aligned}$$

Proof. By $(*)_4$, for any $m, \mathcal{B}$ there are $\mathcal{B}' \succeq \mathcal{B}$ and a $\mathcal{B}'' \subseteq \mathcal{B}'$ such that ρ_m is $BE_\flat$-constant on $\mathcal{B}' \setminus \mathcal{B}''$ and $\mu(\mathcal{B}'') < 1/D_\flat^1$. Write $\widehat{\mathbf{B}}_\star^0(\mathcal{B}, m) = \mathcal{B}'$ and $\widehat{\mathbf{B}}_\star^{0,-}(\mathcal{B}, m) = \mathcal{B}''$. Similarly, for any $q, \mathcal{B}$ there are $\mathcal{B}' \succeq \mathcal{B}$ and a $\mathcal{B}'' \subseteq \mathcal{B}'$ such that λ_r is $BE_\flat$-constant on $\mathcal{B}' \setminus \mathcal{B}''$ and $\mu(\mathcal{B}'') < 1/D_\flat^1$. Write $\widehat{\mathbf{B}}_\star^1(\mathcal{B}, q) = \mathcal{B}'$ and $\widehat{\mathbf{B}}_\star^{1,-}(\mathcal{B}, q) = \mathcal{B}''$.

We plan to use Lemma 15. We need to define $\widehat{\mathbf{B}}_0^0(n_0, p_0, \widehat{\mathbf{k}}_0, \widehat{\mathbf{r}}_0, \mathcal{B}_0)$, $\widehat{\mathbf{B}}_0^1(n_0, p_0, \widehat{\mathbf{k}}_0, \widehat{\mathbf{r}}_0, \mathcal{B}_0)$, $\widehat{\mathbf{m}}_0(n_0, p_0, \widehat{\mathbf{k}}_0, \widehat{\mathbf{r}}_0, \mathcal{B}_0)$, $\widehat{\mathbf{q}}_0(n_0, p_0, \widehat{\mathbf{k}}_0, \widehat{\mathbf{r}}_0, \mathcal{B}_0)$, $\widehat{\mathbf{L}}_0(n_0, p_0, \widehat{\mathbf{k}}_0, \widehat{\mathbf{r}}_0, \mathcal{B}_0)$, and $\widehat{\mathbf{S}}_0(n_0, p_0, \widehat{\mathbf{k}}_0, \widehat{\mathbf{r}}_0, \mathcal{B}_0)$.

On input $n_0, p_0, \widehat{\mathbf{k}}_0, \widehat{\mathbf{r}}_0, \mathcal{B}_0$, which we abbreviate $\dagger$, we proceed as follows. We plan to apply Lemma 6, so we define $\widehat{\mathbf{m}}_\dagger(m_\dagger)$.

Let input $m_\dagger$ be given; we abbreviate $\dagger, m_\dagger$ by $\ddagger$. We plan to apply Lemma 6 again, so we define $\widehat{\mathbf{q}}_\ddagger(q_\ddagger)$.

Let $q_\ddagger$ be given. We define

- $\widehat{\mathbf{k}}_{\ddagger,q_\ddagger}(\widehat{\mathbf{l}}, \widehat{\mathbf{s}}) = \widehat{\mathbf{k}}_0(m_\dagger, q_\ddagger, \widehat{\mathbf{l}}, \widehat{\mathbf{s}}, \widehat{\mathbf{B}}^0_\star(\mathcal{B}_0, m_\dagger), \widehat{\mathbf{B}}^1_\star(\mathcal{B}_0, q_\ddagger))$,
- $\widehat{\mathbf{r}}_{\ddagger,q_\ddagger}(\widehat{\mathbf{l}}, \widehat{\mathbf{s}}) = \widehat{\mathbf{r}}_0(m_\dagger, q_\ddagger, \widehat{\mathbf{l}}, \widehat{\mathbf{s}}, \widehat{\mathbf{B}}^0_\star(\mathcal{B}_0, m_\dagger), \widehat{\mathbf{B}}^1_\star(\mathcal{B}_0, q_\ddagger))$,
- $\widehat{\mathbf{l}}_{\ddagger,q_\ddagger} = \widehat{\mathbf{L}}_\flat(m_\dagger, q_\ddagger, \widehat{\mathbf{k}}_{\ddagger,q_\ddagger}, \widehat{\mathbf{r}}_{\ddagger,q_\ddagger})$,
- $\widehat{\mathbf{s}}_{\ddagger,q_\ddagger} = \widehat{\mathbf{S}}_\flat(m_\dagger, q_\ddagger, \widehat{\mathbf{k}}_{\ddagger,q_\ddagger}, \widehat{\mathbf{r}}_{\ddagger,q_\ddagger})$,
- $\widehat{\mathbf{q}}_\ddagger(q_\ddagger) = \widehat{\mathbf{s}}_{\ddagger,q_\ddagger}(\widehat{\mathbf{k}}_{\ddagger,q_\ddagger}(\widehat{\mathbf{l}}_{\ddagger,q_\ddagger}, \widehat{\mathbf{s}}_{\ddagger,q_\ddagger}), \widehat{\mathbf{r}}_{\ddagger,q_\ddagger}(\widehat{\mathbf{l}}_{\ddagger,q_\ddagger}, \widehat{\mathbf{s}}_{\ddagger,q_\ddagger}))$.

By Lemma 6 applied to $|\mathcal{B}_0|BD^0_\flat E_\flat$, $3D^0_\flat$, $\widehat{\mathbf{q}}_\ddagger$, $\max\{p_\flat, p_0\}$, we obtain $q_\ddagger \geq \max\{p_\flat, p_0\}$ and a set $\mathcal{B}^{1,=} \subseteq \mathcal{B}_0$ so that $\mu(\mathcal{B}^{1,=}) < 1/3D^0_\flat$ and for each $q, q' \in [q_\ddagger, \widehat{\mathbf{q}}_\ddagger(q_\ddagger)]$ and each $\sigma \in \mathcal{B}_0 \setminus \mathcal{B}^{1,=}$, $|\lambda_q(\sigma) - \lambda_{q'}(\sigma)| < 1/|\mathcal{B}_0|BD^0_\flat E_\flat$.

We define

- $\widehat{\mathbf{m}}_\dagger(m_\dagger) = \widehat{\mathbf{l}}_{\ddagger,q_\ddagger}(\widehat{\mathbf{k}}_{\ddagger,q_\ddagger}(\widehat{\mathbf{l}}_{\ddagger,q_\ddagger}, \widehat{\mathbf{s}}_{\ddagger,q_\ddagger}), \widehat{\mathbf{r}}_{\ddagger,q_\ddagger}(\widehat{\mathbf{l}}_{\ddagger,q_\ddagger}, \widehat{\mathbf{s}}_{\ddagger,q_\ddagger}))$.

By Lemma 6 applied to $|\mathcal{B}_0|BD^0_\flat E_\flat$, $3D^0_\flat$, $\widehat{\mathbf{m}}_\dagger$, $\max\{n_\flat, n_0\}$, we obtain $m_\dagger \geq \max\{n_\flat, n_0\}$ and a set $\mathcal{B}^{0,=} \subseteq \mathcal{B}_0$ so that $\mu(\mathcal{B}^{0,=}) < 1/3D^0_\flat$ and for each $m, m' \in [m_\dagger, \widehat{\mathbf{m}}_\dagger(m_\dagger)]$ and each $\sigma \in \mathcal{B}_0 \setminus \mathcal{B}^{0,=}$, $|\rho_m(\sigma) - \rho_{m'}(\sigma)| < 1/|\mathcal{B}_0|BD^0_\flat E_\flat$.

We can now set

- $\widehat{\mathbf{B}}^0_0(n_0, p_0, \widehat{\mathbf{k}}_0, \widehat{\mathbf{r}}_0, \mathcal{B}_0) = \widehat{\mathbf{B}}^0_\star(\mathcal{B}_0, m_\dagger)$,
- $\widehat{\mathbf{B}}^1_0(n_0, p_0, \widehat{\mathbf{k}}_0, \widehat{\mathbf{r}}_0, \mathcal{B}_0) = \widehat{\mathbf{B}}^1_\star(\mathcal{B}_0, q_\ddagger)$,
- $\widehat{\mathbf{q}}_0(n_0, p_0, \widehat{\mathbf{k}}_0, \widehat{\mathbf{r}}_0, \mathcal{B}_0) = q_\ddagger$,
- $\widehat{\mathbf{m}}_0(n_0, p_0, \widehat{\mathbf{k}}_0, \widehat{\mathbf{r}}_0, \mathcal{B}_0) = m_\dagger$,
- $\widehat{\mathbf{L}}_0(n_0, p_0, \widehat{\mathbf{k}}_0, \widehat{\mathbf{r}}_0, \mathcal{B}_0) = \widehat{\mathbf{l}}_{\ddagger,q_\ddagger}$,
- $\widehat{\mathbf{S}}_0(n_0, p_0, \widehat{\mathbf{k}}_0, \widehat{\mathbf{r}}_0, \mathcal{B}_0) = \widehat{\mathbf{s}}_{\ddagger,q_\ddagger}$.

We apply Lemma 15 to $BE_\flat$, $D^1_\flat$, $\widehat{\mathbf{B}}^0_0$, $\widehat{\mathbf{B}}^1_0$, $\widehat{\mathbf{m}}_0$, $\widehat{\mathbf{L}}_0$, $\widehat{\mathbf{q}}_0$, and $\widehat{\mathbf{S}}_0$. We obtain $n_0, p_0, \mathcal{B}_0, \widehat{\mathbf{k}}_0, \widehat{\mathbf{r}}_0$. We set $m_\sharp = m_\dagger$, $q_\sharp = q_\ddagger$, $\widehat{\mathbf{k}}_\sharp = \widehat{\mathbf{k}}_{\ddagger,q_\ddagger}$, and $\widehat{\mathbf{r}}_\sharp = \widehat{\mathbf{r}}_{\ddagger,q_\ddagger}$.

Set

- $\widehat{\mathbf{l}}_\flat = \widehat{\mathbf{L}}_\flat(m_\sharp, q_\sharp, \widehat{\mathbf{k}}_\sharp, \widehat{\mathbf{r}}_\sharp)$,
- $\widehat{\mathbf{s}}_\flat = \widehat{\mathbf{S}}_\flat(m_\sharp, q_\sharp, \widehat{\mathbf{k}}_\sharp, \widehat{\mathbf{r}}_\sharp)$,
- $k_\sharp = \widehat{\mathbf{k}}_\sharp(\widehat{\mathbf{l}}_\flat, \widehat{\mathbf{s}}_\flat)$, and
- $r_\sharp = \widehat{\mathbf{r}}_\sharp(\widehat{\mathbf{l}}_\flat, \widehat{\mathbf{s}}_\flat)$.

Note that $k_\sharp \geq m_\dagger = m_\sharp$ and $r_\sharp \geq q_\dagger = q_\sharp$. Suppose $\widehat{\mathbf{l}}_\flat(k_\sharp, r_\sharp) \geq k_\sharp$ and

$\widehat{s}_\flat(k_\sharp, r_\sharp) \geq r_\sharp$ and let $s \in [r_\sharp, \widehat{s}_\flat(k_\sharp, r_\sharp)]$ and $l \in [k_\sharp, \widehat{1}_\flat(k_\sharp, r_\sharp)]$ be given.

Observe that $\widehat{1}_\flat = \widehat{L}_\flat(m_\dagger, q_\ddagger, \widehat{k}_{\ddagger,q_\ddagger}, \widehat{r}_{\ddagger,q_\ddagger}) = \widehat{1}_{\ddagger,q_\ddagger}$ and similarly, $\widehat{s}_\flat = \widehat{s}_{\ddagger,q_\ddagger}$. Therefore

$$r_\sharp = \widehat{r}_{\ddagger,q_\ddagger}(\widehat{1}_\flat, \widehat{s}_\flat) = \widehat{r}_0(m_\sharp, q_\sharp, \widehat{1}_{\ddagger,q_\ddagger}, \widehat{s}_{\ddagger,q_\ddagger}, \widehat{B}^0_\star(\mathcal{B}_0, m_\dagger), \widehat{B}^1_\star(\mathcal{B}_0, q_\ddagger))$$

and similarly, for $k_\sharp$. Therefore, by our application of Lemma 15, since $m_\sharp \geq n_0$ and $q_\sharp \geq p_0$, we have $k_\sharp \geq m_\sharp$, $r_\sharp \geq q_\sharp$, for every $l \in [k_\sharp, \widehat{1}_\flat(k_\sharp, r_\sharp)]$, $\mu(\mathfrak{D}_{BE_\flat,\mathcal{B}_0,l}(\widehat{B}^0_\star(\mathcal{B}_0, m_\dagger))) < 1/D^1_\flat$, and for every $s \in [r_\sharp, \widehat{s}_\flat(k_\sharp, r_\sharp)]$,

$$\mu(\mathfrak{D}_{BE_\flat,\mathcal{B}_0,s}(\widehat{B}^1_\star(\mathcal{B}_0, q_\ddagger))) < 1/D^1_\flat.$$

Fix some $l \in [k_\sharp, \widehat{1}_\flat(k_\sharp, r_\sharp)]$, $s \in [r_\sharp, \widehat{s}_\flat(k_\sharp, r_\sharp)]$. Set $\mathcal{B}^0_\flat = \widehat{B}^0_\star(\mathcal{B}_0, m_\dagger)$, $\mathcal{B}^1_\flat = \widehat{B}^1_\star(\mathcal{B}_0, q_\ddagger)$, $\mathcal{B}^{0,-} = \mathfrak{D}_{BE_\flat,\mathcal{B}_0,l}(\mathcal{B}^0_\flat) \cup \widehat{B}^{0,-}_\star(\mathcal{B}_0, m_\dagger)$, and $\mathcal{B}^{1,-} = \mathfrak{D}_{BE_\flat,\mathcal{B}_0,s}(\mathcal{B}^1_\flat) \cup \widehat{B}^{1,-}_\star(\mathcal{B}_0, q_\ddagger)$. Therefore $\mu(\mathcal{B}^{0,-}), \mu(\mathcal{B}^{1,-}) < 2/D^1_\flat$.

By Lemma 14,

$$\sum_{\sigma \in (\mathcal{B}_0)_{\lambda_s \leq BD^0_\flat}} |(\rho_{m_\sharp} \lambda_s)(\sigma) - (\rho_{m_\sharp} * \lambda_s)(\sigma)| < \frac{2}{E_\flat} + BD^0_\flat |\rho_{m_\sharp}|(\mathcal{B}^{0,-})$$

$$+ |\rho_{m_\sharp} \lambda_s|(\mathcal{B}^{0,-})$$

and

$$\sum_{\sigma \in (\mathcal{B}_0)_{\rho_l \leq BD^0_\flat}} |(\rho_l \lambda_{q_\sharp})(\sigma) - (\rho_l * \lambda_{q_\sharp})(\sigma)| < \frac{2}{E_\flat} + BD^0_\flat |\lambda_{q_\sharp}|(\mathcal{B}^{1,-})$$

$$+ |\rho_l \lambda_{q_\sharp}|(\mathcal{B}^{1,-}).$$

If $\sigma \in (\mathcal{B}_0)_{\lambda_s \leq BD^0_\flat, \rho_l \leq BD^0_\flat} \setminus (\mathcal{B}^{0,-} \cup \mathcal{B}^{1,-})$ then we have

$$\begin{aligned}
|(\rho_{m_\sharp} * \lambda_s)(\sigma) - (\rho_l * \lambda_{q_\sharp})(\sigma)| &\leq |(\rho_{m_\sharp} * \lambda_s)(\sigma) - (\rho_l * \lambda_s)(\sigma)| \\
&\quad + |(\rho_l * \lambda_s)(\sigma) - (\rho_l * \lambda_{q_\sharp})(\sigma)| \\
&\leq |\delta_{\lambda_s}(\sigma)| \cdot |\rho_{m_\sharp}(\sigma) - \rho_l(\sigma)| \\
&\quad + |\delta_{\rho_l}(\sigma)| \cdot |\lambda_s(\sigma) - \lambda_{q_\sharp}(\sigma)| \\
&\leq BD^0_\flat \cdot \frac{1}{|\mathcal{B}_0|BD^0_\flat E_\flat} \\
&\quad + BD^0_\flat \cdot \frac{1}{|\mathcal{B}_0|BD^0_\flat E_\flat} \\
&= \frac{2}{|\mathcal{B}_0|E_\flat}.
\end{aligned}$$

Let $\mathcal{B}^- = \mathcal{B}^{0,-} \cup \mathcal{B}^{1,-} \cup (\mathcal{B}_0)_{\rho_l > BD^0_\flat} \cup (\mathcal{B}_0)_{\lambda_s > BD^0_\flat}$ and $\mathcal{B}^+ = \mathcal{B}_0 \setminus \mathcal{B}^-$.

We have

$$
\begin{aligned}
|(\rho_{m_\sharp}\lambda_s)(\Omega) - (\rho_l\lambda_{q_\sharp})(\Omega)| &= |(\rho_{m_\sharp}\lambda_s)(\mathcal{B}_0) - (\rho_l\lambda_{q_\sharp})(\mathcal{B}_0)| \\
&\leq |(\rho_{m_\sharp}\lambda_s)(\mathcal{B}^+) - (\rho_l\lambda_{q_\sharp})(\mathcal{B}^+)| \\
&\quad + |\rho_{m_\sharp}\lambda_s|(\mathcal{B}^-) + |\rho_l\lambda_{q_\sharp}|(\mathcal{B}^-) \\
&< |(\rho_{m_\sharp} * \lambda_s)(\mathcal{B}^+) - (\rho_l * \lambda_{q_\sharp})(\mathcal{B}^+)| \\
&\quad + |\rho_{m_\sharp}\lambda_s|(\mathcal{B}^-) + |\rho_l\lambda_{q_\sharp}|(\mathcal{B}^-) \\
&\quad + BD_\flat^0|\rho_{m_\sharp}|(\mathcal{B}^{0,-}) + |\rho_{m_\sharp}\lambda_s|(\mathcal{B}^{0,-}) \\
&\quad + BD_\flat^0|\lambda_{q_\sharp}|(\mathcal{B}^{1,-}) + |\rho_l\lambda_{q_\sharp}|(\mathcal{B}^{1,-}) \\
&\quad + \frac{4}{E_\flat} \\
&\leq |\rho_{m_\sharp}\lambda_s|(\mathcal{B}^-) + |\rho_l\lambda_{q_\sharp}|(\mathcal{B}^-) \\
&\quad + BD_\flat^0|\rho_{m_\sharp}|(\mathcal{B}^{0,-}) + |\rho_{m_\sharp}\lambda_s|(\mathcal{B}^{0,-}) \\
&\quad + BD_\flat^0|\lambda_{q_\sharp}|(\mathcal{B}^{1,-}) + |\rho_l\lambda_{q_\sharp}|(\mathcal{B}^{1,-}) \\
&\quad + \frac{6}{E_\flat}
\end{aligned}
$$

$\dashv$

What remains is a series of lemma in which we choose $D_\flat^0, D_\flat^1$ large enough to bound the various terms.

Lemma 17. *Suppose* $(*)$ *holds. Then for every* $E_\flat, D_\flat^0 \leq D_\flat^1, p_\flat, n_\flat,$ $\widehat{\mathbf{L}}_\flat, \widehat{\mathbf{S}}_\flat$ *there are* $m_\sharp \geq n_\flat,$ $q_\sharp \geq p_\flat,$ $\widehat{\mathbf{k}}_\sharp, \widehat{\mathbf{r}}_\sharp$ *so that, setting*

- $\widehat{\mathbf{l}}_\flat = \widehat{\mathbf{L}}_\flat(m_\sharp, q_\sharp, \widehat{\mathbf{k}}_\sharp, \widehat{\mathbf{r}}_\sharp),$

- $\widehat{\mathbf{s}}_\flat = \widehat{\mathbf{S}}_\flat(m_\sharp, q_\sharp, \widehat{\mathbf{k}}_\sharp, \widehat{\mathbf{r}}_\sharp),$

- $k_\sharp = \widehat{\mathbf{k}}_\sharp(\widehat{\mathbf{l}}_\flat, \widehat{\mathbf{s}}_\flat),$

- $r_\sharp = \widehat{\mathbf{r}}_\sharp(\widehat{\mathbf{l}}_\flat, \widehat{\mathbf{s}}_\flat),$

- $l_\flat = \widehat{\mathbf{l}}_\flat(k_\sharp, r_\sharp),$

- $s_\flat = \widehat{\mathbf{s}}_\flat(k_\sharp, r_\sharp),$

we have $k_\sharp \geq m_\sharp,$ $r_\sharp \geq q_\sharp,$ *and if* $l_\flat \geq k_\sharp$ *and* $s_\flat \geq r_\sharp$ *then for any* $s \in [r_\sharp, s_\flat]$ *and* $l \in [k_\sharp, l_\flat]$ *there are sets* $\mathcal{B}^-, \mathcal{B}^{0,-},$ *and* $\mathcal{B}^{1,-}$ *with* $\mu(\mathcal{B}^-) <$

$4/D_\flat^0$, $\mu(\mathcal{B}^{0,-}) < 2/D_\flat^0$, and $\mu(\mathcal{B}^{1,-}) < 2/D_\flat^1$ so that

$$\left|(\rho_{m_\sharp}\lambda_s)(\Omega) - (\rho_l\lambda_{q_\sharp})(\Omega)\right| \leq |\rho_{m_\sharp}\lambda_s|(\mathcal{B}^-) + |\rho_l\lambda_{q_\sharp}|(\mathcal{B}^-)$$
$$+ |\rho_{m_\sharp}\lambda_s|(\mathcal{B}^{0,-}) + BD_\flat^0|\lambda_{q_\sharp}|(\mathcal{B}^{1,-})$$
$$+ |\rho_l\lambda_{q_\sharp}|(\mathcal{B}^{1,-}) + \frac{7}{E_\flat}.$$

Proof. Towards the use of uniform continuity, we define $\widehat{\mathbf{m}}_0(D_0, m_0)$.

Let D_0, m_0, which we abbreviate †, be given. By Lemma 16 applied to $E_\flat, D_\flat^0, \max\{D_0, D_\flat^1\}, p_\flat, m_0, \widehat{\mathbf{L}}_\flat, \widehat{\mathbf{S}}_\flat$ we obtain $m_\dagger, q_\dagger, \widehat{\mathbf{k}}_\dagger, \widehat{\mathbf{r}}_\dagger$. We define

$$\widehat{\mathbf{m}}_0(D_0, m_0) = m_\dagger.$$

By Theorem 4 applied to $4BD_\flat^0 E_\flat, \widehat{\mathbf{m}}_0, n_\flat$ we obtain $D_0, m_0 \geq n_\flat$ so that whenever $m \in [m_0, \widehat{\mathbf{m}}_0(D_0, m_0)]$ and $\mu(\sigma) < 1/D_0$, $|\rho_m(\sigma)| < 1/4BD_\flat^0 E_\flat$.

We set $m_\sharp = m_\dagger$, $q_\sharp = q_\dagger$, $\widehat{\mathbf{k}}_\sharp = \widehat{\mathbf{k}}_\dagger$, and $\widehat{\mathbf{r}}_\sharp = \widehat{\mathbf{r}}_\dagger$. Let $\widehat{\mathbf{l}}_\flat, \widehat{\mathbf{s}}_\flat, k_\sharp, r_\sharp$ be as in the statement. Let $s \in [r_\sharp, \widehat{\mathbf{s}}_\flat(k_\sharp, r_\sharp)]$ and $l \in [k_\sharp, \widehat{\mathbf{l}}_\flat(k_\sharp, r_\sharp)]$ be given, so for appropriate $\mathcal{B}^-, \mathcal{B}^{0,-}, \mathcal{B}^{1,-}$,

$$\left|(\rho_{m_\sharp}\lambda_s)(\Omega) - (\rho_l\lambda_{q_\sharp})(\Omega)\right| \leq |\rho_{m_\sharp}\lambda_s|(\mathcal{B}^-) + |\rho_l\lambda_{q_\sharp}|(\mathcal{B}^-)$$
$$+ BD_\flat^0|\rho_{m_\sharp}|(\mathcal{B}^{0,-}) + |\rho_{m_\sharp}\lambda_s|(\mathcal{B}^{0,-})$$
$$+ BD_\flat^0|\lambda_{q_\sharp}|(\mathcal{B}^{1,-}) + |\rho_l\lambda_{q_\sharp}|(\mathcal{B}^{1,-}) + \frac{6}{E_\flat}.$$

Since $\mu(\mathcal{B}^{0,-}) < 2/D_0$, $|\rho_{m_\sharp}(\mathcal{B}^{0,-})| < 1/BD_\flat^0 E_\flat$, and therefore

$$\left|(\rho_{m_\sharp}\lambda_s)(\Omega) - (\rho_l\lambda_{q_\sharp})(\Omega)\right| \leq |\rho_{m_\sharp}\lambda_s|(\mathcal{B}^-) + |\rho_l\lambda_{q_\sharp}|(\mathcal{B}^-)$$
$$+ |\rho_{m_\sharp}\lambda_s|(\mathcal{B}^{0,-}) + BD_\flat^0|\lambda_{q_\sharp}|(\mathcal{B}^{1,-})$$
$$+ |\rho_l\lambda_{q_\sharp}|(\mathcal{B}^{1,-}) + \frac{7}{E_\flat}$$

as desired. ⊣

Lemma 18. *Suppose* (∗) *holds. Then for every* $E_\flat, D_\flat, p_\flat, n_\flat, \widehat{\mathbf{L}}_\flat, \widehat{\mathbf{S}}_\flat$ *there are* $m_\sharp \geq n_\flat$, $q_\sharp \geq p_\flat$, $\widehat{\mathbf{k}}_\sharp, \widehat{\mathbf{r}}_\sharp$ *so that, setting*

- $\widehat{\mathbf{l}}_\flat = \widehat{\mathbf{L}}_\flat(m_\sharp, q_\sharp, \widehat{\mathbf{k}}_\sharp, \widehat{\mathbf{r}}_\sharp)$,
- $\widehat{\mathbf{s}}_\flat = \widehat{\mathbf{S}}_\flat(m_\sharp, q_\sharp, \widehat{\mathbf{k}}_\sharp, \widehat{\mathbf{r}}_\sharp)$,
- $k_\sharp = \widehat{\mathbf{k}}_\sharp(\widehat{\mathbf{l}}_\flat, \widehat{\mathbf{s}}_\flat)$,

- $r_\sharp = \widehat{\mathbf{r}}_\sharp(\widehat{\mathbf{l}}_\flat, \widehat{\mathbf{s}}_\flat)$,
- $l_\flat = \widehat{\mathbf{l}}_\flat(k_\sharp, r_\sharp)$,
- $s_\flat = \widehat{\mathbf{s}}_\flat(k_\sharp, r_\sharp)$,

we have $k_\sharp \geq m_\sharp$, $r_\sharp \geq q_\sharp$, and if $l_\flat \geq k_\sharp$ and $s_\flat \geq r_\sharp$ then for any $s \in [r_\sharp, s_\flat]$ and $l \in [k_\sharp, l_\flat]$ there are sets $\mathcal{B}^-, \mathcal{B}^{0,-}$, and $\mathcal{B}^{1,-}$ with $\mu(\mathcal{B}^-) < 4/D_\flat$, $\mu(\mathcal{B}^{0,-}) < 2/D_\flat$, and $\mu(\mathcal{B}^{1,-}) < 2/D_\flat$ so that

$$\left| (\rho_{m_\sharp} \lambda_s)(\Omega) - (\rho_l \lambda_{q_\sharp})(\Omega) \right| \leq |\rho_{m_\sharp} \lambda_s|(\mathcal{B}^-) + |\rho_l \lambda_{q_\sharp}|(\mathcal{B}^-)$$
$$+ |\rho_{m_\sharp} \lambda_s|(\mathcal{B}^{0,-}) + |\rho_l \lambda_{q_\sharp}|(\mathcal{B}^{1,-}) + \frac{8}{E_\flat}.$$

Proof. Towards the use of uniform continuity, we define $\widehat{\mathbf{q}}_0(D_0, q_0)$.

Let D_0, q_0, which we abbreviate †, be given.

By Lemma 17 applied to $E_\flat, D_\flat, \max\{D_\flat, D_0\}, n_\flat, q_0, \widehat{\mathbf{L}}_\flat, \widehat{\mathbf{S}}_\flat$ we obtain $m_\dagger, q_\dagger, \widehat{\mathbf{k}}_\dagger, \widehat{\mathbf{r}}_\dagger$. We define

$$\widehat{\mathbf{q}}_0(D_0, q_0) = q_\dagger.$$

By Theorem 4 applied to $4BD_\flat E_\flat, \widehat{\mathbf{q}}_0, p_\flat$ we obtain $D_0 \geq D_\flat, q_0 \geq p_\flat$.

We set $m_\sharp = m_\dagger$, $q_\sharp = q_\dagger$, $\widehat{\mathbf{k}}_\sharp = \widehat{\mathbf{k}}_\dagger$, and $\widehat{\mathbf{r}}_\sharp = \widehat{\mathbf{r}}_\dagger$, and let $\widehat{\mathbf{l}}_\flat, \widehat{\mathbf{s}}_\flat, k_\sharp, r_\sharp$ be as in the statement, so also $\widehat{\mathbf{l}}_\flat = \widehat{\mathbf{l}}_\dagger$, $\widehat{\mathbf{s}}_\flat = \widehat{\mathbf{s}}_\dagger$, $k_\sharp = k_\dagger$, and $r_\sharp = r_\dagger$. Let $s \in [r_\sharp, \widehat{\mathbf{s}}_\flat(k_\sharp, r_\sharp)]$ and $l \in [k_\sharp, \widehat{\mathbf{l}}_\flat(k_\sharp, r_\sharp)]$ be given, so for appropriate $\mathcal{B}^-, \mathcal{B}^{0,-}, \mathcal{B}^{1,-}$,

$$\left| (\rho_{m_\sharp} \lambda_s)(\Omega) - (\rho_l \lambda_{q_\sharp})(\Omega) \right| \leq |\rho_{m_\sharp} \lambda_s|(\mathcal{B}^-) + |\rho_l \lambda_{q_\sharp}|(\mathcal{B}^-)$$
$$+ |\rho_{m_\sharp} \lambda_s|(\mathcal{B}^{0,-}) + BD_\flat |\lambda_{q_\sharp}|(\mathcal{B}^{1,-})$$
$$+ |\rho_l \lambda_{q_\sharp}|(\mathcal{B}^{1,-}) + \frac{7}{E_\flat}.$$

Since $\mu(\mathcal{B}^{1,-}) < 2/D_\flat$, we have $|\lambda_{q_\sharp}(\mathcal{B}^{1,-})| < 1/BD_\flat E_\flat$, and therefore

$$\left| (\rho_{m_\sharp} \lambda_s)(\Omega) - (\rho_l \lambda_{q_\sharp})(\Omega) \right| \leq |\rho_{m_\sharp} \lambda_s|(\mathcal{B}^-) + |\rho_l \lambda_{q_\sharp}|(\mathcal{B}^-)$$
$$+ |\rho_{m_\sharp} \lambda_s|(\mathcal{B}^{0,-}) + |\rho_l \lambda_{q_\sharp}|(\mathcal{B}^{1,-}) + \frac{8}{E_\flat}$$

as desired. ⊣

Lemma 19. *Suppose* $(*)$ *holds. Then for every* $E_\flat, D_\flat, p_\flat, n_\flat, \widehat{\mathbf{L}}_\flat, \widehat{\mathbf{r}}_\flat$ *there are* $m_\sharp \geq n_\flat$, $q_\sharp \geq p_\flat$, $\widehat{\mathbf{k}}_\sharp, \widehat{\mathbf{s}}_\sharp$ *so that, setting*

- $\widehat{\mathbf{l}}_\flat = \widehat{\mathbf{L}}_\flat(m_\sharp, q_\sharp, \widehat{\mathbf{k}}_\sharp, \widehat{\mathbf{s}}_\sharp)$,
- $r_\flat = \widehat{\mathbf{r}}_\flat(m_\sharp, q_\sharp, \widehat{\mathbf{k}}_\sharp, \widehat{\mathbf{s}}_\sharp)$,
- $k_\sharp = \widehat{\mathbf{k}}_\sharp(\widehat{\mathbf{l}}_\flat, r_\flat)$,
- $s_\sharp = \widehat{\mathbf{s}}_\sharp(\widehat{\mathbf{l}}_\flat, r_\flat)$,
- $l_\flat = \widehat{\mathbf{l}}_\flat(k_\sharp, s_\sharp)$,

we have $k_\sharp \geq m_\sharp$, if $r_\flat \geq q_\sharp$ then $s_\sharp \geq r_\flat$, and for any $l \in [k_\sharp, l_\flat]$ there are sets $\mathcal{B}^-, \mathcal{B}^{0,-}$, and $\mathcal{B}^{1,-}$ with $\mu(\mathcal{B}^-) < 4/D_\flat$, $\mu(\mathcal{B}^{0,-}) < 2/D_\flat$, and $\mu(\mathcal{B}^{1,-}) < 2/D_\flat$ so that

$$\left| (\rho_{m_\sharp} \lambda_{s_\sharp})(\Omega) - (\rho_l \lambda_{q_\sharp})(\Omega) \right| \leq |\rho_l \lambda_{q_\sharp}|(\mathcal{B}^-) + |\rho_l \lambda_{q_\sharp}|(\mathcal{B}^{1,-}) + \frac{20}{E_\flat}.$$

Proof. Towards the use of $n_{/p}$-metastable uniform continuity, we define functions $\widehat{\mathbf{m}}_0(D_0, m_0, p_0, \widehat{\mathbf{r}}_0)$ and $\widehat{\mathbf{q}}_0(D_0, m_0, p_0, \widehat{\mathbf{r}}_0)$.

Let $D_0, m_0, p_0, \widehat{\mathbf{r}}_0$, which we abbreviate †, be given. Without loss of generality, we assume $D_0 \geq D_\flat$. In order to apply Lemma 18 we define

- $\widehat{\mathbf{s}}_{m_\dagger, r}(k, r') = \widehat{\mathbf{r}}_0(m_\dagger, \max\{r, r', p_0\})$,
- $\widehat{\mathbf{l}}_{m_\dagger, \widehat{\mathbf{l}}, r}(k, r') = \widehat{\mathbf{l}}(k, \widehat{\mathbf{r}}_0(m_\dagger, \max\{r, r', p_0\}))$,
- $\widehat{\mathbf{k}}_{m_\dagger, \widehat{\mathbf{k}}_\dagger}(\widehat{\mathbf{l}}, r) = \widehat{\mathbf{k}}_\dagger(\widehat{\mathbf{l}}_{m_\dagger, \widehat{\mathbf{l}}, r}, \widehat{\mathbf{s}}_{m_\dagger, r})$,
- $\widehat{\mathbf{s}}_{m_\dagger, \widehat{\mathbf{r}}_\dagger}(\widehat{\mathbf{l}}, r) = \widehat{\mathbf{r}}_0(m_\dagger, \max\{r, \widehat{\mathbf{r}}_\dagger(\widehat{\mathbf{l}}_{m_\dagger, \widehat{\mathbf{l}}, r}, \widehat{\mathbf{s}}_{m_\dagger, r}), p_0\})$,
- $\widehat{\mathbf{L}}_\dagger(m_\dagger, q_\dagger, \widehat{\mathbf{k}}_\dagger, \widehat{\mathbf{r}}_\dagger) = \widehat{\mathbf{l}}_{m_\dagger, \widehat{\mathbf{L}}_\flat(m_\dagger, q_\dagger, \widehat{\mathbf{k}}_{m_\dagger, \widehat{\mathbf{k}}_\dagger}, \widehat{\mathbf{s}}_{m_\dagger, r}), \widehat{\mathbf{r}}_\flat(m_\dagger, q_\dagger, \widehat{\mathbf{k}}_{m_\dagger, \widehat{\mathbf{k}}_\dagger}, \widehat{\mathbf{s}}_{m_\dagger, \widehat{\mathbf{r}}_\dagger})}$
- $\widehat{\mathbf{S}}_\dagger(m_\dagger, q_\dagger, \widehat{\mathbf{k}}_\dagger, \widehat{\mathbf{r}}_\dagger) = \widehat{\mathbf{s}}_{m_\dagger, \widehat{\mathbf{r}}_\flat(m_\dagger, q_\dagger, \widehat{\mathbf{k}}_{m_\dagger, \widehat{\mathbf{k}}_\dagger}, \widehat{\mathbf{s}}_{m_\dagger, \widehat{\mathbf{r}}_\dagger})}$.

We apply Lemma 18 to $E_\flat, D_0, \max\{p_\flat, p_0\}, n_\flat, \widehat{\mathbf{L}}_\dagger, \widehat{\mathbf{S}}_\dagger$ in order to obtain $m_\dagger, q_\dagger, \widehat{\mathbf{k}}_\dagger, \widehat{\mathbf{r}}_\dagger$. Set $\widehat{\mathbf{l}}_\dagger = \widehat{\mathbf{L}}_\dagger(m_\dagger, q_\dagger, \widehat{\mathbf{k}}_\dagger, \widehat{\mathbf{r}}_\dagger)$, $\widehat{\mathbf{s}}_\dagger = \widehat{\mathbf{S}}_\dagger(m_\dagger, q_\dagger, \widehat{\mathbf{k}}_\dagger, \widehat{\mathbf{r}}_\dagger)$, $k_\dagger = \widehat{\mathbf{k}}_\dagger(\widehat{\mathbf{l}}_\dagger, \widehat{\mathbf{s}}_\dagger)$, and $r_\dagger = \widehat{\mathbf{r}}_\dagger(\widehat{\mathbf{l}}_\dagger, \widehat{\mathbf{s}}_\dagger)$.

We set

- $\widehat{\mathbf{m}}_0(D_0, m_0, p_0, \widehat{\mathbf{r}}_0) = m_\dagger$, and
- $\widehat{\mathbf{q}}_0(D_0, m_0, p_0, \widehat{\mathbf{r}}_0) = \max\{\widehat{\mathbf{r}}_\flat(m_\dagger, q_\dagger, \widehat{\mathbf{k}}_{m_\dagger, \widehat{\mathbf{k}}_\dagger}, \widehat{\mathbf{s}}_{m_\dagger, \widehat{\mathbf{r}}_\dagger}), r_\dagger, p_0\}$.

By $n_{/p}$-metastable uniform continuity we obtain $D_0, m_0, p_0, \widehat{\mathbf{r}}_0$. Set $q_0 = \widehat{\mathbf{q}}_0(D_0, m_0, p_0, \widehat{\mathbf{r}}_0)$ and $r_0 = \widehat{\mathbf{r}}_0(m_\dagger, q_0)$. Since $q_0 \geq p_0$, we have $r_0 \geq q_0$ and when $\mu(\mathcal{B}) < 1/D_0$, $|\rho_{m_\dagger} \lambda_{r_0}|(\mathcal{B}) < 1/E_\flat$.

We set $m_\sharp = m_\dagger$, $q_\sharp = q_\dagger$, $\widehat{\mathbf{k}}_\sharp = \widehat{\mathbf{k}}_{m_\dagger, \widehat{\mathbf{k}}_\dagger}$, and $\widehat{\mathbf{s}}_\sharp = \widehat{\mathbf{s}}_{m_\dagger, \widehat{\mathbf{r}}_\dagger}$. Set $r_\flat = \widehat{\mathbf{r}}_\flat(m_\sharp, q_\sharp, \widehat{\mathbf{k}}_\sharp, \widehat{\mathbf{s}}_\sharp)$, $\widehat{\mathbf{l}}_\flat = \widehat{\mathbf{L}}_\flat(m_\sharp, q_\sharp, \widehat{\mathbf{k}}_\sharp, \widehat{\mathbf{s}}_\sharp)$, $s_\sharp = \widehat{\mathbf{s}}_\sharp(\widehat{\mathbf{l}}_\flat, r_\flat)$, $k_\sharp = \widehat{\mathbf{k}}_\sharp(\widehat{\mathbf{l}}_\flat, r_\flat)$, and $l_\flat = \widehat{\mathbf{l}}_\flat(k_\sharp, s_\sharp)$.

Observe that

- $r_\flat = \widehat{r}_\flat(m_\dagger, q_\dagger, \widehat{k}_{m_\dagger, \widehat{k}_\dagger}, \widehat{s}_{m_\dagger, \widehat{r}_\dagger})$,
- $\widehat{l}_\dagger = \widehat{l}_{m_\dagger, \widehat{l}_\flat, r_\flat}$,
- $\widehat{s}_\dagger = \widehat{s}_{m_\dagger, r_\flat}$,
- $r_\dagger = \widehat{r}_\dagger(\widehat{l}_\dagger, \widehat{s}_\dagger) = \widehat{r}_\dagger(\widehat{l}_{m_\dagger, \widehat{l}_\flat, r_\flat}, \widehat{s}_{m_\dagger, r_\flat})$,
- $q_0 = \max\{r_\flat, r_\dagger, p_0\}$,
- $s_\sharp = \widehat{s}_{m_\dagger, \widehat{r}_\dagger}(\widehat{l}_\flat, r_\flat) = \widehat{r}_0(m_\dagger, \max\{r_\flat, r_\dagger, p_0\}) = \widehat{r}_0(m_\dagger, q_0)$
  $= \widehat{s}_{m_\dagger, r_\flat}(k_\dagger, r_\dagger) = \widehat{s}_\dagger(k_\dagger, r_\dagger)$,
- $k_\sharp = \widehat{k}_{m_\dagger, \widehat{k}_\dagger}(\widehat{l}_\flat, r_\flat) = \widehat{k}_\dagger(\widehat{l}_\dagger, \widehat{s}_\dagger) = k_\dagger$,
- $l_\flat = \widehat{l}_\flat(k_\sharp, s_\sharp) = \widehat{l}_\flat(k_\dagger, \widehat{r}_0(m_\dagger, q_0)) = \widehat{l}_{m_\dagger, \widehat{l}_\flat, r_\flat}(k_\dagger, r_\dagger) = \widehat{l}_\dagger(k_\dagger, r_\dagger)$.

We have $k_\sharp \geq m_\sharp$. Suppose $r_\flat \geq q_\sharp$; then $s_\sharp = \widehat{r}_0(m_\dagger, q_0) \geq q_0 \geq r_\flat$. Let $l \in [k_\sharp, l_\flat]$ be given. Since $l \in [k_\dagger, \widehat{l}_\dagger(k_\dagger, r_\dagger)]$ and $s_\sharp \in [r_\dagger, \widehat{s}_\dagger(k_\dagger, r_\dagger)]$, there are sets $\mathcal{B}^-, \mathcal{B}^{0,-}, \mathcal{B}^{1,-}$ with $\mu(\mathcal{B}^-) < 4/D_0$, $\mu(\mathcal{B}^{0,-}) < 2/D_0$, and $\mu(\mathcal{B}^{1,-}) < 2/D_0$ so that

$$\left|(\rho_{m_\sharp}\lambda_{s_\sharp})(\Omega) - (\rho_l \lambda_{q_\sharp})(\Omega)\right| \leq |\rho_{m_\sharp}\lambda_{s_\sharp}|(\mathcal{B}^-) + |\rho_l \lambda_{q_\sharp}|(\mathcal{B}^-)$$
$$+ |\rho_{m_\sharp}\lambda_{s_\sharp}|(\mathcal{B}^{0,-}) + |\rho_l \lambda_{q_\sharp}|(\mathcal{B}^{1,-}) + \frac{8}{E_\flat}.$$

Since $m_\sharp \in [m_0, \widehat{m}_0(D_0, m_0, p_0, \widehat{r}_0)]$, $s_\sharp = \widehat{r}_0(m_\dagger, q_0)$, $\mu(\mathcal{B}^-) < 4/D_0$ and $\mu(\mathcal{B}^{0,-}) < 2/D_0$, we have $|\rho_{m_\sharp}\lambda_{s_\sharp}|(\mathcal{B}^-) + |\rho_{m_\sharp}\lambda_{s_\sharp}|(\mathcal{B}^{0,-}) < 12/E_\flat$, so

$$\left|(\rho_{m_\sharp}\lambda_{s_\sharp})(\Omega) - (\rho_l \lambda_{q_\sharp})(\Omega)\right| \leq |\rho_l \lambda_{q_\sharp}|(\mathcal{B}^-) + |\rho_l \lambda_{q_\sharp}|(\mathcal{B}^{1,-}) + \frac{20}{E_\flat}.$$

$\dashv$

One more application of the same technique eliminates the last extraneous term in the bound, giving the desired result.

Theorem 8. *Suppose* $(*)$ *holds. Then for every* $E_\flat, p_\flat, n_\flat, \widehat{k}_\flat, \widehat{r}_\flat$ *there are* $m_\sharp \geq n_\flat$, $q_\sharp \geq p_\flat$, $\widehat{l}_\sharp, \widehat{s}_\sharp$ *so that, setting*

- $k_\flat = \widehat{k}_\flat(m_\sharp, q_\sharp, \widehat{l}_\sharp, \widehat{s}_\sharp)$,
- $r_\flat = \widehat{r}_\flat(m_\sharp, q_\sharp, \widehat{l}_\sharp, \widehat{s}_\sharp)$,
- $l_\sharp = \widehat{l}_\sharp(k_\flat, r_\flat)$,
- $s_\sharp = \widehat{s}_\sharp(k_\flat, r_\flat)$,

we have

$$\left|(\rho_{m_\sharp}\lambda_{s_\sharp})(\Omega) - (\rho_{l_\sharp}\lambda_{q_\sharp})(\Omega)\right| \leq \frac{32}{E_\flat}.$$

Proof. Towards the use of p/n-metastable uniform continuity, we define functions $\widehat{q}_0(D_0, q_0, n_0, \widehat{k}_0)$ and $\widehat{m}_0(D_0, q_0, n_0, \widehat{k}_0)$.

Let $D_0, q_0, n_0, \widehat{\mathbf{k}}_0$, which we abbreviate †, be given. In order to apply Lemma 19 we define

- $\widehat{\mathbf{l}}_{\star,q,k}(k', r) = \widehat{\mathbf{k}}_0(q, \max\{k, k'\})$,
- $\widehat{\mathbf{l}}_{\star,q,\widehat{\mathbf{k}}}(k, r) = \widehat{\mathbf{k}}_0(q, \max\{k, \widehat{\mathbf{k}}(r, \widehat{\mathbf{l}}_{\star,q,k})\})$,
- $\widehat{\mathbf{s}}_{\star,q,\widehat{\mathbf{k}},\widehat{\mathbf{s}}}(k, r) = \widehat{\mathbf{s}}(r, \widehat{\mathbf{l}}_{\star,q,\widehat{\mathbf{k}}})$,
- $\widehat{\mathbf{r}}_\dagger(m_\dagger, q_\dagger, \widehat{\mathbf{s}}_\dagger, \widehat{\mathbf{k}}_\dagger) = \widehat{\mathbf{r}}_\flat(m_\dagger, q_\dagger, \widehat{\mathbf{s}}_{\star,q_\dagger,\widehat{\mathbf{k}}_\dagger,\widehat{\mathbf{s}}_\dagger}, \widehat{\mathbf{l}}_{\star,q_\dagger,\widehat{\mathbf{k}}_\dagger})$,
- $\widehat{\mathbf{L}}_\dagger(m_\dagger, q_\dagger, \widehat{\mathbf{s}}_\dagger, \widehat{\mathbf{k}}_\dagger) = \widehat{\mathbf{l}}_{\star,q_\dagger,\widehat{\mathbf{k}}_\flat(m_\dagger,q_\dagger,\widehat{\mathbf{s}}_{\star,q,\widehat{\mathbf{k}}_\dagger,\widehat{\mathbf{s}}_\dagger},\widehat{\mathbf{l}}_{\star,q_\dagger,\widehat{\mathbf{k}}_\dagger})}$,

By Lemma 19 we obtain $m_\dagger, q_\dagger, \widehat{\mathbf{s}}_\dagger, \widehat{\mathbf{k}}_\dagger$. It is convenient to define

- $\widehat{\mathbf{s}}_\star = \widehat{\mathbf{s}}_{\star,q_\dagger,\widehat{\mathbf{k}}_\dagger,\widehat{\mathbf{s}}_\dagger}$,
- $\widehat{\mathbf{l}}_\star = \widehat{\mathbf{l}}_{\star,q_\dagger,\widehat{\mathbf{k}}_\dagger}$,
- $r_\dagger = \widehat{\mathbf{r}}_\flat(m_\dagger, q_\dagger, \widehat{\mathbf{s}}_\star, \widehat{\mathbf{l}}_\star)$,
- $\widehat{\mathbf{l}}_\dagger = \widehat{\mathbf{l}}_{\star,q_\dagger,\widehat{\mathbf{k}}_\flat(m_\dagger,q_\dagger,\widehat{\mathbf{s}}_\star,\widehat{\mathbf{l}}_\star)}$,
- $s_\dagger = \widehat{\mathbf{s}}_\dagger(r_\dagger, \widehat{\mathbf{l}}_\dagger)$,
- $k_\dagger = \widehat{\mathbf{k}}_\dagger(r_\dagger, \widehat{\mathbf{l}}_\dagger)$.

We may then set

- $\widehat{\mathbf{q}}_0(D_0, q_0, n_0, \widehat{\mathbf{k}}_0) = q_\dagger$, and
- $\widehat{\mathbf{m}}_0(D_0, q_0, n_0, \widehat{\mathbf{k}}_0) = \max\{\widehat{\mathbf{k}}_\flat(m_\dagger, q_\dagger, \widehat{\mathbf{s}}_\star, \widehat{\mathbf{l}}_\star), k_\dagger\}$.

By metastable uniform continuity we obtain $D_0, q_0, n_0, \widehat{\mathbf{k}}_0$. Set $m_0 = \widehat{\mathbf{m}}_0(D_0, q_0, n_0, \widehat{\mathbf{k}}_0)$ and $k_0 = \widehat{\mathbf{k}}_0(q_\dagger, m_0)$. Since $m_0 \geq n_0$, we have $k_0 \geq m_0$ and when $\mu(\mathcal{B}') < 1/D_0$, $|\rho_{k_0}\lambda_{q_\dagger}|(\mathcal{B}') < 1/E_\flat$.

We set $m_\sharp = m_\dagger, q_\sharp = q_\dagger, \widehat{\mathbf{l}}_\sharp = \widehat{\mathbf{l}}_\star, \widehat{\mathbf{s}}_\sharp = \widehat{\mathbf{s}}_\star$. Let $k_\flat, r_\flat, l_\sharp, s_\sharp$ be as in the statement. Then since $l_\sharp = \widehat{\mathbf{l}}_\star(k_\flat, r_\flat) = \widehat{\mathbf{k}}_0(q_\dagger, \max\{k_\flat, k_\dagger\}) = \widehat{\mathbf{l}}_{\star,k_\flat}(k_\dagger, s_\dagger) = \widehat{\mathbf{l}}_\dagger(k_\dagger, s_\dagger)$ and $l_\sharp \geq k_\dagger$, we have

$$\left|(\rho_{m_\sharp}\lambda_{s_\sharp})(\Omega) - (\rho_{l_\sharp}\lambda_{q_\sharp})(\Omega)\right| \leq |\rho_{l_\sharp}\lambda_{q_\sharp}|(\mathcal{B}^-) + |\rho_{l_\sharp}\lambda_{q_\sharp}|(\mathcal{B}^{1,-}) + \frac{20}{E_\flat}$$

where $\mu(\mathcal{B}^-) < 4/D_\flat$ and $\mu(\mathcal{B}^{1,-}) < 2/D_\flat$. Therefore $|\rho_{l_\sharp}\lambda_{q_\sharp}|(\mathcal{B}^-) + |\rho_{l_\sharp}\lambda_{q_\sharp}|(\mathcal{B}^{1,-}) < 12/E_\flat$, and so

$$\left|(\rho_{m_\sharp}\lambda_{s_\sharp})(\Omega) - (\rho_{l_\sharp}\lambda_{q_\sharp})(\Omega)\right| \leq \frac{32}{E_\flat}$$

⊣

7   Quantitative Bounds

In this section we work out a concrete case of the bounds given by the work above, in essentially the simplest non-trivial case. This illustrates just how large the bounds above get, and is the result needed in [18].

The simplest meaningful case of 8 is to swap the order of the indices – that is, to show that, for every ϵ, there exist $s > m$ and $l > q$ so that

$$|(\rho_m \lambda_s)(\Omega) - (\rho_l \lambda_q)(\Omega)| < \epsilon.$$

Our goal in this section is to obtain a bound on l and s under some assumptions about the sequences ρ_n and λ_p.

Throughout this section, we make the following assumptions (we use ν_d to stand in for either some ρ_n or λ_p), which are essentially quantitative versions of the assumptions $(*)$ above:

$(*)_1^Q$ Each sequence $(\rho_n \lambda_p)_p$ (for fixed n) and $(\rho_n \lambda_p)_n$ (for fixed p) has bounded fluctuations with the uniform bound $8B^2 E^2$,

$(*)_2^Q$ For all ρ_m, $\omega_{\rho_m}(E) \leq E2^m$ and for all λ_p, $\omega_{\lambda_p}(E) \leq E2^p$,

$(*)_3^Q$ $||\nu_d||_{L^1} \leq B$ for all ν_d,

$(*)_4^Q$ For any E, D, any ν_d, and any $\mathcal{B}$, there is a $\mathcal{B}' \succeq \mathcal{B}$ and a $\mathcal{B}'' \subseteq \mathcal{B}$ such that ν_d is E-constant on $\mathcal{B}'$, $\mu(\mathcal{B}'') < 1/D$, and $|\mathcal{B}'| \leq 2BDE|\mathcal{B}|$.

The last condition is exactly what we would expect if ν_d were the Radon-Nikodym derivative of an actual L^1 function—we just take $\mathcal{B}''$ to consist of the points where the underlying function has large absolute value, and $\mathcal{B}'$ to consist of approximate level sets. We refer to these four assumptions collectively as $(*)^Q$.

We will say that a function $f(x)$ is:

- Polynomial if there are a, b, c so that $f(x) \leq ax^b + c$ for all x,
- Exponential if there are a, b, c, d so that $f(x) \leq a^{bx^c + d}$ for all x,
- Double exponential if there are a, b, c, d so that $f(x) \leq a^{a^{bx^c+d}}$ for all x.

We say a function of multiple inputs, $f(x_1, \ldots, x_n)$ is polynomial (resp. exponential, double exponential) if there is a polynomial (resp. exponential, double exponential) function $f'(x)$ so that for all $x_1, \ldots, x_n$, $f(x_1, \ldots, x_n) \leq f'(\prod_i x_i)$.

We will briefly need to keep track of functions which are polynomial in some inputs and exponential in others; we say $f(x; y)$ is poly-exp if there are a, b, c, d so that for all x, y, $f(x; y) \leq (ax)^{by^c + d}$. We say a function of multiple inputs, $f(x_1, \ldots, x_n; y_1, \ldots, y_m)$ is poly-exp if there is a poly-exp function $f'(x; y)$ so that for all $x_1, \ldots, x_n, y_1, \ldots, y_m$, $f(x_1, \ldots, x_n; y_1, \ldots, y_m) \leq f'(\prod_i x_i; \prod_j y_j)$.

Lemma 20. *[Bounds for Lemma 6] Suppose* $(*)^Q$ *holds. Then there is an exponential function* $\mathfrak{a}_0(B, D_\flat, E_\flat)$ *so that for every* $D_\flat, E_\flat, \mathcal{B}_\flat, \widehat{\mathfrak{m}}_\flat$, $n_\flat$ *there is an* $m_\sharp \in [n_\flat, \widehat{\mathfrak{m}}_\flat^{\mathfrak{a}_0(B, D_\flat, E_\flat)}(n_\flat)]$ *such that, taking*

$$\mathcal{B} = \{\sigma \in \mathcal{B}_\flat \mid \text{for every } m, m' \in [m_\sharp, \widehat{\mathfrak{m}}_\flat(m_\sharp)], \ |\nu_m - \nu_{m'}|(\sigma) < 1/E_\flat\},$$

we have $\mu(\mathcal{B}) \geq (1 - /D_\flat)\mu(\mathcal{B}_\flat)$.

Proof. Set $\mathfrak{a}_0(B, D_\flat, E_\flat) = [2^5 B^2 E_\flat^2]^{\frac{-\ln D_\flat}{\ln(1 - \frac{1}{4B^2 E_\flat^2})}}$.

In the proof, $V_\sharp = 2^5 B^2 E_\flat^2$ and

$$k = \lceil \frac{\ln(1/D_\flat)}{\ln(1 - 1/4B^2 E_\flat^2)} \rceil.$$

One can check that $(2^5 B^2 E_\flat^2)^{-1/\ln(1 - 1/4B^2 E_\flat^2)}$ grows less quickly than $e^{(2^5 B^2 E_\flat^2)^2}$, so we can bound $\mathfrak{a}_0(B, D_\flat, E_\flat)$ by $2^{2^{10} B^4 E_\flat^4 \ln D_\flat}$, which has the specified bounds. $\dashv$

We will often need the quantity $\mathfrak{a}_0(B, 3D_\flat, 3E_\flat) + 1$, which we abbreviate $\mathfrak{a}(B, D_\flat, E_\flat)$. $\mathfrak{a}$ is also exponential.

7.1 Bounds on Regularity

The next few lemmas give bounds on (cases of) the various forms of the the one-dimensional regularity lemma, culminating in bounds on a special case of Lemma 7.

We first note the function

- $\mathfrak{b}_0(B, D_\flat, E_\flat) = 2^{10} D_\flat^3 E_\flat^2 B^2 + 2^9 D_\flat^2 E_\flat^2 B^2$.

which appears in the proof of Lemma 6; this is essentially the bound on the number of iterations needed in that argument. It turns out we will mostly need

- $\mathfrak{b}(B, D_\flat, E_\flat) = \mathfrak{b}(B, 3D_\flat, 3E_\flat)$.

Both of these functions are polynomial.

There are two major simplifications in this special case which will make it much easier to find bounds on the various sequential versions. The first is that we will fix, throughout this subsection, a constant z_0 so that all the functions $\widehat{B}_\flat$, $\widehat{B}_\flat^0$, and $\widehat{B}_\flat^1$ will satisfy the bound $|\widehat{B}_\flat(\cdots, \mathcal{B})| \leq z_0|\mathcal{B}|$, independently of other parameters. Along with this, we will assume that we always apply our lemmas with $|\mathcal{A}_\flat| \leq z_0^{(\mathfrak{b}(B, D_\flat, E_\flat) + 1)(\mathfrak{b}(B, D_\flat, E_\flat))}$, and that all partitions $\mathcal{B}$ appearing in the proofs in this subsection satisfy $|\mathcal{B}| \leq z_0^{(\mathfrak{b}(B, D_\flat, E_\flat) + 2)\mathfrak{b}(B, D_\flat, E_\flat)}$.

The second simplification is that we have a fixed function $\widehat{\mathbf{u}} : \mathbb{N} \to \mathbb{N}$, and all the functions we deal with will be bounded by, roughly, iterations of $\widehat{\mathbf{u}}$ above some base value. Abstractly, we say:

- A function $\widehat{\mathbf{m}}(n)$ is $\widehat{\mathbf{u}}$-bounded by c above n_- if
  $\widehat{\mathbf{m}}(n) \leq \widehat{\mathbf{u}}^c(\max\{n, n_-\})$,

- A function $\widehat{\mathbf{m}}(n, \widehat{\mathbf{k}})$ is $\widehat{\mathbf{u}}$-bounded by f above n_- if whenever $\widehat{\mathbf{k}}$ is $\widehat{\mathbf{u}}$-bounded by c above n_-, $\widehat{\mathbf{m}}(n, \widehat{\mathbf{k}}) \leq \widehat{\mathbf{u}}^{f(c)}(\max\{n, n_-\})$.

Note that this second clause allows us to iterate if we interpret c as being itself a function and f as a higher order functional. We try to limit the quantity of special cases we need to deal with, but it will be helpful to generalize the second clause to the case where there are two input functions:

- A function $\widehat{\mathbf{m}}(n, \widehat{\mathbf{k}}, \widehat{\mathbf{r}})$ is $\widehat{\mathbf{u}}$-bounded by f above n_- if whenever $\widehat{\mathbf{k}}$ is $\widehat{\mathbf{u}}$-bounded by c above n_- and $\widehat{\mathbf{r}}$ is $\widehat{\mathbf{u}}$-bounded by d above n_-, $\widehat{\mathbf{m}}(n, \widehat{\mathbf{k}}, \widehat{\mathbf{r}}) \leq \widehat{\mathbf{u}}^{f(c,d)}(\max\{n, n_-\})$.

For consistency with the later arguments, we find bounds for Lemma 6 with the values $3D_\flat, 3E_\flat$.

Lemma 21. *[Bounds on 6] Suppose that $(*)^Q$ holds. There exist poly-exp functions $\mathfrak{c}(a; B, D_\flat, E_\flat)$ and $\mathfrak{d}(a; B, D_\flat, E_\flat)$ so that whenever z_0, n_-, a, b, $\mathcal{A}_\flat$, $E_\flat$, $D_\flat$, $\widehat{\mathbf{m}}_\flat$, and $\widehat{\mathbf{B}}_\flat$ are given such that:*

- $|\mathcal{A}_\flat| \leq z_0^{(\mathfrak{b}(B, D_\flat, E_\flat)+1)(\mathfrak{b}(B, D_\flat, E_\flat))}$,

- *For all $\widehat{\mathbf{m}}, \mathcal{B}$, $|\widehat{\mathbf{B}}_\flat(\widehat{\mathbf{m}}, \mathcal{B})| \leq z_0|\mathcal{B}|$,*

- *The function $(n, \widehat{\mathbf{k}}) \mapsto \sup_{\mathcal{B}_0, \mathcal{B}_1, \mathcal{B}_2} \widehat{\mathbf{m}}_\flat(n, \widehat{\mathbf{k}}, \mathcal{B}_0, \mathcal{B}_1)$ is $\widehat{\mathbf{u}}$-bounded by $\widehat{\mathbf{f}}(x) = ax + b$ above n_- (where the supremum ranges over partitions refining $\mathcal{A}_\flat$ and satisfying $|\mathcal{B}_i| \leq z_0^{(\mathfrak{b}(B, D_\flat, E_\flat)+2)(\mathfrak{b}(B, D_\flat, E_\flat))}$),*

there exist $\mathcal{B}_\sharp \succeq \mathcal{A}_\flat$, $n_\sharp$, and $\widehat{\mathbf{k}}_\sharp$ such that:

- $|\mathcal{B}_\sharp| \leq z_0^{\mathfrak{b}(B, D_\flat, E_\flat)}|\mathcal{A}_\flat|$,

- $n_\sharp \leq \widehat{\mathbf{u}}^{\mathfrak{d}(a; B, D_\flat, E_\flat)\mathfrak{b}}(n_-)$,

- *The function $n \mapsto \sup_\mathcal{B} \widehat{\mathbf{k}}_\sharp(n, \mathcal{B})$ is $\widehat{\mathbf{u}}$-bounded by $\mathfrak{c}(a; B, D_\flat, E_\flat)\mathfrak{b}$ above $n_\sharp$ (with the supremum ranging over the same partitions as above),*

- *Whenever $\mathcal{B}_\sharp \preceq \mathcal{B} \preceq \mathcal{B}' \preceq \widehat{\mathbf{B}}_\flat(\widehat{\mathbf{m}}_\sharp, \mathcal{B}_\sharp)$, setting $m_\flat = \widehat{\mathbf{m}}_\flat(n_\sharp, \widehat{\mathbf{k}}_\sharp, \mathcal{B}_\sharp, \mathcal{B}, \mathcal{B}')$ and $k_\sharp = \widehat{\mathbf{k}}_\sharp(m_\flat, \mathcal{B}')$, if $m_\flat \geq n_\sharp$ then we have $k_\sharp \geq m_\flat$ and $\mu(\mathfrak{D}_{3E_\flat, \mathcal{B}, k_\sharp}(\mathcal{B}')) < 1/3D_\flat$.*

Proof. If we ignore the bounds, this is essentially Lemma 6 applied to $\mathcal{A}_\flat, 3E_\flat, 3D_\flat, \widehat{\mathbf{m}}_\flat, \widehat{\mathbf{B}}_\flat$. We obtain bounds by examining the proof of Lemma 6.

The main step in the proof is the construction of the sequence of functions $\widehat{\mathbf{k}}_i$, with $\widehat{\mathbf{k}}_\sharp$ bounded by $\widehat{\mathbf{k}}_{\mathfrak{b}(B,D_\flat,E_\flat)}$. First, note that when $\mathcal{A}_0 = \mathcal{A}_\flat$, each element in the sequence of partitions constructed in the proof satisfies $\mathcal{A}_{i+1} \preceq \widehat{\mathbf{B}}_\flat(m_0, \widehat{\mathbf{k}}_i, \mathcal{A}_i)$, so we have $|\mathcal{A}_i| \leq z_0^i |\mathcal{A}_\flat|$, and so $|\mathcal{B}_\sharp| \leq z_0^{\mathfrak{b}(B,D_\flat,E_\flat)} |\mathcal{A}_\flat|$.

We show inductively that $\widehat{\mathbf{k}}_i(m, \mathcal{B})$ is $\widehat{\mathbf{u}}$-bounded by $(a+2)^i b$ above n_-.

Clearly $\widehat{\mathbf{k}}_0(m, \mathcal{B}) = m = \widehat{\mathbf{u}}^0(m)$. When $m_0 \geq n_-$, we have

$$\begin{aligned}
\widehat{\mathbf{k}}_{i+1}(m_0, \mathcal{B}) &= \widehat{\mathbf{k}}_i(\widehat{\mathbf{m}}_\flat(m_0, \widehat{\mathbf{k}}_i, \mathcal{A}_d, \mathcal{B}, \mathcal{B}')) \\
&\leq \widehat{\mathbf{u}}^{(a+2)^i b}(\widehat{\mathbf{n}}^{a(a+2)^i b + b}(m_0)) \\
&= \widehat{\mathbf{u}}^{(a+2)^i b + a(a+2)^i b + b}(m_0) \\
&= \widehat{\mathbf{u}}^{(a+2)^i b(a+1)+b}(m_0) \\
&\leq \widehat{\mathbf{u}}^{(a+2)^{i+1} b}(m_0).
\end{aligned}$$

It suffices, in the last step, to work with $i = \mathfrak{b}(B, D_\flat, E_\flat)$ and $m_0 = n_-$, so $\widehat{\mathbf{k}}_\sharp = \widehat{\mathbf{k}}_{\mathfrak{b}(D_\flat,E_\flat)}$ is $\widehat{\mathbf{u}}$-bounded by $(a+2)^{\mathfrak{b}(B,D_\flat,E_\flat)} b$ above n_-; we see that this bound is poly-exp of the right form. For the largest value that might be used for $n_\sharp$, take the values $n_0 = n_-$, $n_{i+1} = \widehat{\mathbf{m}}_\flat(n_i, \widehat{\mathbf{k}}_{\mathfrak{b}(B,D_\flat,3E_\flat)-i}, \mathcal{A}_d, \mathcal{B}, \mathcal{B}')$, and observe that $n_\sharp = n_i$ for some i. Since $n_i \leq \widehat{\mathbf{u}}^{\sum_{j \leq i} a(a+2)^{\mathfrak{b}(B,D_\flat,3E_\flat)-i} b + b}(n_-)$, we see that the bound $\mathfrak{d}(a, b; B, D_\flat, E_\flat) = \sum_{j \leq i} a(a+2)^{\mathfrak{b}(B,D_\flat,3E_\flat)-i} b + b$ also has the promised size. ⊣

Lemma 22. *[Bounds on* γ*] Suppose* $(*)^Q$ *holds. Then there exists a poly-exp function* $\mathfrak{e}(a; B, D_\flat, E_\flat)$ *so that whenever* $z_0, n_-, \mathcal{A}_\flat, E_\flat, D_\flat$, $\widehat{\mathbf{B}}_\flat, \widehat{\mathbf{n}}_\flat, \widehat{\mathbf{L}}_\flat$ *are given such that:*

- $|\mathcal{A}_\flat| \leq z_0^{(\mathfrak{b}(B,D_\flat,E_\flat)+1)(\mathfrak{b}(B,D_\flat,E_\flat))}$,
- *For all* $\widehat{\mathbf{m}}, \mathcal{B}$, $|\widehat{\mathbf{B}}_\flat(\widehat{\mathbf{m}}, \mathcal{B})| \leq z_0 |\mathcal{B}|$,
- *For all* $n, \widehat{\mathbf{k}}$ *and all* $\mathcal{B} \succeq \mathcal{A}_\flat$ *with* $|\mathcal{B}| \leq z_0^{\mathfrak{b}(B,D_\flat,E_\flat)} |\mathcal{A}_\flat|$*, the function* $k \mapsto \widehat{\mathbf{L}}_\flat(n, \widehat{\mathbf{k}}, \mathcal{B})(k)$ *is* $\widehat{\mathbf{u}}$-bounded by 1 above $\max\{n_-, n, \widehat{\mathbf{m}}_\flat(n, \widehat{\mathbf{k}}, \mathcal{B})\}$,
- *The function* $(n, \widehat{\mathbf{k}}) \mapsto \sup_{\mathcal{B}} \widehat{\mathbf{m}}_\flat(n, \widehat{\mathbf{k}}, \mathcal{B})$ *is* $\widehat{\mathbf{u}}$-bounded by $\widehat{\mathbf{f}}(f) = af(1)+b$ above n_- *(where the supremum is over suitable* $\mathcal{B}$ *as above),*

there are $n_\sharp, \mathcal{B}_\sharp$ *and* $\widehat{\mathbf{m}}_\sharp$ *such that:*

- $|\mathcal{B}_\sharp| \leq z_0^{\mathfrak{b}(B,D_\flat,E_\flat)} |\mathcal{A}_\flat|$,
- $n_\sharp \leq \widehat{\mathbf{u}}^{\mathfrak{e}(a;B,D_\flat,E_\flat)b}(n_-)$,
- $\widehat{\mathbf{k}}_\sharp$ *is* $\widehat{\mathbf{u}}$-bounded by $f(x) = \mathfrak{a}(B, D_\flat, E_\flat)x + \mathfrak{e}(a; B, D_\flat, E_\flat)b$ *above* $n_\sharp$,

- *Setting $m_\flat = \widehat{\mathbf{m}}_\flat(n_\sharp, \widehat{\mathbf{k}}_\sharp, \mathcal{B}_\sharp)$ and $\widehat{\mathbf{l}}_\flat = \widehat{\mathbf{L}}_\flat(n, \widehat{\mathbf{m}}, \mathcal{B})$, if $m_\flat \geq n_\sharp$ then for any $\mathcal{B}_\sharp \preceq \mathcal{B} \preceq \mathcal{B}' \preceq \widehat{\mathbf{B}}_\flat(n_\sharp, \widehat{\mathbf{k}}_\sharp, \mathcal{B}_\sharp)$, setting $k_\sharp = \widehat{\mathbf{k}}_\sharp(m_\flat, \widehat{\mathbf{l}}_\flat, \mathcal{B}, \mathcal{B}')$, we have $k_\sharp \geq m_\flat$ and for every $l \in [k_\sharp, \widehat{\mathbf{l}}_\flat(k_\sharp)]$, $\mu(\mathfrak{D}_{E_\flat, \mathcal{B}, k}(\mathcal{B}')) < 1/D_\flat$.*

Proof. We examine the proof of Lemma 7. Suppose we have been given $n_0, \mathcal{B}_0$, and a $\widehat{\mathbf{k}}_0$ which is $\widehat{\mathbf{u}}$-bounded by y above n_0.

Suppose we have also fixed $\widehat{\mathbf{l}}_\flat$, which is $\widehat{\mathbf{u}}$-bounded by x above

$$\max\{n_0, n_-\}$$

and a $\mathcal{B}' \succeq \mathcal{B} \succeq \mathcal{B}_0$ with $|\mathcal{B}'| \leq z_0^{\flat(B, D_\flat, E_\flat)}|A_\flat|$. The function $m \mapsto \widehat{\mathbf{l}}_\flat(\widehat{\mathbf{k}}_0(m, \mathcal{B}'))$ is $\widehat{\mathbf{u}}$-bounded by $y + x$ above $\max\{n_0, n_-\}$. The function

$$m \mapsto m_{\dagger, m, \widehat{\mathbf{l}}_\flat, \mathcal{B}, \mathcal{B}'}$$

is obtained by applying Lemma 6 with $3D_\flat, 3E_\flat$, and is therefore $\widehat{\mathbf{u}}$-bounded by $\mathfrak{a}_0(B, 3D_\flat, 3E_\flat)(y + x)$ above $\max\{n_0, n_-\}$.

The function $(m, \widehat{\mathbf{l}}_\flat) \mapsto m_{\dagger, m, \widehat{\mathbf{l}}_\flat, \mathcal{B}, \mathcal{B}'}$ is therefore $\widehat{\mathbf{u}}$-bounded by

$$f_y(x) = \mathfrak{a}_0(B, 3D_\flat, 3E_\flat)(y + x)$$

above $\max\{n_0, n_-\}$, and so

$$(m, \widehat{\mathbf{l}}_\flat) \mapsto \sup_{\mathcal{B}, \mathcal{B}'} \widehat{\mathbf{k}}_\dagger(m, \widehat{\mathbf{l}}_\flat, \mathcal{B}, \mathcal{B}')$$

is bounded by $f'_y(x) = \mathfrak{a}_0(B, 3D_\flat, 3E_\flat)(y+x)+y+x = \mathfrak{a}(B, D_\flat, E_\flat)(y+x)$.

The function $(n_0, \widehat{\mathbf{k}}_0) \mapsto \sup_{\mathcal{B}} \widehat{\mathbf{m}}_\flat(n_0, \widehat{\mathbf{k}}_\dagger, \mathcal{B}_0)$ is therefore $\widehat{\mathbf{u}}$-bounded by $af'_y(1) + b = \mathfrak{a}(B, D_\flat, E_\flat)a(y+1) + b$ above n_-. This at last lets us consider the function $\widehat{\mathbf{m}}_0$: it is $\widehat{\mathbf{u}}$-bounded by

$$f''(y) = \mathfrak{a}(B, D_\flat, E_\flat)a(y+1) + b + \mathfrak{a}_0(B, 3D_\flat, 3E_\flat)(y+1)$$
$$\leq \mathfrak{a}(B, D_\flat, E_\flat)(a+1)(y+1) + b$$

above n_-.

We are in the setting of the previous lemma where

$$(n, \widehat{\mathbf{k}}) \mapsto \sup_{\mathcal{B}_0, \mathcal{B}_1, \mathcal{B}_2} \widehat{\mathbf{m}}_0(n, \widehat{\mathbf{k}}, \mathcal{B}_0, \mathcal{B}_1, \mathcal{B}_1)$$

is $\widehat{\mathbf{u}}$-bounded by $f''(y)$; rewriting $f''(y)$ as a linear function,

$$f''(y) = \mathfrak{a}(B, D_\flat, E_\flat)(a+1)y + \mathfrak{a}(B, D_\flat, E_\flat)(a+1) + \flat.$$

Therefore $\widehat{\mathbf{k}}_0$ is $\widehat{\mathbf{u}}$-bounded by

$$\mathfrak{c}(\mathfrak{a}(B, D_\flat, E_\flat)(a+1); B, D_\flat, E_\flat)(\mathfrak{a}(B, D_\flat, E_\flat)(a+1) + b)$$

and

$$n_0 \leq \widehat{\mathbf{u}}^{\partial(\mathfrak{a}(B, D_\flat, E_\flat)(a+1); D_\flat, E_\flat)(\mathfrak{a}(B, D_\flat, E_\flat)(a+1)+b)}(n_-).$$

The function $\widehat{\mathbf{k}}_\sharp$ is therefore bounded by

$$f'_{\mathfrak{c}(\mathfrak{a}(B,D_\flat,E_\flat)(a+1);D_\flat,E_\flat)(\mathfrak{a}(B,D_\flat,E_\flat)(a+1)+b)}(x),$$

which has the specified form (in particular, it is linear in b, polynomial in a and exponential in $B, D_\flat, E_\flat$).

Similarly $n_\sharp = n_0$ is bounded by

$$\widehat{\mathbf{u}}^{\mathfrak{d}(\mathfrak{a}(B,D_\flat,E_\flat)(a+1);D_\flat,E_\flat)(\mathfrak{a}(B,D_\flat,E_\flat)(a+1)+b)}(n_-),$$

which also has the specified bounds. ⊣

7.2 Controlling Intervals

In this subsection we obtain bounds on a special case of Lemma 16. We continue to work with a fixed function $\widehat{\mathbf{u}}$.

Lemma 23. *[Bounds on 15] Suppose* $(*)^Q$ *holds. There are poly-exp functions* $\mathfrak{f}(\rho; B, D_\flat, E_\flat)$ *and* $\mathfrak{g}(\rho; B, D_\flat, E_\flat)$ *so that whenever* z, ρ, n_-, $D_\flat$, $E_\flat$, $\widehat{\mathbf{B}}_\flat^0$, $\widehat{\mathbf{B}}_\flat^1$, $\widehat{\mathbf{m}}_\flat$, $\widehat{\mathbf{L}}_\flat$, $\widehat{\mathbf{q}}_\flat$, *and* $\widehat{\mathbf{S}}_\flat$ *are given such that:*

- *For all* $n, p, \mathcal{B}, \widehat{\mathbf{k}}, \widehat{\mathbf{r}}$ *and* $i \in \{0,1\}$, $|\widehat{\mathbf{B}}_\flat^i(n, p, \widehat{\mathbf{k}}, \widehat{\mathbf{r}}, \mathcal{B})| \leq z|\mathcal{B}|$,
- $\widehat{\mathbf{m}}_\flat(n, p, \widehat{\mathbf{k}}, \widehat{\mathbf{r}}, \mathcal{B})$ *is* $\widehat{\mathbf{u}}$-*bounded by* $\rho(f(1,1) + g(1,1) + 1)$ *above* n_-,
- $\widehat{\mathbf{q}}_\flat(n, p, \widehat{\mathbf{k}}, \widehat{\mathbf{r}}, \mathcal{B})$ *is* $\widehat{\mathbf{u}}$-*bounded by* $\rho(f(1,1) + g(1,1) + 1)$ *above* n_-,
- *For any* $n, p, \mathcal{B}, \widehat{\mathbf{k}}, \widehat{\mathbf{r}}$, *the function* $m \mapsto \widehat{\mathbf{L}}_\flat(n, p, \widehat{\mathbf{k}}, \widehat{\mathbf{r}}, \mathcal{B})(m, q)$ *does not depend on* q *and is* $\widehat{\mathbf{u}}$-*bounded by* 1 *above*

$$\max\{n_-, n, p, \widehat{\mathbf{m}}_\flat(n, p, \widehat{\mathbf{k}}, \widehat{\mathbf{r}}, \mathcal{B}), \widehat{\mathbf{q}}_\flat(n, p, \widehat{\mathbf{k}}, \widehat{\mathbf{r}}, \mathcal{B}) + 1\},$$

- *For any* $n, p, \mathcal{B}, \widehat{\mathbf{k}}, \widehat{\mathbf{r}}$, *the function* $q \mapsto \widehat{\mathbf{S}}_\flat(n, p, \widehat{\mathbf{k}}, \widehat{\mathbf{r}}, \mathcal{B})(m, q)$ *does not depend on* m *and is* $\widehat{\mathbf{u}}$-*bounded by* 1 *above*

$$\max\{n_-, n, p, \widehat{\mathbf{m}}_\flat(n, p, \widehat{\mathbf{k}}, \widehat{\mathbf{r}}, \mathcal{B}) + 1, \widehat{\mathbf{q}}_\flat(n, p, \widehat{\mathbf{k}}, \widehat{\mathbf{r}}, \mathcal{B})\},$$

then there exist $\mathcal{B}_\sharp$, $n_\sharp$, $p_\sharp$, $\widehat{\mathbf{k}}_\sharp$, *and* $\widehat{\mathbf{r}}_\sharp$ *such that:*

- $|\mathcal{B}_0| \leq z^{\mathfrak{b}(B,D_\flat,E_\flat)(\mathfrak{b}(B,D_\flat,E_\flat)+2)}$,
- $n_\sharp \leq \widehat{\mathbf{u}}^{\mathfrak{f}(\rho;B,D_\flat,E_\flat)}(n_-)$,
- $p_\sharp \leq \widehat{\mathbf{u}}^{\mathfrak{g}(\rho;B,D_\flat,E_\flat)}(n_-)$,
- $\widehat{\mathbf{k}}_\sharp$ *is* $\widehat{\mathbf{u}}$-*bounded by* $\mathfrak{a}(B, D_\flat, E_\flat)x + \mathfrak{f}(\rho; B, D_\flat, E_\flat)$ *above* $\max\{n_\sharp, p_\sharp\}$,
- $\widehat{\mathbf{r}}_\sharp$ *is* $\widehat{\mathbf{u}}$-*bounded by* $\mathfrak{a}(B, D_\flat, E_\flat)y + \mathfrak{g}(\rho; B, D_\flat, E_\flat)$ *above* $\max\{n_\sharp, p_\sharp\}$,

Proof. We need to first analyze the inner application of Lemma 7. Suppose we have fixed $\mathcal{B}_0, n_0,$ and $\widehat{\mathbf{k}}_0$ so that $|\mathcal{B}_0| \leq z^{\mathfrak{b}(B,D_\flat,E_\flat)(\mathfrak{b}(B,D_\flat,E_\flat)+1)}$ and $\widehat{\mathbf{k}}_0$ is $\widehat{\mathbf{u}}$-bounded by $h_1(x)$ above $\max\{n_-, n_0\}$.

Suppose we are given $\mathcal{B}_\dagger, p_\dagger, \widehat{\mathbf{r}}_\dagger$ with $|\mathcal{B}_\dagger| \leq z^{\mathfrak{b}(B,D_\flat,E_\flat)}|\mathcal{B}_0|$ and $\widehat{\mathbf{r}}_\dagger$ is $\widehat{\mathbf{u}}$-bounded by $h_2(y)$ above $\max\{n_-, n_0, p_\dagger\}$.

We note bounds on the various helper functions under the assumption (as will turn out to be the case) that $\widehat{\mathsf{l}}(m,q)$ does not depend on q and is $\widehat{\mathsf{u}}$-bounded by 1 above $\max\{n_-, n_0, p_\dagger\}$, and that $\widehat{\mathsf{s}}(m,q)$ does not depend on m and is $\widehat{\mathsf{u}}$-bounded by 1 above $\max\{n_-, n_0, p_\dagger\}$.

- $\widehat{\mathsf{l}}_{\widehat{\mathsf{l}},r}$ does not depend on r and is also $\widehat{\mathsf{u}}$-bounded by 1 above $\max\{n_-, n_0, p_\dagger\}$,
- $\widehat{\mathsf{k}}_{\star,r}$ does not depend on q and is $\widehat{\mathsf{u}}$-bounded by $h_1(1)$ above $\max\{n_-, n_0, p_\dagger\}$,
- $\widehat{\mathsf{s}}_{\star,\mathcal{B}^0,\mathcal{B}^1,m,q,\widehat{\mathsf{l}},\widehat{\mathsf{s}}}$ is also $\widehat{\mathsf{u}}$-bounded by 1 above $\max\{n_-, n_0, p_\dagger\}$,
- $\widehat{\mathsf{r}}_\star$ is $\widehat{\mathsf{u}}$-bounded by $h_2(y)$ above $\max\{n_-, n_0, p_\dagger\}$,
- $\widehat{\mathsf{k}}_\star$ is $\widehat{\mathsf{u}}$-bounded by $h_1(x)$ above $\max\{n_-, n_0, p_\dagger\}$,
- $\widehat{\mathsf{l}}_\star(m,q)$ does not depend on q and is $\widehat{\mathsf{u}}$-bounded by 1 above $\max\{n_-, n_0, p_\dagger\}$,
- $\widehat{\mathsf{s}}_\star(m,q)$ does not depend on m and is $\widehat{\mathsf{u}}$-bounded by 1 above
$$\max\{n_-, n_0, p_\dagger\},$$
- $|\mathcal{B}^0_\star|, |\mathcal{B}^1_\star| \leq z|\mathcal{B}|$,
- $m_\star \leq \widehat{\mathsf{u}}^{\rho(h_1(1)+h_2(1)+1)}(\max\{n_-, n_0, p_\dagger\})$,
- $q_\star \leq \widehat{\mathsf{u}}^{\rho(h_1(1)+h_2(1)+1)}(\max\{n_-, n_0, p_\dagger\})$.

We now see that $\widehat{\mathsf{q}}_\dagger$ is $\widehat{\mathsf{u}}$-bounded by $\mathfrak{g}_{h_1}(h_2) = \rho(h_1(1) + h_2(1) + 1)$ above $\max\{n_-, n_0\}$ and
$$r \mapsto \widehat{\mathsf{S}}_\dagger(p_\dagger, \mathcal{B}_\dagger, \widehat{\mathsf{q}}_\dagger)(r)$$
is $\widehat{\mathsf{u}}$-bounded by 1 above $\max\{n_-, n_0, p_\dagger\}$.

When we apply the previous lemma to $z, \mathcal{B}_0, E_\flat, D_\flat, \widehat{\mathsf{B}}_\dagger, \widehat{\mathsf{q}}_\dagger, \widehat{\mathsf{S}}_\dagger$, we obtain $\mathcal{B}_\dagger, p_\dagger, \widehat{\mathsf{r}}_\dagger$ such that:

- $|\mathcal{B}_\dagger| \leq z^{\mathfrak{b}(B, D_\flat, E_\flat)}|\mathcal{B}_0|$,
- $p_\dagger \leq \widehat{\mathsf{u}}^{\mathfrak{e}(\rho; B, D_\flat, E_\flat)\rho(h_1(1)+1)}(n_0)$,
- $\widehat{\mathsf{r}}_\dagger$ is $\widehat{\mathsf{u}}$-bounded by $\mathfrak{a}(B, D_\flat, E_\flat)y + \mathfrak{e}(\rho; B, D_\flat, E_\flat)\rho(h_1(1)+1)$ above $p_\dagger$.

We now turn to the outer application of Lemma 7. $\widehat{\mathsf{m}}_0$ is $\widehat{\mathsf{u}}$-bounded by
$$\rho(h_1(1) + \mathfrak{a}(B, D_\flat, E_\flat)1 + \mathfrak{e}(\rho; B, D_\flat, E_\flat)\rho(h_1(1) + 1) + 1)$$
$$=\rho h_1(1) + \rho\mathfrak{a}(B, D_\flat, E_\flat) + \rho\mathfrak{e}(\rho; B, D_\flat, E_\flat)\rho(h_1(1) + 1) + \rho$$
$$=\rho\mathfrak{a}(B, D_\flat, E_\flat) + \rho^2\mathfrak{e}(\rho; B, D_\flat, E_\flat) + \rho + (\rho + \rho^2\mathfrak{e}(\rho; B, D_\flat, E_\flat))h_1(1).$$

For any $n_0, \widehat{\mathsf{k}}_0$, the function $k \mapsto \widehat{\mathsf{L}}_0(n_0, \widehat{\mathsf{k}}_0, \mathcal{B}_0)(k)$ is $\widehat{\mathsf{u}}$-bounded by 1 above $\max\{n_-, n_0, p_\dagger\}$. Therefore by the previous lemma applied with $z_0 = z^{\mathfrak{b}(B, D_\flat, E_\flat)+1}$, we obtain $\mathcal{B}_0, n_0, \widehat{\mathsf{k}}_0$ such that:

- $n_0 \leq \widehat{\mathbf{u}}^{\mathfrak{e}(\rho\mathfrak{a}(B,D_\flat,E_\flat)+\rho^2\,\mathfrak{e}(\rho;B,D_\flat,E_\flat)+\rho;B,D_\flat,E_\flat)(\rho+\rho^2\,\mathfrak{e}(\rho;B,D_\flat,E_\flat))}(n_-)$,

- $\widehat{\mathbf{k}}_\sharp$ is $\widehat{\mathbf{u}}$-bounded by

$$\mathfrak{a}(B,D_\flat,E_\flat)x + \mathfrak{e}(\rho\mathfrak{a}(B,D_\flat,E_\flat)+\gamma;B,D_\flat,E_\flat)\gamma$$

above $\max\{n_-,n_0\}$, where $\gamma = \rho + \rho^2\mathfrak{e}(\rho;B,D_\flat,E_\flat)$.

We obtain final bounds with:

- $|\mathcal{B}_\sharp| \leq z^{\mathfrak{b}(B,D_\flat,E_\flat)(\mathfrak{b}(B,D_\flat,E_\flat)+2)}$,

- $n_\sharp = n_0 \leq \widehat{\mathbf{u}}^{\mathfrak{f}(\rho;B,D_\flat;E_\flat)}(n_-)$,

- $p_\sharp = p_\dagger \leq \widehat{\mathbf{u}}^{\mathfrak{g}(\rho;B,D_\flat;E_\flat)}(n_-)$,

- $\widehat{\mathbf{k}}_\sharp$ is $\widehat{\mathbf{u}}$-bounded by $\mathfrak{a}(B,D_\flat,E_\flat)x + \mathfrak{f}(\rho;D_\flat;E_\flat)$ above $\max\{n_-,n_\sharp,p_\sharp\}$,

- $\widehat{\mathbf{r}}_\sharp$ is $\widehat{\mathbf{u}}$-bounded by $\mathfrak{a}(B,D_\flat,E_\flat)y + \mathfrak{g}(\rho;D_\flat;E_\flat)$ above $\max\{n_-,n_\sharp,p_\sharp\}$.

$\dashv$

Lemma 24. *[Bounds on 16] Suppose* $(*)^Q$ *holds. There is a double exponential function* $\mathfrak{i}(B,D_\flat^0,D_\flat^1,E_\flat)$ *so that for any* $n_\flat,p_\flat,D_\flat^0,D_\flat^1,E_\flat,\widehat{\mathbf{u}}$ *such that:*

- $\widehat{\mathbf{u}}(m) > m$ *for all* m,

- *For any* $m,q,\widehat{\mathbf{k}},\widehat{\mathbf{r}}$, *the function* $k \mapsto \widehat{\mathbf{L}}_\flat(m,q,\widehat{\mathbf{k}},\widehat{\mathbf{r}})(k,r)$ *does not depend on* r *and is* $\widehat{\mathbf{u}}$-*bounded by* 1 *above* $\max\{n_\flat,p_\flat,m,q+1\}$,

- *For any* $m,q,\widehat{\mathbf{k}},\widehat{\mathbf{r}}$, *the function* $r \mapsto \widehat{\mathbf{S}}_\flat(m,q,\widehat{\mathbf{k}},\widehat{\mathbf{r}})(k,r)$ *does not depend on* k *and is* $\widehat{\mathbf{u}}$-*bounded by* 1 *above* $\max\{n_\flat,p_\flat,m+1,q\}$,

there are $m_\sharp \geq n_\flat$, $q_\sharp \geq p_\flat$, $\widehat{\mathbf{k}}_\sharp$, *and* $\widehat{\mathbf{r}}_\sharp$ *such that:*

- $m_\sharp \leq \widehat{\mathbf{u}}^{\mathfrak{i}(B,D_\flat^0,D_\flat^1,E_\flat)}(\max\{n_\flat,p_\flat\})$,

- $q_\sharp \leq \widehat{\mathbf{u}}^{\mathfrak{i}(B,D_\flat^0,D_\flat^1,E_\flat)}(\max\{n_\flat,p_\flat\})$,

- $\widehat{\mathbf{k}}_\sharp$ *is* $\widehat{\mathbf{u}}$-*bounded by* $\mathfrak{a}(B,D_\flat^1,BE_\flat)x + \mathfrak{i}(B,D_\flat^0,D_\flat^1,E_\flat)$ *above* $\max\{m_\sharp,q_\sharp\}$,

- $\widehat{\mathbf{r}}_\sharp$ *is* $\widehat{\mathbf{u}}$-*bounded by* $\mathfrak{a}(B,D_\flat^1,BE_\flat)x + \mathfrak{i}(B,D_\flat^0,D_\flat^1,E_\flat)$ *above* $\max\{m_\sharp,q_\sharp\}$,

- *Setting* $\widehat{\mathbf{l}}_\flat = \widehat{\mathbf{L}}_\flat(m_\sharp,q_\sharp,\widehat{\mathbf{k}}_\sharp,\widehat{\mathbf{r}}_\sharp)$, $\widehat{\mathbf{s}}_\flat = \widehat{\mathbf{S}}_\flat(m_\sharp,q_\sharp,\widehat{\mathbf{k}}_\sharp,\widehat{\mathbf{r}}_\sharp)$, $k_\sharp = \widehat{\mathbf{k}}_\sharp(\widehat{\mathbf{l}}_\flat,\widehat{\mathbf{s}}_\flat)$ *and* $r_\sharp = \widehat{\mathbf{r}}_\sharp(\widehat{\mathbf{l}}_\flat,\widehat{\mathbf{s}}_\flat)$, *we have* $k_\sharp \geq m_\sharp$, $r_\sharp \geq q_\sharp$, *and if* $l_\flat \geq k_\sharp$ *and* $s_\flat \geq r_\sharp$ *then for any* $s \in [r_\sharp,\widehat{\mathbf{s}}_\flat(k_\sharp,r_\sharp)]$ *and* $l \in [k_\sharp,\widehat{\mathbf{l}}_\flat(k_\sharp,r_\sharp)]$, *there are sets* $\mathcal{B}^-$, $\mathcal{B}^{0,-}$, *and* $\mathcal{B}^{1,-}$ *with* $\mu(\mathcal{B}^-) < 4/D_\flat^0$, $\mu(\mathcal{B}^{0,-}) < 2/D_\flat^0$, *and* $\mu(\mathcal{B}^{1,-}) < 2/D_\flat^1$ *so that*

$$\left|(\rho_{m_\sharp}\lambda_s)(\Omega) - (\rho_l\lambda_{q_\sharp})(\Omega)\right| \leq |\rho_{m_\sharp}\lambda_s|(\mathcal{B}^-) + |\rho_l\lambda_{q_\sharp}|(\mathcal{B}^-)$$
$$+ BD_\flat^0|\rho_{m_\sharp}|(\mathcal{B}^{0,-}) + |\rho_{m_\sharp}\lambda_s|(\mathcal{B}^{0,-})$$
$$+ BD_\flat^0|\lambda_{q_\sharp}|(\mathcal{B}^{1,-}) + |\rho_l\lambda_{q_\sharp}|(\mathcal{B}^{1,-})$$
$$+ \frac{6}{E_\flat}.$$

Proof. We examine the proof of Lemma 16.

The functions $\widehat{\mathbf{B}}_\star^i$ are particularly simple: $|\widehat{\mathbf{B}}_\star^i(\mathcal{B},\cdot)| \leq 2D_\flat^1 B^2 E_\flat|\mathcal{B}|$ no matter what the second input is. This means that when we apply the previous lemma, we will do so with $z = 2D_\flat^1 B^2 E_\flat$, which means that the set $\mathcal{B}_0$ we ultimately consider will have

$$|\mathcal{B}_0| \leq (2D_\flat^1 B^2 E_\flat)^{\mathfrak{b}(B,D_\flat^1,BE_\flat)(\mathfrak{b}(B,D_\flat^1,BE_\flat)+2)}.$$

We write $\mathfrak{h}_0(B, D_\flat^1, E_\flat) = (2D_\flat^1 B^2 E_\flat)^{\mathfrak{b}(B,D_\flat^1,BE_\flat)(\mathfrak{b}(B,D_\flat^1,BE_\flat)+2)}$.

Suppose that, as in the proof, we are given $n_0, p_0, \widehat{\mathbf{k}}_0, \widehat{\mathbf{r}}_0, \mathcal{B}_0$ with $|\mathcal{B}_0| \leq \mathfrak{h}_0(B, D_\flat^1, E_\flat)$, $\widehat{\mathbf{k}}_0$ is $\widehat{\mathbf{u}}$-bounded by $h_1(x,y)$ above $\max\{n_\flat, p_\flat, n_0, p_0\}$, and $\widehat{\mathbf{r}}_0$ is $\widehat{\mathbf{u}}$-bounded by $h_2(x,y)$ above $\max\{n_\flat, p_\flat, n_0, p_0\}$.

Suppose we are further given the values $m_\dagger$ and $q_\ddagger$. Then

- $\widehat{\mathbf{l}}_{\ddagger,q_\ddagger} = \widehat{\mathbf{L}}_\flat(m_\dagger, q_\ddagger, \widehat{\mathbf{k}}_{\ddagger,q_\ddagger}, \widehat{\mathbf{r}}_{\ddagger,q_\ddagger})$ is $\widehat{\mathbf{u}}$-bounded by 1 above
$$\max\{n_\flat, p_\flat, n_0, p_0, m_\dagger, q_\ddagger + 1\},$$

- $\widehat{\mathbf{s}}_{\ddagger,q_\ddagger} = \widehat{\mathbf{S}}_\flat(m_\dagger, q_\ddagger, \widehat{\mathbf{k}}_{\ddagger,q_\ddagger}, \widehat{\mathbf{r}}_{\ddagger,q_\ddagger})$ is $\widehat{\mathbf{u}}$-bounded by 1 above
$$\max\{n_\flat, p_\flat, n_0, p_0, m_\dagger + 1, q_\ddagger\},$$

- $\widehat{\mathbf{k}}_{\ddagger,q_\ddagger}(\widehat{\mathbf{l}}_{\ddagger,q_\ddagger}, \widehat{\mathbf{s}}_{\ddagger,q_\ddagger}) = \widehat{\mathbf{k}}_0(m_\dagger, q_\ddagger, \widehat{\mathbf{l}}_{\ddagger,q_\ddagger}, \widehat{\mathbf{s}}_{\ddagger,q_\ddagger}, \dots)$ is bounded by
$$\widehat{\mathbf{u}}^{h_1(1,1)}(\max\{n_\flat, p_\flat, n_0, p_0, m_\dagger + 1, q_\ddagger + 1\}),$$

- $\widehat{\mathbf{r}}_{\ddagger,q_\ddagger}(\widehat{\mathbf{l}}_{\ddagger,q_\ddagger}, \widehat{\mathbf{s}}_{\ddagger,q_\ddagger}) = \widehat{\mathbf{r}}_0(m_\dagger, q_\ddagger, \widehat{\mathbf{l}}_{\ddagger,q_\ddagger}, \widehat{\mathbf{s}}_{\ddagger,q_\ddagger}, \cdots)$ is bounded by
$$\widehat{\mathbf{u}}^{h_2(1,1)}(\max\{n_\flat, p_\flat, n_0, p_0, m_\dagger + 1, q_\ddagger + 1\}),$$

- $\widehat{\mathbf{q}}_\ddagger$ is $\widehat{\mathbf{u}}$-bounded by $1 + h_2(1,1)$ above $\max\{n_\flat, p_\flat, n_0, p_0, m_\dagger + 1\}$.

Since the value $q_\ddagger$ is obtained by an application of Lemma 6 to $\widehat{\mathbf{q}}_\ddagger$, it follows that $q_\ddagger$ is bounded by

$$\widehat{\mathbf{u}}^{\mathfrak{a}_0(B,3D_\flat^0,|\mathcal{B}_0|BD_\flat^0 E_\flat)(1+h_2(1,1))}(\max\{n_\flat, p_\flat, n_0, p_0, m_\dagger + 1\}).$$

We write θ for $\mathfrak{a}_0(B, 3D_\flat^0, \mathfrak{h}_0(B, D_\flat^1, E_\flat)BD_\flat^0 E_\flat)$; note that θ is exponential in $D_\flat^0$ and double exponential in B, $D_\flat^1$ and $E_\flat$, and we have $q_\ddagger \leq \widehat{\mathbf{u}}^{\theta(1+h_2(1,1))}(\max\{n_\flat, p_\flat, n_0, p_0, m_\dagger + 1\})$.

In particular, the function $m_\dagger \mapsto q_\ddagger$ is $\widehat{\mathbf{u}}$-bounded by $\theta(1+h_2(1,1))+1$ above
$\max\{n_\flat, p_\flat, n_0, p_0\}$.

Given $m_\dagger$, the value $\widehat{\mathbf{k}}_{\ddagger,q_\ddagger}(\widehat{\mathbf{l}}_{\ddagger,q_\ddagger}, \widehat{\mathbf{r}}_{\ddagger,q_\ddagger})$ is bounded by

$$\widehat{\mathbf{u}}^{h_1(1,1)+\theta(1+h_2(1,1))}(\max\{n_\flat, p_\flat, n_0, p_0, m_\dagger + 1\}),$$

so $\widehat{\mathbf{m}}_\dagger$ is $\widehat{\mathbf{u}}$-bounded by $2 + h_1(1,1) + \theta(1 + h_2(1,1))$ above
$\max\{n_\flat, p_\flat, n_0, p_0\}$. Therefore $m_\dagger$ is bounded by

$$\widehat{\mathbf{u}}^{\theta(2+h_1(1,1)+\theta(1+h_2(1,1)))}(\max\{n_\flat, p_\flat, n_0, p_0\}).$$

We now prepare to apply the previous lemma:

- Each $\widehat{\mathbf{B}}_0^i$ satisfies $|\widehat{\mathbf{B}}_0^i(n_0, p_0, \widehat{\mathbf{k}}_0, \widehat{\mathbf{r}}_0, \mathcal{B}_0)| \leq z|\mathcal{B}_0|$,
- $\widehat{\mathbf{m}}_0(n_0, p_0, \widehat{\mathbf{k}}_0, \widehat{\mathbf{r}}_0, \mathcal{B}_0)$ is $\widehat{\mathbf{u}}$-bounded by $\theta(2+h_1(1,1)+\theta(1+h_2(1,1)))$ above $\max\{n_\flat, p_\flat\}$,
- $\widehat{\mathbf{q}}_0(n_0, p_0, \widehat{\mathbf{k}}_0, \widehat{\mathbf{r}}_0, \mathcal{B}_0)$ is $\widehat{\mathbf{u}}$-bounded by $\theta(2+h_1(1,1)+\theta(1+h_2(1,1)))+\theta(1 + h_2(1,1))$ above $\max\{n_\flat, p_\flat\}$,
- For any $n_0, p_0, \widehat{\mathbf{k}}_0, \widehat{\mathbf{r}}_0, \mathcal{B}_0$, the function
$$m \mapsto \widehat{\mathbf{L}}_0(n_0, p_0, \widehat{\mathbf{k}}_0, \widehat{\mathbf{r}}_0, \mathcal{B}_0)(m, q)$$
does not depend on q and is $\widehat{\mathbf{u}}$-bounded by 1 above
$$\max\{n_\flat, p_\flat, n_0, p_0, m_\dagger, q_\ddagger + 1\},$$
- For any $n_0, p_0, \widehat{\mathbf{k}}_0, \widehat{\mathbf{r}}_0, \mathcal{B}_0$, the function
$$q \mapsto \widehat{\mathbf{S}}_0(n_0, p_0, \widehat{\mathbf{k}}_0, \widehat{\mathbf{r}}_0, \mathcal{B}_0)(m, q)$$
does not depend on m and is $\widehat{\mathbf{u}}$-bounded by 1 above
$$\max\{n_\flat, p_\flat, n_0, p_0, m_\dagger + 1, q_\ddagger\}.$$

This puts us in the setting of the previous lemma with $\rho = \theta^2+2\theta+1$,
so we obtain

- $n_0 \leq \widehat{\mathbf{u}}^{f(\rho; B, D_\flat^1, BE_\flat)}(\max\{n_\flat, p_\flat\})$,
- $p_0 \leq \widehat{\mathbf{u}}^{g(\rho; B, D_\flat^1, BE_\flat)}(\max\{n_\flat, p_\flat\})$,
- $\widehat{\mathbf{k}}_0$ is $\widehat{\mathbf{u}}$-bounded by $\mathfrak{a}(B, D_\flat^1, BE_\flat)x + \mathfrak{f}(\rho; B, D_\flat^1, BE_\flat)$ above $\max\{n_\flat, p_\flat, n_0, p_0\}$,
- $\widehat{\mathbf{r}}_0$ is $\widehat{\mathbf{u}}$-bounded by $\mathfrak{a}(B, D_\flat^1, BE_\flat)y + \mathfrak{g}(\rho; B, D_\flat^1, BE_\flat)$ above $\max\{n_\flat, p_\flat, n_0, p_0\}$.

Therefore, plugging $h_1(x,y) = \mathfrak{a}(B, D_\flat^1, BE_\flat)x + \mathfrak{f}(\rho; B, D_\flat^1, BE_\flat)$ and $h_2(x,y) = \mathfrak{a}(B, D_\flat^1, BE_\flat)y + \mathfrak{g}(\rho; B, D_\flat^1, BE_\flat)$ in to the equations above, we obtain the desired bounds:

- $m_\sharp \leq \widehat{\mathbf{u}}^{i(B, D_\flat^0, D_\flat^1, E_\flat)}(\max\{n_\flat, p_\flat\})$,

- $q_\sharp \leq \widehat{u}^{i(B,D_\flat^0,D_\flat^1,E_\flat)}(\max\{n_\flat, p_\flat\})$,

- $\widehat{k}_\sharp$ is $\widehat{u}$-bounded by $a(B, D_\flat^1, BE_\flat)x + i(B, D_\flat^0, D_\flat^1, E_\flat)$ above $\max\{m_\sharp, q_\sharp\}$,

- $\widehat{r}_\sharp$ is $\widehat{u}$-bounded by $a(B, D_\flat^1, BE_\flat)x + i(B, D_\flat^0, D_\flat^1, E_\flat)$ above $\max\{m_\sharp, q_\sharp\}$.

⊣

7.3 Fast Growing Functions

At this point our bounds start growing much more rapidly. Suppose we have fixed a function $\widehat{u}$. We define:

- $C = 2^{29} B^6 E_\flat^6$,

- $\widehat{w}_{0,\widehat{u},B,E}(m) = \widehat{u}(m)$,

- $\widehat{w}_{j+1,\widehat{u},B,E}(m) = \widehat{w}_{j,\widehat{u},E}^{a(B,E2^{m+5},BE)C^2+i(B,E2^{m+5},E2^{m+5},E)C}(m)$.

We will ultimately be interested in the case where $\widehat{u} = \text{suc}$ where $\text{suc}(m) = m+1$. Observe that in this case, $\widehat{w}_{1,\text{suc},E}$ is triply exponential, $\widehat{w}_{2,\text{suc},E}(m)$ is a tower of exponents of size roughly triply exponential in m. To express the bounds more generally, recall the fast-growing hierarchy:

- $f_0(m) = m + 1$,

- $f_{j+1}(m) = f_j^m(m)$.

Then we have

Lemma 25. *There is a c so that for all j, m, E, B, $\widehat{w}_{j,\text{suc},E}(m) \leq f_{2j+1}(m + E + B + c)$.*

Proof. Since $a(B, E2^{m+5}, BE)C^2 + i(B, E2^{m+5}, E2^{m+5}, E)C$ is triply exponential in m while f_3 is tower exponential, there is a c so that for all m, E, B, $a(B, E2^{m+5}, BE)C^2 + i(B, E2^{m+5}, E2^{m+5}, E)C \leq f_3(m + E + B + c)$. (Indeed, $c = 5$ suffices.)

When $j = 0$ the statement is immediate, since $\widehat{w}_{0,\text{suc},E}(m) = m+1 \leq f_3(m + c)$.

Suppose the claim holds for j. Then

$$\widehat{w}_{j+1,\text{suc},E}(m) = \widehat{w}_{j,\widehat{u},E}^{a(B,E2^{m+5},BE)C^2+i(B,E2^{m+5},E2^{m+5},E)C}(m).$$

Therefore

$$\widehat{\mathbf{w}}_{j+1,\text{suc},E}(m) = \widehat{\mathbf{w}}_{j,\widehat{\mathbf{u}},E}^{\mathfrak{a}(B,E2^{m+5},BE)C^2+\mathfrak{i}(B,E2^{m+5},E2^{m+5},E)C}(m)$$

$$\leq \widehat{\mathbf{w}}_{j,\widehat{\mathbf{u}},E}^{f_3(m+c)}(m)$$

$$\leq f_{2j+1}^{f_3(m+c)}(m+E+B+c)$$

$$\leq f_{2j+1}^{f_{2j+1}(m+c)}(m+E+B+c)$$

$$\leq f_{2j+2}^{m+c}(m+E+B+c)$$

$$= f_{2j+3}(m+E+B+c).$$

$\dashv$

7.4 Bounds on Uniform Continuity

Before continuing, we need bounds on Theorem 4 and Theorem 9.

Lemma 26. *[Bounds on Theorem 4] Suppose* $(*)^Q$ *holds. Let* $E_\flat$, $D_\flat$, $n_\flat$ *be given. Suppose* $\widehat{\mathbf{m}}_\flat(E_\flat 2^{m_0+5}, m_0) \leq \widehat{\mathbf{v}}(m_0)$ *for all* $m_0 \geq \max\{n_\flat, \ln D_\flat\}$. *Then there is an* $m_\sharp \leq \widehat{\mathbf{v}}^{2^7 B^2 E_\flat^2}(\max\{n_\flat, \ln D_\flat\})$ *such that, setting* $D_\sharp = 2^{m_\sharp} \geq D_\flat$, *for each* $m \in [m_\sharp, \widehat{\mathbf{m}}(D_\sharp, m_\sharp)]$, *whenever* $\mu(\sigma) < 1/D_\sharp$, $|\nu_m(\sigma)| < 1/E_\flat$.

Proof. Examining the proof of Theorem 4, the function $\widehat{\mathbf{m}}(m_0)$ is just $\widehat{\mathbf{m}}_\flat(E_\flat 2^{m_0+5}, m_0)$, which is bounded by $\widehat{\mathbf{v}}(m_0)$. The sequence of values m_i are given by $m_0 = \max\{n_\flat, \ln D_\flat\}$ (to ensure that $E_\flat 2^{m_0+5} \geq D_\flat$), $m_{i+1} \leq \widehat{\mathbf{m}}_\flat(E_\flat 2^{m_i+5}, m_i) \leq \widehat{\mathbf{v}}(m_i)$. Since we have bounded fluctuations, we need only consider $m_{2^7 B^2 E_\flat^2}$, so

$$m_\sharp \leq \widehat{\mathbf{v}}^{2^7 B^2 E_\flat^2}(\max\{n_\flat, \ln D_\flat\})$$

as claimed.

$\dashv$

Lemma 27. *[Bounds on Lemma 8] Suppose* $(*)^Q$ *holds. For any* $d, E_\flat$, $\widehat{\mathbf{m}}_\flat$, $\widehat{\mathbf{q}}_\flat$, $n_\flat$, $p_\flat$ *such that:*

- *For any* $D, m, q, \widehat{\mathbf{r}}$ *such that* $\widehat{\mathbf{r}}$ *is* $\widehat{\mathbf{w}}_{j,\widehat{\mathbf{u}},E_\flat}$-*bounded by* C *above* $\max\{m, q, \ln D, n_\flat, p_\flat\}$,

$$\widehat{\mathbf{m}}_\flat(D, m, q, \widehat{\mathbf{r}}) \leq \widehat{\mathbf{w}}_{j+d,\widehat{\mathbf{u}},E_\flat}^{2^{12}B^4 E_\flat^4}(\max\{m, q, \ln D, n_\flat, p_\flat\}),$$

- *For any* $D, m, q, \widehat{\mathbf{r}}$ *such that* $\widehat{\mathbf{r}}$ *is* $\widehat{\mathbf{w}}_{j,\widehat{\mathbf{u}},E_\flat}$-*bounded by* C *above* $\max\{m, q, \ln D, n_\flat, p_\flat\}$,

$$\widehat{\mathbf{q}}_\flat(D, m, q, \widehat{\mathbf{r}}) \leq \widehat{\mathbf{w}}_{j+d,\widehat{\mathbf{u}},E_\flat}^{2^{12}B^4 E_\flat^4}(\max\{m, q, \ln D, n_\flat, p_\flat\}),$$

there exist $D_\sharp, m_\sharp \geq n_\flat, q_\sharp \geq p_\flat, \widehat{\mathbf{r}}_\sharp$ *such that:*

- $m_\sharp \leq \widehat{\mathbf{w}}_{d(2^9 B^2 E_\flat^2),\widehat{\mathbf{u}},E_\flat}^C(\max\{n_\flat, p_\flat\})$

- $q_\sharp \leq \widehat{\mathbf{w}}^C_{d(2^9 B^2 E_\flat^2),\widehat{\mathbf{u}},E_\flat}(\max\{n_\flat, p_\flat\})$,
- $D_\sharp = E_\flat 2^{q_\sharp + 5}$,
- $\widehat{\mathbf{r}}_\sharp$ is $\widehat{\mathbf{w}}^C_{d(2^9 B^2 E_\flat^2),\widehat{\mathbf{u}},E_\flat}$-bounded by 1 above $\max\{m_\sharp, q_\sharp, n_\flat, p_\flat\}$,

and, setting $m_\flat = \widehat{\mathbf{m}}_\flat(D_\sharp, m_\sharp, q_\sharp, \widehat{\mathbf{r}}_\sharp)$, if $q_\flat = \widehat{\mathbf{q}}_\flat(D_\sharp, m_\sharp, q_\sharp, \widehat{\mathbf{r}}_\sharp) \geq q_\sharp$ then $r_\sharp = \widehat{\mathbf{r}}_\sharp(m_\flat, q_\flat) \geq q_\flat$, there is a σ_0 such that whenever $\mu(\sigma_0 \triangle \sigma) < 1/D_\sharp$, $|(\rho_{m_\sharp} \lambda_{r_\sharp})(\sigma) - (\rho_{m_\flat} \lambda_{r_\sharp})(\sigma)| < 4/E_\flat$, and whenever $\mu(\sigma) < 2/D_\sharp$, $|(\rho_{m_\sharp} \lambda_{r_\sharp})(\sigma)| < 1/16E_\flat$.

Proof. We need the following inductive hypothesis: for each i, D, n, p there is a function $\widehat{\mathbf{r}}_{i,D,n,p}$ which is $\widehat{\mathbf{w}}_{di,\widehat{\mathbf{u}},E_\flat}$-bounded by C above

$$\max\{n, p, \ln D, n_\flat, p_\flat\}$$

so that for each m, q, either:

- There exist $D_\sharp, m_\sharp, q_\sharp, \widehat{\mathbf{r}}_\sharp$ satisfying the conclusion of Lemma 4.10 with:
  - $m_\sharp \leq \widehat{\mathbf{w}}^{C+1}_{di,\widehat{\mathbf{u}},E_\flat}(\max\{m, q, n, p, n_\flat, p_\flat\})$
  - $q_\sharp \leq \widehat{\mathbf{w}}^{C+1}_{di,\widehat{\mathbf{u}},E_\flat}(\max\{m, q, n, p, n_\flat, p_\flat\})$,
  - $D_\sharp = E_\flat 2^{q_\sharp + 5}$,
  - $\widehat{\mathbf{r}}_\sharp$ is $\widehat{\mathbf{w}}^C_{di,\widehat{\mathbf{u}},E_\flat}$-bounded by 1 above $\max\{m_\sharp, q_\sharp, m, q, n, p, n_\flat, p_\flat\}$,

  or,
- one of the other cases in the proof Lemma 4.10 holds.

For $\widehat{\mathbf{r}}_{0,D,n,p}(m, q) = q$ this is immediate since $q \leq \widehat{\mathbf{u}}(q)$ for all q.

Suppose we have shown that for every D, n, p, $\widehat{\mathbf{r}}_{i,D,n,p}$ is $\widehat{\mathbf{w}}_{di,\widehat{\mathbf{u}},E_\flat}$-bounded by C above $\max\{n, p, \ln D, n_\flat, p_\flat\}$. We now attempt to bound $\widehat{\mathbf{r}}_{i+1,D,n,p}$. Let m, q be given. Observe that

$$\widehat{\mathbf{m}}_\flat(D^*, m, q^*, \widehat{\mathbf{r}}_{i,D^*,m,q^*}) \leq \widehat{\mathbf{w}}^{2^{12}B^4 E_\flat^4}_{d(i+1),\widehat{\mathbf{u}},E_\flat}(u)$$

where $u = \max\{q^*, \ln D^*, m, q, n, p, n_\flat, p_\flat\}$ and similarly for $\widehat{\mathbf{q}}_\flat$, so

$$\widehat{\mathbf{r}}^*(D^*, q^*) \leq \widehat{\mathbf{w}}^{2^{12}B^4 E_\flat^4+1}_{d(i+1),\widehat{\mathbf{u}},E_\flat}(u).$$

In particular,

$$\widehat{\mathbf{r}}^*(E_\flat 2^{q^*+5}, q^*) \leq \widehat{\mathbf{w}}^{(2^{12}B^4 E_\flat^4)+1}_{d(i+1),\widehat{\mathbf{u}},E_\flat}(u')$$

where $u' = \max\{q^* + \ln E_\flat + 1, m, q, n, p, \ln D + \ln 2, n_\flat, p_\flat\}$, so by Lemma 26 applied to $16E_\flat$ we obtain

$$q^* \leq \widehat{\mathbf{w}}^{((2^{12}B^4 E_\flat^4)+2)2^{15}B^2 E_\flat^2}_{d(i+1),\widehat{\mathbf{u}},E_\flat}(\max\{m, q, n, p, \ln D, n_\flat, p_\flat\})$$

and $D^* = E_\flat 2^{q^*+5}$. Then

$$\widehat{\mathbf{r}}_{i+1,D,n,p}(m, q) \leq \widehat{\mathbf{w}}^C_{d(i+1),\widehat{\mathbf{u}},E_\flat}(\max\{m, q, n, p, \ln D, n_\flat, p_\flat\}).$$

as required.

We also need to bound the possible witnesses $D_\sharp, m_\sharp, q_\sharp, \widehat{\mathbf{r}}_\sharp$. We take $m' = \widehat{\mathbf{m}}_\flat(D^*, m, q^*, \widehat{\mathbf{r}}_{i,D^*,m,q^*})$ and $q' = \widehat{\mathbf{q}}_\flat(D^*, m, q^*, \widehat{\mathbf{r}}_{i,D^*,m,q^*})$, and have

$$m', q' \leq \widehat{\mathbf{w}}^C_{d(i+1),\widehat{\mathbf{u}},E_\flat}(\max\{m, q, n, p, \ln D, n_\flat, p_\flat\}).$$

In particular, when we apply the inductive hypothesis to $\widehat{\mathbf{r}}_{i,D^*,m,q^*}(m', q')$, we potentially obtain $D_\sharp, m_\sharp, q_\sharp, \widehat{\mathbf{r}}_\sharp$ with

$$m_\sharp, q_\sharp \leq \widehat{\mathbf{w}}^{C+1}_{di,\widehat{\mathbf{u}},E_\flat}(\max\{m', q', m, q^*, \ln D^*, n_\flat, p_\flat\})$$

$$\leq \widehat{\mathbf{w}}^{C+1}_{d(i+1),\widehat{\mathbf{u}},E_\flat}(\max\{n, p, m, q, \ln D, n_\flat, p_\flat\})$$

which satisfies the promised bounds.

If we choose $m_\sharp = m, q_\sharp = q^*, D_\sharp = D^*, \widehat{\mathbf{r}}_\sharp = \widehat{\mathbf{r}}_{i,D^*,m,q^*}$ then again the promised bounds hold.

In the proof, we work with

$$\widehat{\mathbf{r}}^*(D^*, q^*) = \widehat{\mathbf{r}}_{2^9 B^2 E_\flat^2, D^*, 1, q^*}(\widehat{\mathbf{m}}_\flat(D^*, 1, q^*, \widehat{\mathbf{r}}_{2^9 B^2 E_\flat^2, D^*, 1, q^*}),$$

$$\widehat{\mathbf{q}}_\flat(D^*, 1, q^*, \widehat{\mathbf{r}}_{2^9 B^2 E_\flat^2, D^*, 1, q^*})).$$

In particular, as above

$$\widehat{\mathbf{r}}^*(E_\flat 2^{q^*+5}, q^*) \leq \widehat{\mathbf{w}}^{2^{12} B^4 E_\flat^4 + 1}_{d(2^9 B^2 E_\flat^2 + 1),\widehat{\mathbf{u}},E_\flat}(\max\{q^* + \ln E_\flat + 1, n_\flat, p_\flat\})$$

and therefore when we apply Lemma 26 to $16E_\flat$, we obtain $q^* \leq \widehat{\mathbf{w}}^{2^{28} B^6 E_\flat^6}_{d(2^9 B^2 E_\flat^2 + 1),\widehat{\mathbf{u}},E_\flat}(\max\{n_\flat, p_\flat\})$. In particular, whatever case we are in, we have the specified bounds. $\dashv$

Lemma 28. *[Bounds on Lemma 9] Suppose $(*)^Q$ holds. Let $d, E_\flat, n_\flat, p_\flat, \widehat{\mathbf{m}}_\flat, \widehat{\mathbf{q}}_\flat, d$ be given so that:*

- *For any $D, m, q, \widehat{\mathbf{r}}$ such that $\widehat{\mathbf{r}}$ is $\widehat{\mathbf{w}}_{j,\widehat{\mathbf{u}},E_\flat}$-bounded by C above $\max\{m, q, \ln D, n_\flat, p_\flat\}$,*

$$\widehat{\mathbf{m}}_\flat(D, m, q, \widehat{\mathbf{r}}) \leq \widehat{\mathbf{w}}^{2^{12} B^4 E_\flat^4}_{j+d,\widehat{\mathbf{u}},E_\flat}(\max\{m, q, \ln D+, n_\flat, p_\flat\}),$$

- *For any $D, m, q, \widehat{\mathbf{r}}$ such that $\widehat{\mathbf{r}}$ is $\widehat{\mathbf{w}}_{j,\widehat{\mathbf{u}},E_\flat}$-bounded by C above $\max\{m, q, \ln D, n_\flat, p_\flat\}$,*

$$\widehat{\mathbf{q}}_\flat(D, m, q, \widehat{\mathbf{r}}) \leq \widehat{\mathbf{w}}^{2^{12} B^4 E_\flat^4}_{j+d,\widehat{\mathbf{u}},E_\flat}(\max\{m, q, \ln D, n_\flat, p_\flat\}),$$

Then there are $D_\sharp, m_\sharp, p_\sharp, \widehat{\mathbf{r}}_\sharp$ so that:

- $m_\sharp \leq \widehat{\mathbf{w}}^C_{d(2^9 B^2 E_\flat^2),\widehat{\mathbf{u}},E_\flat}(\max\{n_\flat, p_\flat\})$
- $q_\sharp \leq \widehat{\mathbf{w}}^C_{d(2^9 B^2 E_\flat^2),\widehat{\mathbf{u}},E_\flat}(\max\{n_\flat, p_\flat\})$,
- $D_\sharp = E_\flat 2^{q_\sharp + 5}$,
- $\widehat{\mathbf{r}}_\sharp$ *is $\widehat{\mathbf{w}}^C_{d(2^9 B^2 E_\flat^2),\widehat{\mathbf{u}},E_\flat}$-bounded by 1 above $\max\{m_\sharp, q_\sharp, n_\flat, p_\flat\}$,*

- If $\widehat{\mathbf{m}}_\flat(D_\sharp, m_\sharp, q_\sharp, \widehat{\mathbf{r}}_\sharp) \geq m_\sharp$ and $\widehat{\mathbf{q}}_\flat(D_\sharp, m_\sharp, p_\sharp, \widehat{\mathbf{r}}_\sharp, m) \geq p_\sharp$ then

$$\widehat{\mathbf{r}}_\sharp(\widehat{\mathbf{m}}_\flat(D_\sharp, m_\sharp, q_\sharp, \widehat{\mathbf{r}}_\sharp), \widehat{\mathbf{q}}_\flat(D_\sharp, m_\sharp, p_\sharp, \widehat{\mathbf{r}}_\sharp)) \geq \widehat{\mathbf{q}}_\flat(D_\sharp, m_\sharp, p_\sharp, \widehat{\mathbf{r}}_\sharp)$$

and for any σ with $\mu(\sigma) < 1/D_\sharp$,

$$|(\rho_{\widehat{\mathbf{m}}_\flat(D_\sharp, m_\sharp, q_\sharp, \widehat{\mathbf{r}}_\sharp)} \lambda_{\widehat{\mathbf{r}}_\sharp(\widehat{\mathbf{m}}_\flat(D_\sharp, m_\sharp, q_\sharp, \widehat{\mathbf{r}}_\sharp), \widehat{\mathbf{q}}_\flat(D_\sharp, m_\sharp, p_\sharp, \widehat{\mathbf{r}}_\sharp))})(\sigma)| < 1/E_\flat.$$

Proof. Theorem 9 follows by applying Lemma 8 to $4E_\flat, \widehat{\mathbf{m}}_\flat, \widehat{\mathbf{p}}_\flat, 0, 0$. Lemma 27 is already the $4E_\flat$ case, so we simply apply the previous lemma to obtain the desired bounds. $\dashv$

7.5 Refining the Bounds

Lemma 29. *[Bounds on 17] Suppose* $(*)^Q$ *holds. Let* j, $E_\flat$, $D_\flat^0 \leq D_\flat^1$, $n_\flat$, $p_\flat$, $\widehat{\mathbf{L}}_\flat$, $\widehat{\mathbf{S}}_\flat$ *be given so that:*

- *For any* $m, q, \widehat{\mathbf{k}}, \widehat{\mathbf{r}}$, *the function* $k \mapsto \widehat{\mathbf{L}}_\flat(m, q, \widehat{\mathbf{k}}, \widehat{\mathbf{r}})(k, r)$ *does not depend on* r *and is* $\widehat{\mathbf{w}}_{j,\widehat{\mathbf{u}}, E_\flat}$-*bounded by* C *above* $\max\{n_\flat, p_\flat, m, q+1\}$,
- *For any* $m, q, \widehat{\mathbf{k}}, \widehat{\mathbf{r}}$, *the function* $r \mapsto \widehat{\mathbf{S}}_\flat(m, q, \widehat{\mathbf{k}}, \widehat{\mathbf{r}})(k, r)$ *does not depend on* k *and is* $\widehat{\mathbf{w}}_{j,\widehat{\mathbf{u}}, E_\flat}$-*bounded by* C *above* $\max\{n_\flat, p_\flat, m+1, q\}$.

Then there are $m_\sharp \geq n_\flat$, $q_\sharp \geq p_\flat$, $\widehat{\mathbf{k}}_\sharp$, *and* $\widehat{\mathbf{r}}_\sharp$ *such that, setting* $c_\sharp = \widehat{\mathbf{w}}_{j+1,\widehat{\mathbf{u}}, E_\flat}^{2^7 B^2 E_\flat^2}(\max\{n_\flat, p_\flat, \ln D_\flat^1\})$:

- $m_\sharp \leq \widehat{\mathbf{w}}_{j+1,\widehat{\mathbf{u}}, E_\flat}(c_\sharp)$,
- $q_\sharp \leq \widehat{\mathbf{w}}_{j+1,\widehat{\mathbf{u}}, E_\flat}(c_\sharp)$,
- $\widehat{\mathbf{k}}_\sharp$ *is* $\widehat{\mathbf{w}}_{j,\widehat{\mathbf{u}}, E_\flat}$-*bounded by*

$$\mathfrak{a}(B, E_\flat 2^{c_\sharp+5}, BE_\flat)Cx + \mathfrak{i}(B, D_\flat^0, E_\flat 2^{c_\sharp+5}, E_\flat)C$$

above $\max\{m_\sharp, q_\sharp\}$,

- $\widehat{\mathbf{r}}_\sharp$ *is* $\widehat{\mathbf{w}}_{j,\widehat{\mathbf{u}}, E_\flat}$-*bounded by*

$$\mathfrak{a}(B, E_\flat 2^{c_\sharp+5}, BE_\flat)Cx + \mathfrak{i}(B, D_\flat^0, E_\flat 2^{c_\sharp+5}, E_\flat)C$$

above $\max\{m_\sharp, q_\sharp\}$,

- *Setting* $\widehat{\mathbf{l}}_\flat = \widehat{\mathbf{L}}_\flat(m_\sharp, q_\sharp, \widehat{\mathbf{k}}_\sharp, \widehat{\mathbf{r}}_\sharp)$, $\widehat{\mathbf{s}}_\flat = \widehat{\mathbf{S}}_\flat(m_\sharp, q_\sharp, \widehat{\mathbf{k}}_\sharp, \widehat{\mathbf{r}}_\sharp)$, $k_\sharp = \widehat{\mathbf{k}}_\sharp(\widehat{\mathbf{l}}_\flat, \widehat{\mathbf{s}}_\flat)$ *and* $r_\sharp = \widehat{\mathbf{r}}_\sharp(\widehat{\mathbf{l}}_\flat, \widehat{\mathbf{s}}_\flat)$, *we have* $k_\sharp \geq m_\sharp$, $r_\sharp \geq q_\sharp$, *and for any* $s \in [r_\sharp, \widehat{\mathbf{s}}_\flat(k_\sharp, r_\sharp)]$ *and* $l \in [k_\sharp, \widehat{\mathbf{l}}_\flat(k_\sharp, r_\sharp)]$, *there are sets* $\mathcal{B}^-$, $\mathcal{B}^{0,-}$, *and* $\mathcal{B}^{1,-}$ *with* $\mu(\mathcal{B}^-) < 4/D_\flat^0$, $\mu(\mathcal{B}^{0,-}) < 2/D_\flat^0$, *and* $\mu(\mathcal{B}^{1,-}) < 2/D_\flat^1$ *so that*

$$\begin{aligned}
\left|(\rho_{m_\sharp} \lambda_s)(\Omega) - (\rho_l \lambda_{q_\sharp})(\Omega)\right| &\leq |\rho_{m_\sharp} \lambda_s|(\mathcal{B}^-) + |\rho_l \lambda_{q_\sharp}|(\mathcal{B}^-) \\
&+ |\rho_{m_\sharp} \lambda_s|(\mathcal{B}^{0,-}) + BD_\flat^0 |\lambda_{q_\sharp}|(\mathcal{B}^{1,-}) \\
&+ |\rho_l \lambda_{q_\sharp}|(\mathcal{B}^{1,-}) + \frac{7}{E_\flat}.
\end{aligned}$$

Proof. This lemma and the next amount to combining Lemma 16 with Theorem 4.

Note that being $\widehat{\mathbf{w}}_{j,\widehat{\mathbf{u}},E_\flat}$-bounded by C is the same as being $\widehat{\mathbf{w}}^C_{j,\widehat{\mathbf{u}},E_\flat}$-bounded by 1, so the lemmas in the previous subsections apply. Suppose we have fixed $D_0 \geq D_\flat^1$ and $m_0 \geq \max\{n_\flat, p_\flat\}$. Let

$$c_\sharp = \widehat{\mathbf{w}}_{j,\widehat{\mathbf{u}},E_\flat}^{\mathfrak{i}(B,D_\flat^0,D_0,E_\flat)C}(m_0).$$

In particular,

$$\widehat{\mathbf{m}}_0(E_\flat 2^{m_0+5}, m_0) \leq \widehat{\mathbf{w}}_{j,\widehat{\mathbf{u}},E_\flat}^{\mathfrak{i}(B,E_\flat 2^{m_0+5},E_\flat 2^{m_0+2},E_\flat)C}(m_0) \leq \widehat{\mathbf{w}}_{j+1,\widehat{\mathbf{u}},E_\flat}(m_0).$$

Applying Lemma 26, $m_0 \leq \widehat{\mathbf{w}}_{j+1,\widehat{\mathbf{u}},E_\flat}^{32B^2E_\flat^2}(\max\{n_\flat, p_\flat, \ln D_\flat^1\})$. Setting $c_\sharp = \widehat{\mathbf{w}}_{j+1,\widehat{\mathbf{u}},E_\flat}^{32B^2E_\flat^2}(\max\{n_\flat, p_\flat, \ln D_\flat^1\})$, the remaining bounds follow from the previous subsections. ⊣

Lemma 30. *[Bounds on 18] Suppose* $(*)^Q$ *holds. Let* $E_\flat, D_\flat, n_\flat, p_\flat, \widehat{\mathbf{L}}_\flat, \widehat{\mathbf{S}}_\flat, \widehat{\mathbf{u}},$ *be given so that:*

- *For any* $m, q, \widehat{\mathbf{k}}, \widehat{\mathbf{r}}$, *the function* $k \mapsto \widehat{\mathbf{L}}_\flat(m, q, \widehat{\mathbf{k}}, \widehat{\mathbf{r}})(k, r)$ *does not depend on* r *and is* $\widehat{\mathbf{w}}_{j,\widehat{\mathbf{u}},E_\flat}$*-bounded by* C *above* $\max\{n_\flat, p_\flat, m, q+1\}$,

- *For any* $m, q, \widehat{\mathbf{k}}, \widehat{\mathbf{r}}$, *the function* $r \mapsto \widehat{\mathbf{S}}_\flat(m, q, \widehat{\mathbf{k}}, \widehat{\mathbf{r}})(k, r)$ *does not depend on* k *and is* $\widehat{\mathbf{w}}_{j,\widehat{\mathbf{u}},E_\flat}$*-bounded by* C *above* $\max\{n_\flat, p_\flat, m+1, q\}$.

Then there are $m_\sharp \geq n_\flat$, $q_\sharp \geq p_\flat$, $\widehat{\mathbf{k}}_\sharp$, *and* $\widehat{\mathbf{r}}_\sharp$ *such that, setting* $c_\sharp = \widehat{\mathbf{w}}_{j+1,\widehat{\mathbf{u}},E_\flat}^{2^{11}B^4E_\flat^4}(\max\{n_\flat, p_\flat, \ln D_\flat\})$:

- $m_\sharp \leq \widehat{\mathbf{w}}_{j+1,\widehat{\mathbf{u}},E_\flat}^{32B^2E_\flat^2+1}(c_\sharp)$,

- $q_\sharp \leq \widehat{\mathbf{w}}_{j+1,\widehat{\mathbf{u}},E_\flat}^{32B^2E_\flat^2+1}(c_\sharp)$,

- $\widehat{\mathbf{k}}_\sharp$ *is* $\widehat{\mathbf{w}}_{j,\widehat{\mathbf{u}},E_\flat}$*-bounded by*

$$\mathfrak{a}(B, E_\flat 2^{c_\sharp+5}, BE_\flat)Cx + \mathfrak{i}(B, D_\flat, E_\flat 2^{c_\sharp+5}, E_\flat)C$$

above $\max\{m_\sharp, q_\sharp\}$,

- $\widehat{\mathbf{r}}_\sharp$ *is* $\widehat{\mathbf{w}}_{j,\widehat{\mathbf{u}},E_\flat}$*-bounded by*

$$\mathfrak{a}(B, E_\flat 2^{c_\sharp+5}, BE_\flat)Cy + \mathfrak{i}(B, D_\flat, E_\flat 2^{c_\sharp+5}, E_\flat)C$$

above $\max\{m_\sharp, q_\sharp\}$,

- *Setting* $\widehat{\mathbf{l}}_\flat = \widehat{\mathbf{L}}_\flat(m_\sharp, q_\sharp, \widehat{\mathbf{k}}_\sharp, \widehat{\mathbf{r}}_\sharp)$, $\widehat{\mathbf{s}}_\flat = \widehat{\mathbf{S}}_\flat(m_\sharp, q_\sharp, \widehat{\mathbf{k}}_\sharp, \widehat{\mathbf{r}}_\sharp)$, $k_\sharp = \widehat{\mathbf{k}}_\sharp(\widehat{\mathbf{l}}_\flat, \widehat{\mathbf{s}}_\flat)$ *and* $r_\sharp = \widehat{\mathbf{r}}_\sharp(\widehat{\mathbf{l}}_\flat, \widehat{\mathbf{s}}_\flat)$, *so that* $k_\sharp \geq m_\sharp$, $r_\sharp \geq q_\sharp$, *and for any* $s \in [r_\sharp, \widehat{\mathbf{s}}_\flat(k_\sharp, r_\sharp)]$ *and* $l \in [k_\sharp, \widehat{\mathbf{l}}_\flat(k_\sharp, r_\sharp)]$, *there are sets* $\mathcal{B}^-, \mathcal{B}^{0,-},$ *and* $\mathcal{B}^{1,-}$

with $\mu(\mathcal{B}^-) < 4/D_\flat^0$, $\mu(\mathcal{B}^{0,-}) < 2/D_\flat^0$, and $\mu(\mathcal{B}^{1,-}) < 2/D_\flat^1$ so that

$$\left|(\rho_{m_\sharp}\lambda_s)(\Omega) - (\rho_l\lambda_{q_\sharp})(\Omega)\right| \le |\rho_{m_\sharp}\lambda_s|(\mathcal{B}^-) + |\rho_l\lambda_{q_\sharp}|(\mathcal{B}^-)$$
$$+ |\rho_{m_\sharp}\lambda_s|(\mathcal{B}^{0,-}) + |\rho_l\lambda_{q_\sharp}|(\mathcal{B}^{1,-}) + \frac{8}{E_\flat}.$$

Proof. We again combine the previous lemma with Theorem 4.

The function $q_0 \mapsto \widehat{\mathbf{q}}_0(E_\flat 2^{q_0+5}, q_0)$ is bounded by

$$\widehat{\mathbf{w}}_{j+1,\widehat{\mathbf{u}},E_\flat}^{32B^2E_\flat^2+1}(\max\{q_0, \ln(E_\flat 2^{q_0+5}), n_\flat, p_\flat, \ln D_\flat\})$$

$$\le \widehat{\mathbf{w}}_{j+1,\widehat{\mathbf{u}},E_\flat}^{32B^2E_\flat^2+1}(\max\{n_\flat, q_0\}),$$

so by Lemma 26 we have $q_0 \le \widehat{\mathbf{w}}_{j+1,\widehat{\mathbf{u}},E_\flat}^{2^{11}B^4E_\flat^4}(\max\{n_\flat, p_\flat, \ln D_\flat\})$. Again, the remaining bounds follow from the previous subsection. $\dashv$

Lemma 31. *[Bounds on 19] Suppose* $(*)^Q$ *holds. Let* $E_\flat, D_\flat, p_\flat, n_\flat,$ $\widehat{\mathbf{L}}_\flat, \widehat{\mathbf{r}}_\flat$ *be given so that:*

- *For any* $m, q, \widehat{\mathbf{k}}, \widehat{\mathbf{s}}$, *the function* $k \mapsto \widehat{\mathbf{L}}_\flat(m, q, \widehat{\mathbf{k}}, \widehat{\mathbf{s}})(k, r)$ *does not depend on* r *and is* $\widehat{\mathbf{u}}$*-bounded by* 1 *above* $\max\{n_\flat, p_\flat, m, q+1\}$,
- *For any* $m, q, \widehat{\mathbf{k}}, \widehat{\mathbf{s}}$, $\widehat{\mathbf{r}}_\flat(m, q, \widehat{\mathbf{k}}, \widehat{\mathbf{s}}) = m + 1$.

Then there are $m_\sharp \ge n_\flat, q_\sharp \ge p_\flat, \widehat{\mathbf{k}}_\sharp, \widehat{\mathbf{s}}_\sharp$ *such that:*

- $m_\sharp \le \widehat{\mathbf{w}}_{2^9B^2E_\flat^2+1,\widehat{\mathbf{u}},E_\flat}^{2^{12}B^4E_\flat^4+1}(\max\{\ln D_\flat, n_\flat, p_\flat\})$,
- $q_\sharp \le \widehat{\mathbf{w}}_{2^9B^2E_\flat^2+1,\widehat{\mathbf{u}},E_\flat}^{2^{12}B^4E_\flat^4+1}(\max\{\ln D_\flat, n_\flat, p_\flat\})$,
- $\widehat{\mathbf{k}}_\sharp$ *is* $\widehat{\mathbf{w}}_{2^9B^2E_\flat^2+1,\widehat{\mathbf{u}},E_\flat}$*-bounded by*

$$\mathfrak{a}(B, E_\flat 2^{c_\dagger+5}, BE_\flat)Cx + \mathfrak{i}(B, D_\flat, E_\flat 2^{c_\sharp+5}, E_\flat)C$$

above $\max\{m_\sharp, q_\sharp, \ln D_\flat, n_\flat, p_\flat\}$,
- $\widehat{\mathbf{s}}_\sharp$ *is* $\widehat{\mathbf{w}}_{2^9B^2E_\flat^2+1,\widehat{\mathbf{u}},E_\flat}$*-bounded by*

$$\mathfrak{a}(B, E_\flat 2^{c_\dagger+5}, BE_\flat)Cy + \mathfrak{i}(B, D_\flat, E_\flat 2^{c_\sharp+5}, E_\flat)C$$

above $\max\{m_\sharp, q_\sharp, \ln D_\flat, n_\flat, p_\flat\}$, *and*
- *setting* $\widehat{\mathbf{l}}_\flat = \widehat{\mathbf{L}}_\flat(m_\sharp, q_\sharp, \widehat{\mathbf{k}}_\sharp, \widehat{\mathbf{s}}_\sharp)$, $r_\flat = \widehat{\mathbf{r}}_\flat(m_\sharp, q_\sharp, \widehat{\mathbf{k}}_\sharp, \widehat{\mathbf{s}}_\sharp)$, $k_\sharp = \widehat{\mathbf{k}}_\sharp(\widehat{\mathbf{l}}_\flat, r_\flat)$, *and* $s_\sharp = \widehat{\mathbf{s}}_\sharp(\widehat{\mathbf{l}}_\flat, r_\flat)$, *if* $r_\flat \ge q_\sharp$ *then* $s_\sharp \ge r_\flat$ *and if also* $l_\flat \ge k_\sharp$ *then for any* $l \in [k_\sharp, l_\flat]$ *there are sets* $\mathcal{B}^-$ *and* $\mathcal{B}^{1,-}$ *with* $\mu(\mathcal{B}^-) < 4/D_\flat$, $\mu(\mathcal{B}^{1,-}) < 2/D_\flat$ *so that*

$$\left|(\rho_{m_\sharp}\lambda_{s_\sharp})(\Omega) - (\rho_l\lambda_{q_\sharp})(\Omega)\right| \le |\rho_l\lambda_{q_\sharp}|(\mathcal{B}^-) + |\rho_l\lambda_{q_\sharp}|(\mathcal{B}^{1,-}) + \frac{20}{E_\flat}.$$

Proof. As always, we examine the proof.

Suppose $D_0, m_0, p_0, \widehat{\mathbf{r}}_0$ are fixed and that $\widehat{\mathbf{r}}_0$ is $\widehat{\mathbf{w}}_{j,\widehat{\mathbf{u}},E_\flat}$-bounded by C above $\max\{m_0, p_0, \ln D_0, p_\flat, n_\flat, \ln D_\flat\}$. Then:

- $\widehat{\mathbf{s}}_{m_\dagger,r}(k,r') \leq \widehat{\mathbf{w}}^C_{j,\widehat{\mathbf{u}},E_\flat}(\max\{m_\dagger,r,r',m_0,p_0,\ln D_0,p_\flat,n_\flat,\ln D_\flat\})$,
- If $\widehat{\mathbf{l}}$ is $\widehat{\mathbf{u}}$-bounded by 1 above c, $\widehat{\mathbf{l}}_{m_\dagger,\widehat{\mathbf{l}},r}(k,r') \leq \widehat{\mathbf{u}}(\max\{q_\dagger+1,k,c\})$,
- $\widehat{\mathbf{L}}_\dagger(m_\dagger,q_\dagger,\widehat{\mathbf{k}}_\dagger,\widehat{\mathbf{r}}_\dagger)(k,r) \leq \widehat{\mathbf{u}}(\max\{q_\dagger+1,k,m_\dagger,n_\flat,p_\flat\})$,
- $\widehat{\mathbf{S}}_\dagger(m_\dagger,q_\dagger,\widehat{\mathbf{k}}_\dagger,\widehat{\mathbf{r}}_\dagger)(k,r) = \widehat{\mathbf{s}}_{m_\dagger,m_\dagger+1}(k,r') \leq \widehat{\mathbf{w}}^C_{j,\widehat{\mathbf{u}},E_\flat}(\max\{r',m_\dagger+1,m_0,p_0,p_\flat,n_\flat\})$.

In particular, both $\widehat{\mathbf{u}}$ and $\widehat{\mathbf{r}}_0$ are bounded by $\widehat{\mathbf{w}}^C_{j,\widehat{\mathbf{u}},E_\flat}$. We apply Lemma 30 to

$$E_\flat, \max\{D_0,D_\flat\}, \max\{m_0,n_\flat\}, \max\{p_0,p_\flat\}, \widehat{\mathbf{L}}_\dagger, \widehat{\mathbf{s}}_\dagger, \widehat{\mathbf{w}}_{j,\widehat{\mathbf{u}},E_\flat},$$

and therefore obtain the bounds:

- $c_\dagger = \widehat{\mathbf{w}}^{2^{11}B^4E_\flat^4}_{j+1,\widehat{\mathbf{u}},E_\flat}(\max\{m_0,p_0,\ln D_0,n_\flat,p_\flat,\ln D_\flat\})$,
- $m_\dagger \leq \widehat{\mathbf{w}}^{32B^2E_\flat^2+1}_{j+1,\widehat{\mathbf{u}},E_\flat}(c_\dagger)$,
- $q_\dagger \leq \widehat{\mathbf{w}}^{32B^2E_\flat^2+1}_{j+1,\widehat{\mathbf{u}},E_\flat}(c_\dagger)$,
- $\widehat{\mathbf{k}}_\dagger$ is $\widehat{\mathbf{w}}_{j,\widehat{\mathbf{u}},E_\flat}$-bounded by

$$\mathfrak{a}(B,E_\flat 2^{c_\dagger+5},BE_\flat)Cx + \mathfrak{i}(B,\max\{D_0,D_\flat\},E_\flat 2^{c_\sharp+5},E_\flat)C$$

above $\max\{m_\dagger,q_\dagger,m_0,p_0,\ln D_0,n_\flat,p_\flat,\ln D_\flat\}$,

- $\widehat{\mathbf{r}}_\dagger$ is $\widehat{\mathbf{w}}_{j,\widehat{\mathbf{u}},E_\flat}$-bounded by

$$\mathfrak{a}(B,E_\flat 2^{c_\dagger+5},BE_\flat)Cy + \mathfrak{i}(B,\max\{D_0,D_\flat\},E_\flat 2^{c_\dagger+5},E_\flat)C$$

above $\max\{m_\dagger,q_\dagger,m_0,p_0,\ln D_0,n_\flat,p_\flat,\ln D_\flat\}$.

Note that $\widehat{\mathbf{l}}_\dagger = \widehat{\mathbf{L}}_\dagger(m_\dagger,q_\dagger,\widehat{\mathbf{k}}_\dagger,\widehat{\mathbf{r}}_\dagger)(k,r) = \widehat{\mathbf{u}}(\max\{q_\dagger+1,k\})$ while $\widehat{\mathbf{s}}_\dagger = \widehat{\mathbf{S}}_\dagger(m_\dagger,q_\dagger,\widehat{\mathbf{k}}_\dagger,\widehat{\mathbf{r}}_\dagger)$ is $\widehat{\mathbf{w}}_{j,\widehat{\mathbf{u}},E_\flat}$-bounded by C above $\max\{m_\dagger+1,m_0,p_0,\ln D_0,n_\flat,p_\flat,\ln D_\flat\}$, so $r_\dagger = \widehat{\mathbf{r}}_\dagger(\widehat{\mathbf{l}}_\dagger,\widehat{\mathbf{s}}_\dagger)$ is bounded by

$$\widehat{\mathbf{w}}^{\mathfrak{a}(B,E_\flat 2^{c_\dagger+5},BE_\flat)C+\mathfrak{i}(B,\max\{D_0,D_\flat\},E_\flat 2^{c_\dagger+5},E_\flat)C}_{j,\widehat{\mathbf{u}},E_\flat}(u)$$

where $u = \max\{m_\dagger,q_\dagger,m_0,p_0,\ln D_0,n_\flat,p_\flat,\ln D_\flat\}$, which is in turn bounded by

$$\widehat{\mathbf{w}}^{2^{12}B^4E_\flat^4}_{j+1,\widehat{\mathbf{u}},E_\flat}(\max\{m_0,p_0,\ln D_0,n_\flat,p_\flat,\ln D_\flat\}).$$

In particular,

- $\widehat{\mathbf{m}}_0(D_0,m_0,p_0,\widehat{\mathbf{r}}_0) \leq \widehat{\mathbf{w}}^{2^{12}B^4E_\flat^4}_{j+1,\widehat{\mathbf{u}},E_\flat}(\max\{m_0,p_0,\ln D_0,n_\flat,p_\flat,\ln D_\flat\})$,
- $\widehat{\mathbf{q}}_0(D_0,m_0,p_0,\widehat{\mathbf{r}}_0) \leq \widehat{\mathbf{w}}^{2^{12}B^4E_\flat^4}_{j+1,\widehat{\mathbf{u}},E_\flat}(\max\{m_0,p_0,\ln D_0,n_\flat,p_\flat,\ln D_\flat\})$.

This puts us in the setting of Lemma 28 with $d = 1$, so we obtain

- $m_0 \leq \widehat{\mathbf{w}}^C_{2^9B^2E_\flat^2,\widehat{\mathbf{u}},E_\flat}(\max\{n_\flat,p_\flat,\ln D_\flat\})$
- $q_0 \leq \widehat{\mathbf{w}}^C_{2^9D^2E_\flat^2,\widehat{\mathbf{u}},E_\flat}(\max\{n_\flat,p_\flat,\ln D_\flat\})$,

- $D_0 = E_\flat 2^{q_0+5}$,
- $\widehat{\mathbf{r}}_0$ is $\widehat{\mathbf{w}}^C_{2^9 B^2 E_\flat^2, \widehat{\mathbf{u}}, E_\flat}$-bounded by 1 above $\max\{m_0, q_0, n_\flat, p_\flat, \ln D_\flat\}$.

Therefore:

- $m_\sharp \leq \widehat{\mathbf{w}}^{2^{12} B^4 E_\flat^4 + 1}_{2^9 B^2 E_\flat^2 + 1, \widehat{\mathbf{u}}, E_\flat}(\max\{n_\flat, p_\flat, \ln D_\flat\})$,
- $q_\sharp \leq \widehat{\mathbf{w}}^{2^{12} B^4 E_\flat^4 + 1}_{2^9 B^2 E_\flat^2 + 1, \widehat{\mathbf{u}}, E_\flat}(\max\{n_\flat, p_\flat, \ln D_\flat\})$,
- $\widehat{\mathbf{k}}_\sharp$ is $\widehat{\mathbf{w}}_{2^9 B^2 E_\flat^2 + 1, \widehat{\mathbf{u}}, E_\flat}$-bounded by

$$\mathfrak{a}(B, E_\flat 2^{c_\dagger+5}, BE_\flat)Cx + \mathfrak{i}(B, D_0, E_\flat 2^{c_\sharp+5}, E_\flat)C$$

above $\max\{m_\sharp, q_\sharp, n_\flat, p_\flat, \ln D_\flat\}$,

- $\widehat{\mathbf{s}}_\sharp$ is $\widehat{\mathbf{w}}_{2^9 B^2 E_\flat^2 + 1, \widehat{\mathbf{u}}, E_\flat}$-bounded by

$$\mathfrak{a}(B, E_\flat 2^{c_\dagger+5}, BE_\flat)Cy + \mathfrak{i}(B, D_0, E_\flat 2^{c_\sharp+5}, E_\flat)C$$

above $\max\{m_\sharp, q_\sharp, n_\flat, p_\flat, \ln D_\flat\}$.

$\dashv$

At last we obtain actual (large) numeric bounds.

Theorem 9 (Bounds on 8). *Suppose* $(*)^Q$ *holds. Then for every* $E_\flat$ *there exist* $m_\sharp < s_\sharp$ *and* $q_\sharp < l_\sharp$ *such that:*

- $s_\sharp \leq \widehat{\mathbf{w}}^{C+2}_{2^{20} B^4 E_\flat^4, \mathrm{suc}, E_\flat}(1)$,
- $l_\sharp \leq \widehat{\mathbf{w}}^{C+2}_{2^{20} B^4 E_\flat^4, \mathrm{suc}, E_\flat}(1)$,
- $|(\rho_{m_\sharp} \lambda_{s_\sharp})(\Omega) - (\rho_{l_\sharp} \lambda_{q_\sharp})(\Omega)| < \frac{20}{E_\flat}$.

Proof. We apply Lemma 8 with $\widehat{\mathbf{k}}_\flat(m_\sharp, q_\sharp, \widehat{\mathbf{l}}_\sharp, \widehat{\mathbf{s}}_\sharp) = q_\sharp + 1$ and $\widehat{\mathbf{r}}_\flat(m_\sharp, q_\sharp, \widehat{\mathbf{l}}_\sharp, \widehat{\mathbf{s}}_\sharp) = m_\sharp + 1$.

Suppose we are given D_0, q_0, n_0, and $\widehat{\mathbf{k}}_0$ which is $\widehat{\mathbf{w}}^C_{j, \mathrm{suc}, E_\flat}$-bounded by C above $\max\{n_0, q_0, \ln D_0\}$. Then

- $\widehat{\mathbf{l}}_{\star, q, k}(k', r) \leq \widehat{\mathbf{w}}^C_{j, \mathrm{suc}, E_\flat}(\max\{q, k, k'\})$,
- $\widehat{\mathbf{r}}_\dagger(m_\dagger, q_\dagger, \widehat{\mathbf{s}}_\dagger, \widehat{\mathbf{k}}_\dagger) = m_\dagger + 1$,
- $\widehat{\mathbf{L}}_\dagger(m_\dagger, q_\dagger, \widehat{\mathbf{s}}_\dagger, \widehat{\mathbf{k}}_\dagger)(k', r) = \widehat{\mathbf{l}}_{\star, q_\dagger, q_\dagger + 1} \leq \widehat{\mathbf{w}}^C_{j, \mathrm{suc}, E_\flat}(\max\{q_\dagger + 1, k'\})$.

Therefore by the preceding lemma we obtain $m_\dagger, q_\dagger, \widehat{\mathbf{s}}_\dagger, \widehat{\mathbf{k}}_\dagger$ with:

- $m_\dagger \leq \widehat{\mathbf{w}}^{2^{12} B^4 E_\flat^4 + 1}_{2^9 B^2 E_\flat^2 + j + 2, \mathrm{suc}, E_\flat}(\max\{n_0, q_0, \ln D_0\})$,
- $q_\dagger \leq \widehat{\mathbf{w}}^{2^{12} B^4 E_\flat^4 + 1}_{2^9 B^2 E_\flat^2 + j + 2, \mathrm{suc}, E_\flat}(\max\{n_0, q_0, \ln D_0\})$,
- $\widehat{\mathbf{k}}_\dagger$ is $\widehat{\mathbf{w}}_{2^9 B^2 E_\flat^2 + j + 2, \mathrm{suc}, E_\flat}$-bounded by

$$\mathfrak{a}(B, E_\flat 2^{q_\dagger+5}, BE_\flat)Cx + \mathfrak{i}(B, D_0, E_\flat 2^{q_\dagger+5}, E_\flat)C$$

above $\max\{m_\dagger, q_\dagger, n_0, q_0, \ln D_0\}$,

- $\widehat{\mathbf{s}}_\dagger$ is $\widehat{\mathbf{w}}_{2^9 B^2 E_b^2 + j + 2, \text{suc}, E_b}$-bounded by

$$\mathfrak{a}(B, E_b 2^{q_\dagger + 5}, B E_b) C y + \mathfrak{i}(B, D_0, E_b 2^{q_\dagger + 5}, E_b) C$$

above $\max\{m_\dagger, q_\dagger, n_0, q_0, \ln D_0\}$.

In particular, since $\widehat{\mathbf{l}}_\dagger = \widehat{\mathbf{L}}_\dagger(m_\dagger, q_\dagger, \widehat{\mathbf{s}}_\dagger, \widehat{\mathbf{k}}_\dagger)$ is bounded by $\widehat{\mathbf{w}}_{j,\text{suc},E_b}^C (\max\{q_\dagger + 1, k'\})$, $k_\dagger = \widehat{\mathbf{k}}_\dagger(r_\dagger, \widehat{\mathbf{l}}_\dagger)$ is bounded by

$$\widehat{\mathbf{w}}_{2^9 B^2 E_b^2 + j + 2, \text{suc}, E_b}^{\mathfrak{a}(B, E_b 2^{q_\dagger + 5}, B E_b) C^2 + \mathfrak{i}(B, D_0, E_b 2^{q_\dagger + 5}, E_b) C} (u)$$

where $u = \max\{m_\dagger + 1, q_\dagger + 1, n_0, q_0, \ln D_0\}$, which is bounded by

$$\widehat{\mathbf{w}}_{2^9 B^2 E_b^2 + j + 3, \text{suc}, E_b} (\max\{m_\dagger + 1, q_\dagger + 1, n_0, q_0, \ln D_0\}).$$

This puts us in the setting of Lemma 28 with $d = 2^9 B^2 E_b^2 + 5$, so we ultimately obtain:

- $q_0 \leq \widehat{\mathbf{w}}_{2^{19} B^4 E_b^4, \text{suc}, E_b}^C (1)$,

- $n_0 \leq \widehat{\mathbf{w}}_{2^{19} B^4 E_b^2, \text{suc}, E_b}^C (1)$,

- $D_0 = E_b 2^{q_0 + 5}$,

- $\widehat{\mathbf{k}}_0$ is $\widehat{\mathbf{w}}_{2^{19} B^4 E_b^2, \text{suc}, E_b}^C$-bounded by 1 above $\max\{q_0, n_0\}$.

In particular,

- $m_\sharp = m_\dagger \leq \widehat{\mathbf{w}}_{2^{19} B^4 E_b^2, \text{suc}, E_b}^{C + 2^{12} B^4 E_b^4 + 1} (1)$,

- $q_\sharp = q_\dagger \leq \widehat{\mathbf{w}}_{2^{19} B^4 E_b^2, \text{suc}, E_b}^{C + 2^{12} B^4 E_b^4 + 1} (1)$,

- $k_\dagger \leq \widehat{\mathbf{w}}_{2^{20} B^4 E_b^4, \text{suc}, E_b}^C (\max\{m_\sharp, q_\sharp\})$,

- $r_\dagger = m_\sharp + 1$,

- $l_\sharp \leq \widehat{\mathbf{w}}_{2^{20} B^4 E_b^4, \text{suc}, E_b}^{C + 1} (\max\{m_\sharp, q_\sharp\})$,

- $s_\sharp \leq \widehat{\mathbf{w}}_{2^{20} B^4 E_b^4, \text{suc}, E_b}^{C + 1} (\max\{m_\sharp, q_\sharp\})$.

$\dashv$

In particular, bounds on $s_\sharp$ and $l_\sharp$ as a function of $B \cdot E_b$ are given by

$$B \cdot E_b \mapsto \widehat{\mathbf{w}}_{2^{20}(B E_b)^4, \text{suc}, E_b}^{C + 2} (1) \leq f_{2^{22}(B E_b)^4}(B + E + c).$$

Recall the function $f_\omega(m) = f_m(m)$; then $s_\sharp$ and $l_\sharp$ are bounded by

$$f_\omega(2^{22}(B E_b)^4 + c)$$

for some constant c.

References

[1] Jeremy Avigad and Solomon Feferman. Gödel's functional ("Dialectica") interpretation. In *Handbook of proof theory*, volume 137 of *Stud. Logic Found. Math.*, pages 337–405. North-Holland, Amsterdam, 1998.

[2] Jeremy Avigad, Philipp Gerhardy, and Henry Towsner. Local stability of ergodic averages. *Trans. Amer. Math. Soc.*, 362(1):261–288, 2010.

[3] Ernie Croot. Notes on szemeredi's regularity lemma. http://people.math.gatech.edu/~ecroot/regularity.pdf.

[4] Joseph Diestel. *Sequences and series in Banach spaces*, volume 92 of *Graduate Texts in Mathematics*. Springer-Verlag, New York, 1984.

[5] Jaime Gaspar and Ulrich Kohlenbach. On Tao's "finitary" infinite pigeonhole principle. *J. Symbolic Logic*, 75(1):355–371, 2010.

[6] Philipp Gerhardy and Ulrich Kohlenbach. General logical metatheorems for functional analysis. *Trans. Amer. Math. Soc.*, 360 (5):2615–2660, 2008.

[7] Isaac Goldbring and Henry Towsner. An approximate logic for measures. *Israel Journal of Mathematics*, 199(2):867–913, 2014.

[8] U. Kohlenbach. *Applied proof theory: proof interpretations and their use in mathematics*. Springer Monographs in Mathematics. Springer-Verlag, Berlin, 2008.

[9] U. Kohlenbach and L. Leuştean. On the computational content of convergence proofs via Banach limits. *Philos. Trans. R. Soc. Lond. Ser. A Math. Phys. Eng. Sci.*, 370(1971):3449–3463, 2012.

[10] U. Kohlenbach and P. Oliva. Proof mining: a systematic way of analyzing proofs in mathematics. *Tr. Mat. Inst. Steklova*, 242(Mat. Logika i Algebra):147–175, 2003.

[11] Ulrich Kohlenbach. Effective bounds from ineffective proofs in analysis: an application of functional interpretation and majorization. *J. Symbolic Logic*, 57(4):1239–1273, 1992.

[12] Ulrich Kohlenbach. Analysing proofs in analysis. In *Logic: from foundations to applications (Staffordshire, 1993)*, Oxford Sci. Publ., pages 225–260. Oxford Univ. Press, New York, 1996.

[13] Ulrich Kohlenbach and Angeliki Koutsoukou-Argyraki. Rates of convergence and metastability for abstract Cauchy problems generated by accretive operators. *J. Math. Anal. Appl.*, 423(2):1089–1112, 2015.

[14] Daniel Körnlein and Ulrich Kohlenbach. Rate of metastability for Bruck's iteration of pseudocontractive mappings in Hilbert space. *Numer. Funct. Anal. Optim.*, 35(1):20–31, 2014.

[15] Laurenţiu Leuştean. An application of proof mining to nonlinear iterations. *Ann. Pure Appl. Logic*, 165(9):1484–1500, 2014.

[16] Terence Tao. Norm convergence of multiple ergodic averages for commuting transformations. *Ergodic Theory Dynam. Systems*, 28(2):657–688, 2008.

[17] Terence Tao. Norm convergence of multiple ergodic averages for commuting transformations. *Ergodic Theory Dynam. Systems*, 28(2):657–688, 2008.

[18] Henry Towsner. An inverse Ackermannian lower bound on the local unconditionality constant of the James space. draft, 4 2015.

[19] Henry Towsner. A worked example of the functional interpretation. submitted, 4 2015.

9

Notes on Various Versions of Friedman's Self-Embedding Theorem

KEITA YOKOYAMA

Abstract: In this paper, we will review a typical argument to construct a self-embedding of nonstandard models of arithmetic, and will give several variations of Friedman's self-embedding theorem for subsystems of PA.

In [5], Harvey Friedman showed the famous self-embedding theorem for PA which asserts that every countable nonstandard model of PA has a cut which is isomorphic to itself. Various variations of this theorem are known, see e.g., [9, Section 12], [7, Chapter IV, Section 2(d)] and [14]. In this paper, we will give several "optimal versions" of the self-embedding theorem. There are several optimal versions of the self-embedding theorem, e.g., Dimitracopoulos/Paris [3, Theorem 2.2] and Ressayre [13, Theorems 1.I and 2.I]. Our versions will give characterization of models of subsystems $I\Sigma_n$ and $B\Sigma_n$ of PA.

In this paper, we use ω to denote the set of standard natural number. For a given $M \models I\Sigma_0$ and $\alpha \in M$, code(α) denotes the M-finite set

The author is grateful to Dr. Tin Lok Wong, Prof. Stephen G. Simpson and an anonymous referee for useful comments. His work is partially supported by JSPS Grant-in-Aid for Research Activity Start-up grant number 25887026, JSPS-NUS Bilateral Joint Research Projects J150000618 (PI's: K. Tanaka, C. T. Chong), JSPS fellowship for research abroad, and JSPS Core-to-Core Program (A. Advanced Research Networks).

Studies in Weak Arithmetics, Volume 3.
Patrick Cégielski, Ali Enayat,
Roman Kossak

coded by α (we fix a canonical coding, e.g., by binary expansions), and let $\mathrm{Cod}(M) = \{\mathrm{code}(\alpha) \mid \alpha \in M\}$. If M is nonstandard, the *standard system of M* (denoted by $\mathrm{SSy}(M)$) is defined as $\mathrm{SSy}(M) = \{\mathrm{code}(\alpha) \cap \omega \mid \alpha \in M\}$.

1 $\mathrm{B}\Sigma_n$ and Self-Embedding Theorem

In this section, we study a self-embedding theorem for a countable recursively saturated model of $\mathrm{B}\Sigma_n$ studied, e.g., in [9, Section 12]. The discussions in this section are more or less known folklore, but we will review the arguments.

Throughout this paper, we will use the following theorem freely.

Theorem 1 (Parsons, Paris/Kirby). *Let $n \geq 0$. Then,*

$$\mathrm{I}\Sigma_{n+1} \Rightarrow \mathrm{B}\Sigma_{n+1} \Leftrightarrow \mathrm{B}\Pi_n \Rightarrow \mathrm{I}\Sigma_n.$$

Proof. See, e.g., [12, Theorem A] or [9, Section 7]. ⊣

It is well-known that $\mathrm{B}\Sigma_n$ is closely related to the existence of partial elementary end extension. See, e.g., [12].

Lemma 1. (1) *Let $n \geq 2$. Let M, N be models of $\mathrm{I}\Sigma_0$ such that N is a Σ_n-elementary proper end extension of M. Then, M is a model of $\mathrm{B}\Sigma_n$.*

  (2) *Let $n \geq 1$. Let M be a model of $\mathrm{I}\Sigma_0$ and N be a model of $\mathrm{I}\Sigma_{n-1}$ such that N is a Σ_n-elementary proper end extension of M. Then, M is a model of $\mathrm{B}\Sigma_{n+1}$.*

  (3) *Let M, N be models of $\mathrm{I}\Sigma_0$ such that N is a Σ_0-elementary proper end extension of M. Then, M is a model of $\mathrm{B}\Sigma_1$.*

  (4) *Let $n \geq 0$. Let M, N be models of $\mathrm{I}\Sigma_0$ such that N is a Σ_n-elementary proper end extension of M and $M \equiv N$. Then, M is a model of $\mathrm{B}\Sigma_{n+1}$.*

Proof. For 1, see [12, Theorem B] and for 2, see [2, Proposition 3]. The proof for 3 is similar to the proof of 2. For 4, the case $n = 0, 1$ is trivial by 2 and 3. Let $n \geq 2$. Then, by 1, $M \models \mathrm{B}\Sigma_n$, hence $N \models \mathrm{B}\Sigma_n$ since $M \equiv N$. Thus, $N \models \mathrm{I}\Sigma_{n-1}$, and we have $M \models \mathrm{B}\Sigma_{n+1}$ by 2. ⊣

The following lemma is the key for Friedman's self-embedding theorem, which essentially appears in [9, page 166, Exercise 12.2].

Lemma 2. *Let $n \geq 0$. Let M be a countable short $\Sigma_{n+1} \cup \Pi_{n+1}$-recursively saturated model of $\mathrm{B}\Sigma_{n+1}$ and let N be a countable Σ_{n+1}-recursively saturated model of $\mathrm{B}\Sigma_n$ such that $\mathrm{SSy}(M) = \mathrm{SSy}(N)$. Let $a \in M$ and $b, c \in N$ such that $M \models \exists x \psi(x, a)$ implies $N \models \exists x < b\psi(x, c)$ for any Π_n formulas $\psi(x, y)$. Then, there exists an embedding*

$f : M \to N$ such that $f(M) \subseteq_e N$, $f(M) < b$, $f(a) = c$ and f is an elementary embedding with respect to Π_n formulas.

Proof. We will construct sequences $\{a_i\}_{i\in\omega} = M$ and $\{c_i\}_{i\in\omega} \subseteq_e N_{<b}$ such that $a_0 = a$, $c_0 = c$ and $M \models \exists x \psi(x, \bar{a}_i)$ implies $N \models \exists x < b\psi(x, \bar{c}_i)$ for any Π_n formulas by a back and forth argument, where $\bar{a}_i = (a_0, \ldots, a_i)$ and $\bar{c}_i = (c_0, \ldots, c_i)$. We fix enumerations $M = \{p_k\}_{k\in\omega}$ and $N = \{q_k\}_{k\in\omega}$ such that each element $d \in N$ occurs infinitely often in $\{q_k\}_{k\in\omega}$.

Assume that we have already constructed $\{a_j\}_{j<i}$ and $\{c_j\}_{j<i}$ which satisfy the desired conditions. If $i = 2k + 1$, put $a_i = p_k$. By short $\Sigma_{n+1} \cup \Pi_{n+1}$-recursive saturation, there exists $\alpha \in M$ such that for any $\theta(x) \in \Pi_n$, $\lceil \theta(x) \rceil \in \mathrm{code}(\alpha) \leftrightarrow \exists z\theta(\langle \bar{a}_i, z \rangle)$. Since $\mathrm{SSy}(M) = \mathrm{SSy}(N)$, there exists $\beta \in N$ such that $\mathrm{code}(\alpha) \cap \omega = \mathrm{code}(\beta) \cap \omega$. Then, $q(y) = \{\lceil \theta(x) \rceil \in \mathrm{code}(\beta) \to \exists z < b\theta(\langle \bar{c}_{i-1}, y, z \rangle) \wedge y < b \mid \theta(x) \in \Pi_n\}$ is a Σ_{n+1}-recursive type over N (we can easily check that $q(y)$ is finitely satisfiable). Take a solution c' of $q(y)$ by Σ_{n+1}-recursive saturation, and define $c_i = c'$. Then $\{a_j\}_{j\leq i}$ and $\{c_j\}_{j\leq i}$ satisfy the desired conditions.

If $i = 2k + 2$ and $q_k > \max\{\bar{c}_{i-1}\}$, put $c_i = c_0$ and $a_i = a_0$. If $i = 2k + 2$ and $q_k \leq \max\{\bar{c}_{i-1}\}$, put $c_i = q_k$. By Σ_{n+1}-recursive saturation, there exists $\beta \in N$ such that for any $\theta(x) \in \Sigma_n$, $\lceil \theta(x) \rceil \in \mathrm{code}(\beta) \leftrightarrow \forall z < b\theta(\langle \bar{c}_i, z \rangle)$. Since $\mathrm{SSy}(N) = \mathrm{SSy}(M)$, there exists $\alpha \in M$ such that $\mathrm{code}(\beta) \cap \omega = \mathrm{code}(\alpha) \cap \omega$. Then, $p(x) = \{\lceil \theta(x) \rceil \in \mathrm{code}(\alpha) \to \forall z\theta(\langle \bar{a}_{i-1}, x, z \rangle) \wedge x \leq \max\{\bar{a}_{i-1}\} \mid \theta(x) \in \Sigma_n\}$ is a short Π_{n+1}-recursive type over M. To show that $p(x)$ is finitely satisfiable, let $\theta_0(x), \ldots, \theta_{l-1}(x) \in \Sigma$ such that $N \models \bigwedge_{k<l} \forall z < b\theta_k(\langle \bar{c}_i, z \rangle)$. Then, $N \models \forall y < b\exists x \leq \max\{\bar{c}_{i-1}\} \bigwedge_{k<l} \forall z \leq y\theta_k(\langle \bar{c}_{i-1}, x, z \rangle)$. Since $\{a_j\}_{j<i}$ and $\{c_j\}_{j<i}$ satisfy the desired conditions, we have $M \models \forall y\exists x \leq \max\{\bar{a}_{i-1}\} \bigwedge_{k<l} \forall z \leq y\theta_k(\langle \bar{a}_{i-1}, x, z \rangle)$ (note that there is a Σ_n formula which is equivalent to $\exists x \leq \max\{\bar{u}_{i-1}\} \bigwedge_{k<l} \forall z \leq y\theta_k(\langle \bar{u}_{i-1}, x, z \rangle)$ over $\mathrm{B}\Sigma_n$). Thus, by $M \models \mathrm{B}\Sigma_{n+1}$,

$$M \models \exists x \leq \max\{\bar{a}_{i-1}\} \bigwedge_{k<l} \forall z\theta_k(\langle \bar{a}_{i-1}, x, z \rangle),$$

which means that $p(x)$ is finitely satisfiable. Take a solution a' of $p(x)$ by short Π_{n+1}-recursive saturation, and define $a_i = a'$. Then $\{a_j\}_{j\leq i}$ and $\{c_j\}_{j\leq i}$ satisfy the desired conditions.

Define a function $f : M \to N$ as $f(a_i) = c_i$. Then, we can easily check that f is the desired embedding. ⊣

Theorem 2. *Let $n \geq 0$, and let M be a countable short $\Sigma_{n+1} \cup \Pi_{n+1}$-recursively saturated model of $\mathrm{I}\Sigma_0$. Then, the following are equivalent.*

(1) *M is a model of $\mathrm{B}\Sigma_{n+1}$.*

(2) *There exists a self-embedding* $f : M \to M$ *such that* $f(M) \subsetneq_e M$ *and* f *is an elementary embedding with respect to* Π_n-*formulas.*

Proof. $2 \to 1$ is Lemma 1.4. We will show $1 \to 2$. Note that M is Σ_1-recursively saturated since short Σ_{n+1}-recursive saturation is equivalent to Σ_{n+1}-recursive saturation, which was pointed out by Wong[15]. Let M be a countable recursively saturated model of $B\Sigma_{n+1}$, and let N be a copy of M, *i.e.*, $M \cong N$. Define a recursive type $p(x)$ over M as $p(x) = \{\exists y \theta(y) \to \exists y < x \theta(y) \mid \theta \in \Pi_n\}$. Then, there exists $b \in N$ such that $N \models p(b)$. Define $a = 0 \in M$ and $c = 0 \in N$, then, M, N, a, b, c enjoy the requirements of the previous lemma. ⊣

Remark. The recursive saturation which we assumed in Theorem 2 may be not essentially needed. See [3].

2 $I\Sigma_n$ and Self-Embedding Theorem

In this section, we will slightly improve the argument in the previous section, and give a characterization of countable models of $I\Sigma_n$ by a self-embedding theorem with semi-regularity. Let M be a model of $I\Sigma_0 + \exp$, and let $X \subseteq M$ be a finite (coded) set in M (*i.e.*, $X = \text{code}(\alpha)$ for some $\alpha \in M$). Then, we can define the cardinality $\text{card}(X)$ within M.

Definition 1 (semi-regular cut). *Let* M *be a model of* $I\Sigma_0 + \exp$ *and let* I *be a cut of* M. *Then,* I *is said to be semi-regular if for any finite set* X *in* M *such that* $\text{card}(X) \in I$, $X \cap I$ *is bounded in* I.

Let M be a model of $I\Sigma_0 + \exp$ and let I be a proper cut of M. Then, we define $\text{Cod}(M/I) \subseteq \mathcal{P}(I)$, the coded class on I over M, as $\text{Cod}(M/I) = \{\text{code}(\alpha) \cap I \mid \alpha \in M\}$. We will use the following well-known theorem freely.

Theorem 3. *Let* M *be a model of* $I\Sigma_0 + \exp$ *and let* I *be a cut of* M, *and let* $S = \text{Cod}(M/I)$. *Then the following are equivalent.*

(1) I *is semi-regular in* M.

(2) $(I; S) \models I\Sigma_1(S)$ *as an* $\mathcal{L}_{\text{PA}} \cup S$-*structure.*

(3) $(I, S) \models \text{WKL}_0$ *as an* $\mathcal{L}_2$-*structure.*

Proof. See, e.g., [10, Theorems 7.1.5, 7.1.7]. ⊣

To prove the main theorem, we will prepare a few lemmas.

Lemma 3. *Let* $n \geq 0$. *Let* M *be a nonstandard model of* $I\Sigma_n$ *and let* I *be a* Π_n-*elementary proper cut of* M *which is semi-regular. Then,* $I \models I\Sigma_{n+1}$.

Proof. Let $S = \text{Cod}(M/I)$. Then, by the previous theorem, $(I; S) \models I\Sigma_1(S)$. By $I\Sigma_n$, every bounded Σ_n-definable subset in M is coded by an element in M. Thus, every Σ_n-definable subset in I is in S. Thus, $I \models I\Sigma_{n+1}$.　　　　　　　　　　　　　　　　　　　⊣

Note that the above lemma is closely related to Kaye [8, Lemma 3.4].

Lemma 4. *Let $n \geq 1$. Then, every nonstandard model of $I\Sigma_n$ is Σ_n-recursively saturated and short $\Sigma_n \cup \Pi_n$-recursively saturated.*

Proof. This follows from the fact that the standard cut is not definable by a boolean combination of Σ_n and Π_n-formulas in a nonstandard model of $I\Sigma_n$, see, e.g., [2, Corollary 6]. For Σ_n-recursive saturation, see, e.g., [9, Theorem 11.5]. One can prove short $\Sigma_n \cup \Pi_n$-recursive saturation with the same idea. See also [4, Section 4].　　　　⊣

　　Next, we consider a property of Σ_n-definable trees. Within $I\Sigma_0 + \exp$, let $2^{<\mathbb{N}}$ to be the set of all finite sequences of 0's and 1's, and let $2^{=m}$ be the set of all finite sequences of 0's and 1's of length m. A (definable) set $T \subseteq 2^{<\mathbb{N}}$ is said to be a *tree* if it is closed under initial subsequences. For given a binary sequence $\tau \in 2^{<\mathbb{N}}$ and a number m, $|\tau|$ denotes the length of τ, and $\tau \restriction m$ denotes the initial subsequence of τ of length m. For a finite set X, we define $X[m] = \langle z_j \in 2 \mid j < m, z_j = 1 \leftrightarrow j \in X \rangle$. Note that $X[m]$ is defined even when $m > \max X$. For given a tree $T \subseteq 2^{<\mathbb{N}}$ and a sequence $\tau \in T$, we say that τ is *extendible in T* if there are infinitely many extensions of τ in T. Note that the assertion "a Σ_n-definable tree T is infinite" can be expressed by a Π_n-formula since it is $\forall m \exists \tau \in 2^{=m}(\tau \in T)$. Intuitively, the following lemma says that if a Σ_n-definable tree has an infinite path with its cardinality is bounded, then one can find such a path within $I\Sigma_{n+1}$.

Lemma 5. *Let $n \geq 0$. Let $\psi(x, \tau)$ be a Σ_n-formula. Then, $I\Sigma_{n+1}$ proves the following:*

- *for any u, if $\forall m \exists \tau \in 2^{=m}(\text{card}(\tau) \leq u \wedge \forall x \leq m \psi(x, \tau \restriction x))$, then there exists a finite set A such that $\forall m \psi(m, A[m]) \wedge \text{card}(A) \leq u$.*

Here, $\text{card}(\tau)$ denotes the number of 1's in τ.

Proof. We work within $I\Sigma_{n+1}$. Let u be a number such that $\forall m \exists \tau \in 2^{=m}(\text{card}(\tau) \leq u \wedge \forall x \leq m \psi(x, \tau \restriction x))$. Put $T(k) = \{\tau \in 2^{<\mathbb{N}} \mid \text{card}(\tau) \leq k \wedge \forall w \leq |\tau| \psi(w, \tau \restriction w)\}$. Then, by $B\Sigma_n$, $T(k)$ is a Σ_n-definable tree, and $T(u)$ is infinite by the assumption. ($T(k)$ forms a tree since the condition is closed under subsequence.) By $I\Sigma_{n+1}$, take $v = \min\{k \leq u \mid T(k)$ is infinite$\}$. Then, we can see that there exists $\tau \in T(v)$ such that $\text{card}(\tau) = v$ and τ is extendable in $T(v)$. If $v = 0$, this is trivial since the empty sequence is extendible. Otherwise, $T(v-1)$ is finite, thus take large enough m so that $T(v-1) \cap 2^{=m} = \emptyset$. Since $T(v)$ is infinite, $T(v) \cap 2^{=m} \neq \emptyset$, and for any $\tau \in T(v) \cap 2^{=m}$, $\text{card}(\tau) = v$. Now, by the pigeonhole principle for Σ_n-formulas, which is available within $I\Sigma_n$, there exists $\tau \in T(v) \cap 2^{=m}$ which is extendible in $T(v)$. Then, for any l, $\tau^\frown 0^l \in T(v)$. Thus, $A := \text{set}(\tau) = \{x \mid x < |\tau|, \tau(x) = 1\}$ is the desired finite set. ⊣

Now, we are ready to prove the main lemma.

Lemma 6. *Let $n \geq 0$. Let M and N be countable models of $I\Sigma_{n+1}$ such that $\text{SSy}(M) = \text{SSy}(N)$. Let $a \in M$ and $b, c \in N$ such that $M \models \exists m \psi(m, a)$ implies $N \models \exists m < b \psi(m, c)$ for any Π_n-formulas $\psi(m, z)$. Then, there exists an embedding $f : M \to N$ such that $f(M) \subseteq_e N$, $f(M) < b$, $f(a) = c$, $f(M)$ is a semi-regular cut of N and f is an elementary embedding with respect to Π_n-formulas.*

Proof. We will construct sequences $\{a_i\}_{i \in \omega} = M$, $\{c_i\}_{i \in \omega} \subseteq_e N_{<b}$, $\{A_i\}_{i \in \omega} \subseteq \text{Cod}(M)$ and $\{C_i\}_{i \in \omega} \subseteq \text{Cod}(N)$ such that $a_0 = a$, $c_0 = c$ and $M \models \exists m \psi(m, \bar{a}_i, \bar{A}_i[m])$ implies $N \models \exists m < b \psi(m, \bar{c}_i, \bar{C}_i[m])$ for any Π_n-formulas ψ by a back and forth argument. We fix enumerations $M = \{p_k\}_{k \in \omega}$, $N = \{q_k\}_{k \in \omega}$ and $\text{Cod}(N) = \{Q_k\}_{k \in \omega}$ such that each element $d \in N$ occurs infinitely often in $\{q_k\}_{k \in \omega}$ and each element $D \in \text{Cod}(N)$ occurs infinitely often in $\{Q_k\}_{k \in \omega}$. Note that both of M and N are Σ_{n+1}-recursively saturated and short $\Sigma_{n+1} \cup \Pi_{n+1}$-recursively saturated by Lemma 4.

Assume that we have already constructed $\{a_j\}_{j<i}$, $\{c_j\}_{j<i}$, $\{A_j\}_{j<i}$, $\{C_j\}_{j<i}$ which satisfy the desired conditions. If $i = 3k+1$, put $a_i = p_k$ and $A_i = C_i = \emptyset$. Then, we can find the desired c_i by using the same method of proof of Lemma 2.

If $i = 3k+2$ and $q_k > \max\{\bar{c}_{i-1}\}$, put $c_i = c_0$, $a_i = a_0$ and $A_i = C_i = \emptyset$. If $i = 3k+2$ and $q_k \leq \max\{\bar{c}_{i-1}\}$, put $c_i = q_k$ and $A_i = C_i = \emptyset$. Then, we can find the desired a_i by using the same method of proof of Lemma 2.

If $i = 3k+3$ and $\text{card}(Q_k) > \max\{\bar{c}_{i-1}\}$, put $c_i = c_0$, $a_i = a_0$ and $A_i = C_i = \emptyset$. If $i = 3k+3$ and $\text{card}(Q_k) \leq \max\{\bar{c}_{i-1}\}$, put $c_i = c_0$, $a_i = a_0$ and $C_i = Q_k$. We will find a coded set $A_i \in \text{Cod}(M)$ which satisfies the desired condition. By Σ_{n+1}-recursive saturation, there exists $\beta \in N$ such that for any Π_n-formulas ψ, $\lceil \psi \rceil \in \text{code}(\beta) \leftrightarrow \forall m < b \neg \psi(m, \bar{c}_i, \bar{C}_{i-1}[m], C_i[m])$. Since $\text{SSy}(N) = \text{SSy}(M)$, there exists $\alpha \in M$ such that $\text{SSy}(\beta) = \text{SSy}(\alpha)$. Then, put $p(W) = \{\lceil \psi \rceil \in \text{code}(\alpha) \to \forall m \neg \psi(m, \bar{a}_i, \bar{A}_{i-1}[m], W[m]) \mid \psi \in \Pi_n\}$. We will find a coded set $A \in \text{Cod}(M)$ realizing $p(W)$. Note that Π_{n+1}-recursive saturation is not available in our setting. Thus, to find a solution of $p(W)$, we will directly use overspill instead of recursive saturation.

For a given standard natural number $e_0 \in \omega$, we have

$$N \models \bigwedge_{\lceil \psi \rceil \in \mathrm{code}(\beta)[e_0]} \forall m < b \neg \psi(m, \bar{c}_i, \bar{C}_{i-1}[m], C_i[m]).$$

Then,

$$N \models \forall m < b \exists \tau \in 2^{=m} \left(\mathrm{card}(\tau) \leq \max\{\bar{c}_{i-1}\} \right.$$

$$\left. \wedge \forall w \leq m \bigwedge_{\lceil \psi \rceil \in \mathrm{code}(\beta)[e_0]} \neg \psi(w, \bar{c}_i, \bar{C}_{i-1}[w], \tau \upharpoonright w) \right).$$

We can easily check that this formula is equivalent to a formula of the form $\forall m < b \neg \tilde{\psi}(\bar{c}_{i-1}, \bar{C}_{i-1}[m])$ for some Π_n-formula $\tilde{\psi}$. Thus, we have

$$M \models \forall m \exists \tau \in 2^{=m} \left(\mathrm{card}(\tau) \leq \max\{\bar{a}_{i-1}\} \right.$$

$$\left. \wedge \forall w \leq m \bigwedge_{\lceil \psi \rceil \in \mathrm{code}(\alpha)[e_0]} \neg \psi(, w, \bar{a}_i, \bar{A}_{i-1}, \tau \upharpoonright w) \right) \qquad (\dagger)$$

since $\mathrm{code}(\beta)[e_0] = \mathrm{code}(\alpha)[e_0]$.

Let $\mathrm{Tr}_{\Pi_n}(e, x)$ be the truth definition for Π_n-formulas which is described by a Π_n-formula if $n > 0$ and described by a Σ_1-formula if $n = 0$. Then, $(\dagger)$ can be rephrased as

$$M \models \forall m \exists \tau \in 2^{=m} (\mathrm{card}(\tau) \leq \max\{\bar{a}_{i-1}\} \wedge \forall w \leq m \forall e < e_0$$

$$(e \in \mathrm{code}(\alpha) \to \neg \mathrm{Tr}_{\Pi_n}(e, w, \bar{a}_i, \bar{A}_{i-1}, \tau \upharpoonright w))). \qquad (\dagger\dagger)$$

Now, by overspill for Π_{n+1}-formulas which is available in M, there exists $e_0 \in M \setminus \omega$ such that $(\dagger\dagger)$ holds. (This argument is essentially the same as for proving Σ_{n+1}-saturation from $I\Sigma_{n+1}$.) Thus, by Lemma 5, there exists $A \in \mathrm{Cod}(M)$ such that $M \models \forall m \forall e < e_0 (e \in \mathrm{code}(\alpha) \to \neg \mathrm{Tr}_{\Pi_n}(e, m, \bar{a}_i, \bar{A}_{i-1}[m], A[m]))$, which means that A is a solution of $p(W)$. Define $A_i = A$, then $\{a_j\}_{j \leq i}$, $\{c_j\}_{j \leq i}$, $\{A_j\}_{j \leq i}$ and $\{C_j\}_{j \leq i}$ satisfy the desired conditions.

Define a function $f : M \to N$ as $f(a_i) = c_i$. Then, we can easily check that f is an elementary embedding with respect to Π_n-formulas and $f(M)$ is a cut of N. Let $D \in \mathrm{Cod}(N)$ such that $\mathrm{card}(D) \in f(M)$. Then, there exists $i \in \omega$ such that $C_i = D$. By the construction, for any $x \in M$, $f(x) \in D \Rightarrow x \in A_i \Rightarrow f(x) \leq f(\max A_i)$. Thus, $D \cap f(M)$ is bounded in $f(M)$. This means that $f(M)$ is semi-regular. $\dashv$

Theorem 4. *Let $n \geq 1$, and let M be a nonstandard countable model of $I\Sigma_0 + \exp$. Then, the following are equivalent*

(1) M is a model of $I\Sigma_n$.

(2) There exists a self-embedding $f : M \to M$ such that $f(M) \subsetneq_e M$ is a semi-regular cut and f is an elementary embedding with respect to Π_{n-1}-formulas.

Proof. $2 \to 1$ is a straightforward consequence of Lemmas 1.4 and 3. We can prove $1 \to 2$ by Lemma 6 as in the proof of Theorem 2. ⊣

Remark. The author heard that Richard Kaye proved the case $n = 1$ independently. The case $n = 1$ gives a new proof of Harrington's theorem, which asserts that for every countable model M of $I\Sigma_1$, there exists $S \subseteq \mathcal{P}(M)$ such that $(M, S) \models \mathsf{WKL}_0$. Conversely, we can prove the case $n = 1$ by combining Harrington's theorem and Tanaka's self-embedding theorem for WKL_0 [14]. We will see this in the next section.

3 Some More Variations

As we have mentioned in the previous section, the combination of the self-embedding theorem for WKL_0 plus the extension theorem to be a model of WKL_0 provide self-embeddings with semi-regularity. In this section, we will generalize Theorem 4 with this idea.

In this section, we mainly work within the second-order arithmetic. Here, $I\Sigma_n^0$ and $B\Sigma_n^0$ denote the system of Σ_n^0-induction and Σ_n^0-bounding in the second-order arithmetic, *i.e.* they allow set parameters. For a given second-order structure (M, S) and a cut $I \subseteq_e M$, let $S{\restriction}I := \{X \cap I \mid X \in S\}$. Here are the main tools of this section, Tanaka's self-embedding theorem and the generalization of Harrington's theorem.

Theorem 5 (Tanaka [14], see also Enayat/Wong [4]). *Let (M, S) be a countable nonstandard model of WKL_0, and let $a \in M$ and $A \in S$. Then, there exists $f : (M, S) \to (M, S)$ such that $I := f(M) \subsetneq_e M$, $(I, \mathrm{Cod}(M/I)) = (I, f(S){\restriction}I) \cong (M, S)$ via f, $f(x) = x$ for any $x < a$ and $f(A) = A$.*

Theorem 6 (Harrington, Paris [11], Hájek [6], see also Belanger [1]). *Let M be a countable model of $I\Sigma_0$ and $X \subseteq M$. Let $n \geq 1$.*

(1) *If $(M; X) \models I\Sigma_n(X)$, then there exists $S \subseteq \mathcal{P}(M)$ such that $X \in S$ and $(M, S) \models \mathsf{WKL}_0 + I\Sigma_n^0$.*

(2) *If $(M; X) \models B\Sigma_{n+1}(X)$, there exists $S \subseteq \mathcal{P}(M)$ such that $X \in S$ and $(M, S) \models \mathsf{WKL}_0 + B\Sigma_{n+1}^0$.*

Let $k \geq 1$. A cut $I \subseteq_e M$ is said to be *k-inductive* if $(I, \mathrm{Cod}(M/I))$ is a model of Σ_k^0-induction. A cut $I \subseteq_e M$ is said to be *k-bounding* if $(I, \mathrm{Cod}(M/I))$ is a model of Σ_k^0-bounding. Thus, a 1-inductive cut is a semi-regular cut, and a 2-bounding cut is a regular cut.

Theorem 7. *Let $1 \leq k \leq n$, and let M be a countable nonstandard model of $I\Sigma_0 + \exp$. Then, the following are equivalent.*

(1) *M is a model of $I\Sigma_n$.*

(2) *There exists a self-embedding $f : M \to M$ such that $f(M) \subsetneq_e M$ is a k-inductive cut and f is an elementary embedding with respect to Π_{n-k}-formulas.*

Proof. We will first show $1 \to 2$. Let M be a countable model of $I\Sigma_n$, and let $1 \leq k \leq n$ and $m = n-k$. If $m = 0$, put $X = \emptyset$. Otherwise, define a set $X \subseteq M$ as $(e, x) \in X \Leftrightarrow M \models \mathrm{Tr}_{\Pi_m}(e, x)$, where $\mathrm{Tr}_{\Pi_m}(e, x)$ is the truth definition for Π_m-formulas. Then, one can easily check that $(M; X)$ is a model of $I\Sigma_k(X)$. Now, by Theorem 6.1, there exists a countable set $S \subseteq \mathcal{P}(M)$ such that $X \in S$ and $(M, S) \models \mathsf{WKL}_0 + I\Sigma_k^0$. Thus, by Theorem 5, there exists $f : (M, S) \to (M, S)$ such that $I := f(M) \subsetneq_e M$, $(I, \mathrm{Cod}(M/I)) = (I, f(S){\restriction}I) \cong (M, S)$ via f and $f(X) = X$. Then, I is a k-inductive cut. To show that f is elementary with respect to Π_m-formulas, let $I \models \varphi(a)$ for a Π_m-formula $\varphi(x)$ and $a = f(a_0) \in I$. Then $(M, S) \models \varphi(a_0) \Leftrightarrow (M, S) \models (\lceil\varphi\rceil, a_0) \in X$. Since f is an isomorphism between $(M; X)$ and $(I; X \cap I)$, we have $(\lceil\varphi\rceil, f(a_0)) \in X \cap I$. Thus, $(\lceil\varphi\rceil, a) \in X$, which means $M \models \varphi(a)$.

Next, we show $2 \to 1$. Let $1 \leq k \leq n$, and assume that $M \cong f(M) \subsetneq_e M$ is a k-inductive cut and f is an elementary embedding with respect to Π_{n-k} formulas. Let $m = n - k$, $I = f(M)$, and $S = \mathrm{Cod}(M/I)$. Then, $(I, S) \models I\Sigma_k^0$. Since I is a semi-regular cut and f is elementary with respect to Π_m-formulas, $M \cong I \models I\Sigma_m$ by Theorem 4. We only need to show that $M \cong I \models I\Sigma_n$. Let $\varphi(x)$ be a Σ_n-formula such that $I \models \exists x \varphi(x)$. We will find a minimal $x \in I$ satisfying $\varphi(x)$. Write $\varphi(x) \equiv \exists y_1 \ldots Qy_k \theta(x, y_1, \ldots, y_k)$ such that θ is a Π_m or Σ_m-formula. Take $a \in M \setminus I$. Since $M \models I\Sigma_m$, there exists M-finite set $Z \subseteq M$ such that $\langle x, y_1, \ldots, y_k \rangle \in Z \Leftrightarrow M \models \theta(x, y_1, \ldots, y_k) \wedge \langle x, y_1, \ldots, y_k \rangle < a$. Let $Z' = Z \cap I \in S$. Since I is a Π_m-elementary substructure of M, for any $x, y_1, \ldots, y_k \in I$, we have $I \models \theta(x, y_1, \ldots, y_k) \Leftrightarrow M \models \theta(x, y_1, \ldots, y_k) \Leftrightarrow \langle x, y_1, \ldots, y_k \rangle \in Z'$. Thus, $(I, S) \models \varphi(x) \leftrightarrow \exists y_1 \ldots Qy_k \langle x, y_1, \ldots, y_k \rangle \in Z'$. Since $(I, S) \models I\Sigma_k^0$, there exists a minimal x in I satisfying $\exists y_1 \ldots Qy_k \langle x, y_1, \ldots, y_k \rangle \in Z'$, i.e., satisfying $\varphi(x)$. $\dashv$

Theorem 8. *Let $2 \leq k \leq n$, and let M be a countable nonstandard model of $I\Sigma_0 + \exp$. Then, the following are equivalent.*

(1) *M is a model of $B\Sigma_n$.*

(2) *There exists a self-embedding* $f : M \to M$ *such that* $f(M) \subsetneq_e M$ *is a k-bounding cut and f is an elementary embedding with respect to* Π_{n-k}*-formulas.*

Proof. The same as the proof for Theorem 7 with using Theorem 6.2.

$\dashv$

References

[1] Belanger D. A., Conservation theorems for the cohesiveness principle, 2015, to appear.

[2] Clote P., Partition relations in arithmetic, in *Methods in mathematical logic (Caracas, 1983), Lecture Notes in Math.*, vol. 1130, Springer, Berlin, 1985, pp. 32–68.

[3] Dimitracopoulos C. and Paris J., A note on a theorem of H. Friedman, *Z. Math. Logik Grundlag. Math.*, vol. 34(1), 1988, pp. 13–17.

[4] Enayat A. and Wong T. L., Model theory of WKL$_0^*$, 2015, preprint.

[5] Friedman H., Countable models of set theories, in *Cambridge Summer School in Math. Logic, Lecture Notes in Math.*, vol. 337, 1973, pp. 539–573.

[6] Hájek P., Interpretability and fragments of arithmetic, in P. Clote and J. Krajíček, editors, *Arithmetic, Proof Theory and Computational Complexity*, Oxford, Clarendon Press, 1993, pp. 185–196.

[7] Hájek P. and Pudlák P., *Metamathematics of First-Order Arithmetic*, Springer-Verlag, Berlin, 1993.

[8] Kaye R., Model-theoretic properties characterizing Peano arithmetic, *J. Symbolic Logic*, vol. 56(3), 1991, pp. 949–963.

[9] Kaye R., *Models of Peano Arithmetic*, Oxford Logic Guides 15, Oxford University Press, 1991.

[10] Kossak R. and Schmerl J. H., *The structure of models of Peano arithmetic*, Oxford Logic Guides 50, Oxford University Press, 2006.

[11] Paris J. B., Some conservation results for fragments of arithmetic, in *Model theory and arithmetic (Paris, 1979–1980), Lecture Notes in Math.*, vol. 890, Springer, Berlin-New York, 1981, pp. 251–262.

[12] Paris J. B. and Kirby L. A. S., Σ_n-collection schemas in arithmetic, in *Logic Colloquium 77 (Proc. Conf., Wrocław, 1977), Stud. Logic Foundations Math.*, vol. 96, North-Holland, Amsterdam - New York, 1978, pp. 199–209.

[13] Ressayre J.-P., Nonstandard universes with strong embeddings, and their finite approximations, in *Logic and combinatorics (Arcata, Calif.,*

1985), *Contemp. Math.*, vol. 65, Amer. Math. Soc., Providence, RI, 1987, pp. 333–358.

[14] Tanaka K., The self-embedding theorem of WKL$_0$ and a non-standard method, *Ann. Pure Appl. Logic*, vol. 84(1), 1997, pp. 41–49.

[15] Wong T.L., private communication, March 2016.